Advances in Clinical Diabetes, Obesity, and Metabolic Diseases

Advances in Clinical Diabetes, Obesity, and Metabolic Diseases

Guest Editors

Yuzuru Ohshiro
Kunimasa Yagi
Yasuhiro Maeno

Basel • Beijing • Wuhan • Barcelona • Belgrade • Novi Sad • Cluj • Manchester

Guest Editors

Yuzuru Ohshiro
Department of Internal Medicine
Omoromachi Medical Center
Naha
Japan

Kunimasa Yagi
School of Medicine
Kanazawa Medical University
Uchinada
Japan

Yasuhiro Maeno
Comprehensive Internal Medicine
Shiga University of Medical Science
Higashiomi
Japan

Editorial Office
MDPI AG
Grosspeteranlage 5
4052 Basel, Switzerland

This is a reprint of the Special Issue, published open access by the journal *Medicina* (ISSN 1648-9144), freely accessible at: https://www.mdpi.com/journal/medicina/special_issues/UYHX989072.

For citation purposes, cite each article independently as indicated on the article page online and as indicated below:

Lastname, A.A.; Lastname, B.B. Article Title. *Journal Name* **Year**, *Volume Number*, Page Range.

ISBN 978-3-7258-3830-1 (Hbk)
ISBN 978-3-7258-3829-5 (PDF)
https://doi.org/10.3390/books978-3-7258-3829-5

© 2025 by the authors. Articles in this book are Open Access and distributed under the Creative Commons Attribution (CC BY) license. The book as a whole is distributed by MDPI under the terms and conditions of the Creative Commons Attribution-NonCommercial-NoDerivs (CC BY-NC-ND) license (https://creativecommons.org/licenses/by-nc-nd/4.0/).

Contents

About the Editors . . . . **vii**

Preface . . . . **ix**

Yuzuru Ohshiro, Kunimasa Yagi and Yasuhiro Maeno
Editorial for the Special Issue "Advances in Clinical Diabetes, Obesity, and Metabolic Diseases"
Reprinted from: *Medicina* **2025**, *61*, 595, https://doi.org/10.3390/medicina61040595 1

Kunimasa Yagi, Michiko Inagaki, Yuya Asada, Mako Komatsu, Fuka Ogawa, Tomomi Horiguchi, et al.
Improved Glycemic Control through Robot-Assisted Remote Interview for Outpatients with Type 2 Diabetes: A Pilot Study
Reprinted from: *Medicina* **2024**, *60*, 329, https://doi.org/10.3390/medicina60020329 4

Toshitaka Sawamura, Shigehiro Karashima, Azusa Ohbatake, Takuya Higashitani, Ai Ohmori, Kei Sawada, et al.
Effects of Switching from Degludec to Glargine U300 in Patients with Insulin-Dependent Type 1 Diabetes: A Retrospective Study
Reprinted from: *Medicina* **2024**, *60*, 450, https://doi.org/10.3390/medicina60030450 17

Jawaria Ali Tariq, KaleemUllah Mandokhail, Naheed Sajjad, Abrar Hussain, Humera Javaid, Aamir Rasool, et al.
Effects of Age and Biological Age-Determining Factors on Telomere Length in Type 2 Diabetes Mellitus Patients
Reprinted from: *Medicina* **2024**, *60*, 698, https://doi.org/10.3390/medicina60050698 27

Andra-Elena Nica, Emilia Rusu, Carmen Dobjanschi, Florin Rusu, Claudia Sivu, Oana Andreea Parlițeanu and Gabriela Radulian
The Relationship between the Ewing Test, Sudoscan Cardiovascular Autonomic Neuropathy Score and Cardiovascular Risk Score Calculated with SCORE2-Diabetes
Reprinted from: *Medicina* **2024**, *60*, 828, https://doi.org/10.3390/medicina60050828 43

Nilgun Tan Tabakoglu and Mehmet Celik
Investigation of the Systemic Immune Inflammation (SII) Index as an Indicator of Morbidity and Mortality in Type 2 Diabetic Retinopathy Patients in a 4-Year Follow-Up Period
Reprinted from: *Medicina* **2024**, *60*, 855, https://doi.org/10.3390/medicina60060855 54

Haruna Arakawa and Masashi Inafuku
BCG Vaccination Suppresses Glucose Intolerance Progression in High-Fat-Diet-Fed C57BL/6 Mice
Reprinted from: *Medicina* **2024**, *60*, 866, https://doi.org/10.3390/medicina60060866 67

Eiji Kutoh, Alexandra N. Kuto, Rumiko Okada, Midori Akiyama and Rumi Kurihara
Diverse Strategies for Modulating Insulin Resistance: Causal or Consequential Inference on Metabolic Parameters in Treatment-Naïve Subjects with Type 2 Diabetes
Reprinted from: *Medicina* **2024**, *60*, 991, https://doi.org/10.3390/medicina60060991 80

Karol Graňák, Matej Vnučák, Monika Beliančinová, Patrícia Kleinová, Tímea Blichová, Margaréta Pytliaková and Ivana Dedinská
Regular Physical Activity in the Prevention of Post-Transplant Diabetes Mellitus in Patients after Kidney Transplantation
Reprinted from: *Medicina* **2024**, *60*, 1210, https://doi.org/10.3390/medicina60081210 98

Fulya Bakılan, Sultan Şan Kuşcu, Burcu Ortanca, Fezan Şahin Mutlu, Pınar Yıldız and Onur Armağan
Ultrasonographic Achilles Tendon Measurements and Static and Dynamic Balance in Prediabetes
Reprinted from: *Medicina* 2024, 60, 1349, https://doi.org/10.3390/medicina60081349 . . . . **109**

Khanh Ngoc Nguyen, Van Khanh Tran, Ngoc Lan Nguyen, Thi Bich Ngoc Can, Thi Kim Giang Dang, Thu Ha Nguyen, et al.
Hyperornithinemia–Hyperammonemia–Homocitrullinuria Syndrome in Vietnamese Patients
Reprinted from: *Medicina* 2024, 60, 1877, https://doi.org/10.3390/medicina60111877 . . . . **119**

Miloš Vuković, Igor Nosek, Johannes Slotboom, Milica Medić Stojanoska and Duško Kozić
Neurometabolic Profile in Obese Patients: A Cerebral Multi-Voxel Magnetic Resonance Spectroscopy Study
Reprinted from: *Medicina* 2024, 60, 1880, https://doi.org/10.3390/medicina60111880 . . . . **131**

Emilia Rusu, Eduard Lucian Catrina, Iulian Brezean, Ana Maria Georgescu, Alexandra Vișinescu, Daniel Andrei Vlad Georgescu, et al.
Lower Extremity Amputations Among Patients with Diabetes Mellitus: A Five-Year Analysis in a Clinical Hospital in Bucharest, Romania
Reprinted from: *Medicina* 2024, 60, 2001, https://doi.org/10.3390/medicina60122001 . . . . **149**

Yuya Asada, Tomomi Horiguchi, Kunimasa Yagi, Mako Komatsu, Ayaka Yamashita, Ren Ueta et al.
Robot-Assisted Approach to Diabetes Care Consultations: Enhancing Patient Engagement and Identifying Therapeutic Issues
Reprinted from: *Medicina* 2025, 61, 352, https://doi.org/10.3390/medicina61020352 . . . . **161**

Jung-Eun Han, Jun-Hwan Choi, So-Yeon Yoo, Gwan-Pyo Koh, Sang-Ah Lee, So-Young Lee and Hyun-Jung Lee
Association of Nerve Conduction Study Variables with Hematologic Tests in Patients with Type 2 Diabetes Mellitus
Reprinted from: *Medicina* 2025, 61, 430, https://doi.org/10.3390/medicina61030430 . . . . **176**

About the Editors

Yuzuru Ohshiro

Dr. Yuzuru Ohshiro is the Director of the Department of Diabetes and Endocrinology at Omoromachi Medical Center, and a Visiting Associate Professor in the Department of Internal Medicine at the University of the Ryukyus. He is a board-certified specialist in internal medicine, diabetes, and dialysis therapy. In addition to his clinical practice, he is actively engaged in clinical research in these fields. Dr. Ohshiro also has experience in basic research, having served as a researcher at the Joslin Diabetes Center, Harvard Medical School. His work in both basic and clinical research has been recognized with prestigious honors, including the Overseas Research Award from the Kanae Foundation for the Promotion of Medical Science and the Research Award from the Okinawa Medical Science Foundation.

Kunimasa Yagi

Dr. Kunimasa Yagi is a Professor of Internal Medicine at Kanazawa Medical University School of Medicine and its affiliated hospitals. He is a board-certified specialist in internal medicine, diabetes, endocrinology, lipidology, atherosclerosis and arteriosclerosis, cardiology, and genomic medicine. He has engaged in basic research as a visiting scientist at the Joslin Diabetes Center, Harvard Medical School. Dr. Yagi is dedicated to clinical research and medical practice for pre-heart failure (HF) management in patients with diabetes mellitus. His clinical research interests focus on the fragmented QRS (fQRS) and the fourth cardiac sound (S4) as tools for the early detection of cardiac diastolic dysfunction. He is also engaged in collaborative research with engineering and nursing specialists on using humanoid robots to achieve behavioral change in the pre-heart failure stage, including exercise intervention.

Yasuhiro Maeno

Dr. Yasuhiro Maeno is an Associate Professor of Comprehensive Internal Medicine at Shiga University of Medical Science (SUMS). He is also the Chief of the Diabetes & Endocrinology Department at the National Hospital Organization Higashi-ohmi General Medical Center, the second teaching hospital of SUMS. He is involved in the education of medical students and residents at the hospital and also practices and researches community medicine.

Dr. Maeno is a board-certified specialist and instructor in diabetes, comprehensive internal medicine, and primary care medicine. He is also an academic councilor of the Japan Diabetes Society. His area of expertise is diabetic vascular complications, and he has served as a research fellow at the Joslin Diabetes Center, affiliated with Harvard Medical School. His current interest is in the care of the rapidly growing elderly diabetic population.

Preface

Dear Colleagues,

Diabetes, obesity, and metabolic diseases have emerged as some of the most pressing global health challenges of our time. Their rising prevalence not only burdens individuals and families but also exerts profound pressure on healthcare systems worldwide. These conditions are intricately linked, often overlapping in pathophysiology and complicating clinical management.

In recent years, we have witnessed remarkable breakthroughs in both pharmacological therapies and digital technologies. Novel agents such as weekly insulin, dual and triple incretin receptor agonists, and oral GLP-1 receptor agonists are transforming glycemic control and weight management. At the same time, the integration of digital tools—including AI-assisted monitoring, wearable devices, and telehealth platforms—has begun to redefine the way we deliver care. Yet, despite these promising advances, many critical questions remain. How effective are these treatments in real-world settings? How can technology be best utilized to personalize therapy and improve long-term outcomes? And how do we ensure equitable access to these innovations across diverse global populations?

This Special Issue, entitled "Advances in Clinical Diabetes, Obesity, and Metabolic Diseases", brings together cutting-edge research, clinical insights, and real-world experiences from experts around the world. It aims to provide a comprehensive snapshot of where we stand and where we are heading in the fight against these complex diseases.

We are deeply grateful to the authors and reviewers whose invaluable contributions have made this collection possible. It is our hope that the studies presented here will inspire continued dialogue, collaboration, and innovation in this ever-evolving field.

We now invite you to explore the following Editorial, which offers a comprehensive overview of the significant findings and ongoing challenges highlighted in this Special Issue.

Yuzuru Ohshiro, Kunimasa Yagi, and Yasuhiro Maeno
Guest Editors

Editorial

Editorial for the Special Issue "Advances in Clinical Diabetes, Obesity, and Metabolic Diseases"

Yuzuru Ohshiro [1,*], Kunimasa Yagi [2] and Yasuhiro Maeno [3,4]

1. Department of Internal Medicine, Omoromachi Medical Center, 1-3-1 Uenoya, Naha 900-0011, Okinawa, Japan
2. School of Medicine, Kanazawa Medical University, 1-1 Daigaku, Uchinada 920-0293, Ishikawa, Japan; yagikuni@icloud.com
3. Comprehensive Internal Medicine, Shiga University of Medical Science, 255 Gochi-cho, Higashiomi 527-8505, Shiga, Japan; maeno@belle.shiga-med.ac.jp
4. Diabetes & Endocrinology Department, National Hospital Organization Higashi-ohmi General Medical Center, 255 Gochi-cho, Higashiomi 527-8505, Shiga, Japan
* Correspondence: ooshiro@aol.com

Diabetes, obesity, and metabolic diseases are posing significant challenges to healthcare systems globally. These conditions contribute to increased morbidity and mortality, so continuous advancements in therapeutic strategies and technologies are crucial. Over the past year or so, our Special Issue "Advances in Clinical Diabetes, Obesity, and Metabolic Diseases" has served as a platform for the sharing of novel interventions, innovative technologies, and real-world data that contribute to improved patient care. This Editorial will provide an overview of key developments in this field, highlight the contributions of this Special Issue, and outline future research directions.

Remarkable progress has been made in the management of diabetes and metabolic disorders over the last decade. Novel pharmacological treatments, such as weekly insulin [1], dual glucagon-like peptide (GLP)-1 and gastric inhibitory polypeptide (GIP) receptor agonists [2], and triple-combination therapies (GLP-1, GIP, and glucagon receptor agonists) [3], have been used in clinical practice. The emergence of oral GLP-1 receptor agonists has increased convenience for patients by enhancing their compliance and glycemic control [4]. Despite these advancements, critical gaps remain in our understanding of long-term cardiovascular outcomes, patient adherence, and real-world effectiveness in relation to these novel therapies [5]. Additionally, further evaluations are required regarding the integration of technology, including artificial intelligence and digital health monitoring, into diabetes management to optimize patient outcomes [6].

This Special Issue includes many high-quality studies, several of which stand out for their significant contributions to the field. Yagi et al. [7] investigated the effect of a robot-assisted diabetes self-management monitoring system and demonstrated significant improvements in glycemic control. Their findings highlight the potential of robotic systems in supporting diabetes care professionals and enhancing patient engagement. The number of patients with diabetes continues to increase worldwide, with notable increases in Asia, the Middle East, and Africa. Economic growth in these regions has led to changes in dietary habits and reduced physical activity, contributing to a sharp increase in the prevalence of diabetes. Simultaneously, access to diabetes specialists and appropriate treatment remain insufficient in these areas, leaving many patients without adequate care [8]. The study by Yagi et al. provides valuable insights into addressing this critical issue.

Tariq et al. [9] explored the association between telomere length and aging determinants in patients with type 2 diabetes. Their study challenged the notion that telomere length is the sole marker of biological aging by revealing significant correlations between telomere length and factors such as hypertension, smoking, and stress. These insights underscore the need for multidimensional assessments of age-related diabetes.

In another important study, Graňák et al. [10] examined the role of physical activity in preventing post-transplant diabetes mellitus. Their findings suggest that regular exercise significantly improves glucose tolerance outcomes after kidney transplantation, supporting the incorporation of structured exercise programs into post-transplant care protocols.

In another key study, Vuković et al. [11] investigated the neurometabolic alterations in obesity using cerebral multivoxel magnetic resonance spectroscopy. Their study revealed significant negative correlations between obesity markers and brain metabolites involved in cognitive and emotional processing, with hyperinsulinemia emerging as a critical factor affecting neurometabolic health. These findings emphasize the need for targeted metabolic interventions.

As the research on diabetes, obesity, and metabolic diseases continues to evolve, several key areas will require further investigation. The development of personalized medical approaches will enable more tailored treatment strategies by enabling the consideration of genetic, metabolic, and behavioral factors that influence individual responses to therapy. Additionally, expanding the roles of artificial intelligence, wearable devices, and telemedicine can enhance real-time monitoring and personalized interventions. Further studies are needed to evaluate the long-term cardiovascular outcomes of GLP-1 and GIP receptor agonists beyond glycemic control. Moreover, a deeper understanding of the complex interactions among obesity, insulin resistance, and neurodegeneration is essential in developing targeted therapeutic interventions for metabolic brain dysfunction.

This Special Issue provides valuable insights into the evolving landscapes of diabetes, obesity, and metabolic diseases. The studies featured here contribute to bridging knowledge gaps and advancing clinical applications. However, continued efforts are essential to address remaining challenges and enhance patient outcomes. We extend our gratitude to all the contributing authors, reviewers, and researchers who have enriched this collection with their expertise and dedication. As this field continues to develop, we anticipate that further ground-breaking discoveries will shape the future of metabolic disease management.

Author Contributions: Conceptualization, Y.O.; writing—original draft preparation, Y.O.; writing—review and editing, K.Y. and Y.M. All authors have read and agreed to the published version of the manuscript.

Funding: This research received no external funding.

Conflicts of Interest: The authors declare no conflicts of interest.

Abbreviations

The following abbreviations are used in this manuscript:

GLP glucagon-like peptide
GIP gastric inhibitory polypeptide

References

1. Rosenstock, J.; Bain, S.C.; Gowda, A.; Jódar, E.; Liang, B.; Lingvay, I.; Nishida, T.; Trevisan, R.; Mosenzon, O. Weekly Icodec versus Daily Glargine U100 in Type 2 Diabetes without Previous Insulin. *N. Engl. J. Med.* **2023**, *389*, 297–308. [CrossRef] [PubMed]
2. Salmen, T.; Potcovaru, C.G.; Bica, I.C.; Giglio, R.V.; Patti, A.M.; Stoica, R.A.; Ciaccio, M.; El-Tanani, M.; Janež, A.; Rizzo, M.; et al. Evaluating the Impact of Novel Incretin Therapies on Cardiovascular Outcomes in Type 2 Diabetes: An Early Systematic Review. *Pharmaceuticals* **2024**, *17*, 1322. [CrossRef] [PubMed]
3. Kwon, J.; Thiara, D.; Watanabe, J.H. Oral versus subcutaneous semaglutide weight loss outcomes after two years among patients with type 2 diabetes in a real-world database. *Expert Rev. Endocrinol. Metab.* **2025**, *20*, 163–168. [CrossRef] [PubMed]

4. Jiang, Y.; Zhu, H.; Gong, F. Why does GLP-1 agonist combined with GIP and/or GCG agonist have greater weight loss effect than GLP-1 agonist alone in obese adults without type 2 diabetes? *Diabetes Obes. Metab.* **2025**, *27*, 1079–1095. [CrossRef] [PubMed]
5. An, X.; Sun, W.; Wen, Z.; Duan, L.; Zhang, Y.; Kang, X.; Ji, H.; Sun, Y.; Jiang, L.; Zhao, X.; et al. Comparison of the efficacy and safety of GLP-1 receptor agonists on cardiovascular events and risk factors: A review and network meta-analysis. *Diabetes Obes. Metab.* **2025**. Online ahead of print. [CrossRef]
6. Anwar, A.; Rana, S.; Pathak, P. Artificial intelligence in the management of metabolic disorders: A comprehensive review. *J. Endocrinol. Investig.* **2025**. Online ahead of print. [CrossRef] [PubMed]
7. Yagi, K.; Inagaki, M.; Asada, Y.; Komatsu, M.; Ogawa, F.; Horiguchi, T.; Yamaaki, N.; Shikida, M.; Origasa, H.; Nishio, S. Improved glycemic control through robot-assisted remote interview for outpatients with type 2 diabetes: A pilot study. *Medicina* **2024**, *60*, 329. [CrossRef] [PubMed]
8. Sun, H.; Saeedi, P.; Karuranga, S.; Pinkepank, M.; Ogurtsova, K.; Duncan, B.B.; Stein, C.; Basit, A.; Chan, J.C.N.; Mbanya, J.C.; et al. IDF diabetes atlas: Global, regional and country-level diabetes prevalence estimates for 2021 and projections for 2045. *Diabetes Res. Clin. Pract.* **2022**, *183*, 109119. [CrossRef] [PubMed]
9. Tariq, J.A.; Mandokhail, K.; Sajjad, N.; Hussain, A.; Javaid, H.; Rasool, A.; Sadaf, H.; Javaid, S.; Durrani, A.R. Effects of age and biological age-determining factors on telomere length in type 2 diabetes mellitus patients. *Medicina* **2024**, *60*, 698. [CrossRef] [PubMed]
10. Graňák, K.; Vnučák, M.; Beliančinová, M.; Kleinová, P.; Blichová, T.; Pytliaková, M.; Dedinská, I. Regular physical activity in the prevention of post-transplant diabetes mellitus in patients after kidney transplantation. *Medicina* **2024**, *60*, 1210. [CrossRef] [PubMed]
11. Vuković, M.; Nosek, I.; Slotboom, J.; Medić Stojanoska, M.; Kozić, D. Neurometabolic profile in obese patients: A cerebral multi-voxel magnetic resonance spectroscopy study. *Medicina* **2024**, *60*, 1880. [CrossRef] [PubMed]

Disclaimer/Publisher's Note: The statements, opinions and data contained in all publications are solely those of the individual author(s) and contributor(s) and not of MDPI and/or the editor(s). MDPI and/or the editor(s) disclaim responsibility for any injury to people or property resulting from any ideas, methods, instructions or products referred to in the content.

Article

Improved Glycemic Control through Robot-Assisted Remote Interview for Outpatients with Type 2 Diabetes: A Pilot Study

Kunimasa Yagi [1,2,*], Michiko Inagaki [3], Yuya Asada [3], Mako Komatsu [4], Fuka Ogawa [4], Tomomi Horiguchi [3], Naoto Yamaaki [5], Mikifumi Shikida [4], Hideki Origasa [6] and Shuichi Nishio [7]

1. Department of Internal Medicine, Kanazawa Medical University Hospital, Ishikawa 920-0293, Japan
2. First Department of Internal Medicine, Toyama University Hospital, Toyama 930-0152, Japan
3. Faculty of Health Sciences, Institute of Medical, Pharmaceutical and Health Sciences, Kanazawa University, Ishikawa 920-1192, Japan; ja9xbh@yahoo.co.jp (M.I.); y-asada@staff.kanazawa-u.ac.jp (Y.A.); horiguchi@mhs.mp.kanazawa-u.ac.jp (T.H.)
4. School of Informatics, Kochi University of Technology, Kochi 780-8515, Japan; komatsu.21151@gmail.com (M.K.); 240301e@ugs.kochi-tech.ac.jp (F.O.); shikida.mikifumi@kochi-tech.ac.jp (M.S.)
5. Isobe Clinic, Ishikawa 290-0511, Japan; isobe_dm_clinic@yahoo.co.jp
6. Data Science and AI Innovation Research Promotion Center, Institute of Statistical Mathematics, Shiga University, Shiga 525-0034, Japan; origasahideki@gmail.com
7. Institute for Open and Transdisciplinary Research Initiatives, Osaka University, Osaka 565-0871, Japan; nishio@botransfer.org
* Correspondence: yagikuni@icloud.com

Abstract: *Background and Objectives*: Our research group developed a robot-assisted diabetes self-management monitoring system to support Certified Diabetes Care and Education Specialists (CDCESs) in tracking the health status of patients with type 2 diabetes (T2D). This study aimed to evaluate the impact of this system on glycemic control and to identify suitable candidates for its use. *Materials and Methods*: After obtaining written informed consent from all participants with T2D, the CDCESs conducted remote interviews with the patients using RoBoHoN. All participants completed a questionnaire immediately after the experiment. HbA1c was assessed at the time of the interview and two months later, and glycemic control status was categorized as either "Adequate" or "Inadequate" based on the target HbA1c levels outlined in the guidelines for adult and elderly patients with type 2 diabetes by the Japan Diabetes Society. Patients who changed their medication regimens within the two months following the interview were excluded from the study. *Results*: The clinical characteristics of the 28 eligible patients were as follows: 67.9 ± 14.8 years old, 23 men (69%), body mass index (24.7 ± 4.9 kg/m^2), and HbA1c levels 7.16 ± 1.11% at interview and two months later. Glycemic control status (GCS) was Adequate (A) to Inadequate (I): 1 case; I to A: 7 cases; A to A good: 14 cases; I to I: 6 cases (*p*-value = 0.02862 by Chi-square test). Multiple regression analyses showed that Q1 (Did RoBoHoN speak clearly?) and Q7 (Was RoBoHoN's response natural?) significantly contributed to GCS, indicating that the naturalness of the responses did not impair the robot-assisted interviews. The results suggest that to improve the system in the future, it is more beneficial to focus on the content of the conversation rather than pursuing superficial naturalness in the responses. *Conclusions*: This study demonstrated the efficacy of a robot-assisted diabetes management system that can contribute to improved glycemic control.

Keywords: type 2 diabetes; robot; RoBoHoN; certified diabetes care and education specialists; diabetes self-management; information and communication technology

1. Introduction

The management of type 2 diabetes (T2D) involves multifaceted challenges, particularly in self-management, which includes diet, exercise, and adherence to medication [1,2]. Certified Diabetes Care and Education Specialists (CDCESs) play a pivotal role in this sense [3], employing a comprehensive approach to assess and enhance patients' knowledge, behaviors, and support systems [4]. The effectiveness of CDCES in facilitating behavioral change, enhancing glycemic control, and preventing diabetes-related complications is well established, reflecting their crucial role in improving the quality of life of individuals with T2DM [5].

However, the increasing global prevalence of diabetes, with an estimated 463 million adults living with the condition in 2019 and projections suggesting a rise to 700 million by 2045 [6], has led to a considerable gap in healthcare provision. This burgeoning epidemic is outstripping the availability of skilled CDCES, especially in Japan's super-aged society [7]. The uneven distribution of CDCES, predominantly concentrated in urban medical centers and university hospitals rather than in rural private clinics, exacerbates this shortfall. Additionally, the COVID-19 pandemic has exacerbated this situation by limiting face-to-face interactions in medical settings, thereby hindering traditional methods of diabetes education and care.

In response to these challenges, significant global investment has been directed toward developing information and communication technology (ICT) systems to support diabetes management. Our research group acknowledges that although these technological advancements are promising, the absence of human empathy and interaction in computer-mediated education frequently results in less-than-ideal patient outcomes. To address this, we developed a novel robot-assisted diabetes self-management monitoring system [8–10]. This system, designed to support CDSESs in various tasks such as information gathering and patient education, includes robots that can mimic human expressions. This feature allows for a more engaging and personable approach to providing diabetes nutritional guidance.

Introducing such expressive robots bridges the gap between technological innovation and empathetic patient care. By doing so, we revolutionize patient engagement and transform the landscape of diabetes education. This study evaluated the impact of our system on glycemic management and identified suitable candidates for this technology. Our approach addresses the shortfall in CDSES availability. It challenges the belief that significant investments in medical ICT do not necessarily result in practical medical improvements, potentially leading to considerable decreases in future healthcare expenses and improving the well-being of individuals with T2D.

2. Materials and Methods

2.1. Study Design and Ethical Issues

This is a retrospective cross-sectional observational study of a hospital-based cohort. All procedures followed the ethical standards of the responsible committee on human experimentation and the Helsinki Declaration of 1964, as well as its later amendments. The study protocol was approved by the Ethics Committee of Toyama University Hospital (IRB# R2021083). We obtained written informed consent from all participants, informing them that they could opt out at any time.

2.2. Study Population

This study analyzed data from patients with T2D to assess their glycemic control status (GCS) at the Kitano Internal Medicine Clinic (Kanazawa, Japan), between November 2021 and October 2022. All patients in the study cohort regularly attended the clinic monthly or bimonthly. Even for those attending monthly, the initial HbA1c measurement was confirmed two months after the robotic interview.

The inclusion criteria included patients with T2D who had their HbA1c levels evaluated at baseline and again two months later. The exclusion criteria were as follows: (i) type 1 diabetes, (ii) secondary diabetes, (iii) refractory malignant diseases, (iv) dependency on hemodialysis, (v) renal dysfunction with serum creatine levels over 2.5 mg/dL, (vi) symptomatic coronary artery disease or percutaneous coronary intervention within the past year, (vii) severe hepatic dysfunction (Child–Pugh score $\geq$10), and (viii) patients who had changed their prescriptions within the two months prior to the interview.

2.3. Definitions

T2D diagnoses were established based on the diagnostic criteria outlined by the American Diabetes Association and The Japan Diabetes Society (JDS), involving the presence of HbA1c levels $\geq$6.5% (National Glycohemoglobin Standardization Program), fasting blood glucose concentrations $\geq$126 mg/dL (7.0 mmol/L), random blood glucose concentrations $\geq$200 mg/dL, or the current use of medications for diabetes [11,12].

Diabetic nephropathy is defined by urinary albumin excretion $\geq$30 mg/g creatinine or an estimated glomerular filtration rate < 60 mL/min/1.73 m^2 [11,12].

GCS was classified as Adequate (A) or Inadequate (I) based on HbA1c levels according to the JDS treatment goal guidelines for adult and elderly patients with T2D, considering age and activities of daily living for the elderly [11].

2.4. Robot-Assisted Interview

The development of a robot-based remote medical interview system was motivated by the unique capabilities of robotic technology for providing diabetes care guidance. Detailed information on this innovative system can be found in a separate publication [13]. In brief, this system involves the remote operation of a robot by a CDCES to gather relevant information regarding the patient's diabetes self-management, including knowledge related to diet and exercise, behavioral patterns, and the support environment. All responses provided by the robot are preprogrammed in its control system prior to the interaction, allowing the CDCES to communicate precise questions and acknowledgments to patients with diabetes at the touch of a button. Patient responses are recorded using a camera and microphone embedded in the RoBoHoN (Sharp Corporation, Japan) [14] (Figure 1), enabling the CDCES to retrospectively analyze the conversation. This innovative approach improves the efficiency and effectiveness of remote diabetes management by incorporating robotic technology into the healthcare delivery process.

Figure 1. RoBoHoN image from https://cocorostore.sharp.co.jp/robohon/ (accessed on 1 January 2024). (In Japanese).

2.5. Evaluation of Robot-Assisted Interviews by Certified Diabetes Care and Education Specialists

Two CDCESs, Y.A. and M.I., actively participated in this research endeavor. Their combined experience as CDCESs spans 12 and 38 years, respectively, contributing valuable insights to the system's development, as previously detailed. Beyond their involvement in the system's development, these CDCES independently assessed the self-care behaviors in patients with T2DM.

2.6. Assessment of Robot-Assisted Interviews with Participating Individuals with Diabetes

The effectiveness of the robot-assisted interviews was evaluated by administering questionnaires immediately after the interviews, with the relevant questions categorized under Quality of Care (QC) (Table 1). These questionnaires aimed to assess the perceived usefulness and efficacy of the interview method, offering important perspectives on the patient experience and the overall influence of the robot-assisted approach on healthcare interactions. The specific QC items on the questionnaire addressed various aspects of the interview, highlighting its thoroughness, patient engagement, and potential areas for improvement.

Table 1. Questionnaires evaluating diabetes self-care status, categorized under Quality of Care.

Category 1: Functional quality of RoBoHoN-mediated interview	
QC-01	Did RoBoHoN speak clearly?
QC-02	Did RoBoHoN speak at an appropriate speed to be understood?
QC-03	Do you feel satisfied that you have told RoBoHoN what you wanted to say?
QC-04	Did RoBoHoN ask questions in timely and in natural?
QC-05	Were RoBoHoN's questions difficult to understand?
QC-06	Did RoBoHoN reply in a natural time?
QC-07	Was RoBoHoN's response natural?
Category 2: Impression of RoBoHoN	
QC-08	Was RoBoHoN cute?
QC-09	Did you feel afraid of RoBoHoN (scary, cold, etc.)?
QC-10	Did you feel attached to RoBoHoN while talking with him?
QC-11	Did you feel familiar with the way RoBoHoN speaks?
QC-12	Did you want to talk more with RoBoHoN?
Category 3: Advantages of RoBoHoN-mediated interviews over usual CDCESs'	
QC-13	Did you feel more comfortable talking to RoBoHoN than to a medical professional?
QC-14	Was it easier to talk about difficult things with RoBoHoN than with a medical professional?
QC-15	Did you feel more pressure of tested when talking to a RoBoHoN than when communicating with a medical professional?
QC-16	Did you feel uncomfortable discussing personal matters because of RoBoHoN's childish way of speaking?
QC-17	Did you feel that coming to the clinic was more fun?
Category 4: RoBoHoN's effect on reflection on diabetes self-care	
QC-18	Did your conversation with RoBoHoN make you reflect on your own knowledge about diabetes?
QC-19	Did your conversation with RoBoHoN give you an opportunity to reflect on your own diet and eating habits regarding diabetes?
QC-20	Did your conversation with RoBoHoN give you an opportunity to reflect on your own activity level and exercise habits regarding diabetes?
QC-21	Did your conversation with RoBoHoN cause you to reflect on medications (oral medications and insulin) in relation to diabetes?
QC-22	Did your conversation with RoBoHoN cause you to reflect over your glycemic management?

CDCES, Certified Diabetes Care and Education Specialists; QC, questionnaires categorized under Quality of Care. Each questionnaire should be answered in five degrees: strongly disagree, disagree, neutral, agree somewhat, and strongly agree.

2.7. Statistical Analysis

The sample size was calculated using Lehr's formula, aiming for a power of 80% and a significance level (α) of 0.05, resulting in a required sample size of 32 patients. Continuous variables were expressed as mean $\pm$ SD and median, and categorical variables were expressed as numbers and percentages. Continuous variables were compared using an independent samples *t*-test. A comparison of the categorical variables between the groups was performed using a chi-square test and Kruskal-Wallis test. Multivariate liner regression analysis was conducted to evaluate the predictive effect of robot usage on questionnaire outcomes, adjusting for other potential confounders. Statistical analyses were performed using JMP ver. 16.1.2000. (SAS Institute Inc., Cary, NC, USA), R 4.3.0 GUI 1.79, and R studio ver. 2023.06.0 + 421 (Boston, MA, USA) on a Macintosh computer.

3. Results

We obtained written informed consent from the 33 patients with diabetes. Three participants were unable to complete the follow-up data collection at the two-month mark. Two participants were excluded due to changes in their oral medication regimen. Consequently, the analysis incorporated data from 28 participants, and their clinical characteristics are summarized in Table 1. In summary, the cohort had a mean age of 69.5 $\pm$ 12.7 years, included 20 males (71%), and had a mean body mass index (BMI) of 24.9 $\pm$ 4.9 (Table 2). Pharmacological treatments for diabetes were as follows: three cases with insulin injections, seven cases with GLP-1 analog injections, seven cases with biguanides, ten with insulin secretagogue, eighteen with SGLT2 inhibitors, sixteen with DPP4 inhibitors, and three with thiazolidine, and three with imeglimin. Throughout the study, all patients demonstrated sufficient compliance with the interview process.

Table 2. Baseline characteristics of the participants.

Characteristics	Variables
Number of subjects	28
Age, years old	69.5 $\pm$ 12.7, 70.5 [63.75, 79.25]
Gender (Male/Female)	20/8
HbA1c (0M), %	7.06 $\pm$ 0.69, 7.1 [6.75, 7.45]
HbA1c (2M), %	6.86 $\pm$ 0.62, 6.9 [6.675, 7.125]
ΔHbA1c, %	-0.19 ± 0.29, -0.2 [-0.4, 0]
Glycemic control (0M) Adequate/Inadequate	15/13
Glycemic control (2M) Adequate/Inadequate	21/7
BMI, Kg/m^2	24.9 $\pm$ 4.9, 24.2 [22.25, 27.4]
Number of diabetes medications	2.1 $\pm$ 1.1, 2 [1,3]
Subjects with diabetic complications	5
Treated with insulin injections	3 (11%)
Treated with GLP-1 analog injections	7 (25%)
Treated with biguanide	7 (25%)
Treated with sulfonylureas and glinides	10 (36%)
Treated with SGLT2 inhibitors	18 (64%)
Treated with DPP-4 inhibitors	16 (57%)
Treated with thiazolidine	3 (11%)
Treated with imeglimin	3 (11%)
Grade of diabetes care behavior (6–18)	9.50 [8.00, 12.00]

Table 2. *Cont.*

Characteristics	Variables
Understanding of diabetes therapy (3–9)	4.00 [3.75, 5.00]
Acceptance of diabetes (3–6)	4.00 [4.00, 5.00]
Effectiveness of Robot interview (2–4)	4.00 [4.00, 4.00]
QC-01 (1–5)	5.00 [4.00, 5.00]
QC-02 (1–5)	5.00 [4.00, 5.00]
QC-03 (1–5)	4.00 [4.00, 5.00]
QC-04 (1–5)	4.00 [3.75, 5.00]
QC-05 (1–5)	3.00 [3.00, 4.00]
QC-06 (1–5)	4.00 [3.00, 5.00]
QC-07 (1–5)	4.00 [3.00, 5.00]
QC-08 (1–5)	4.00 [3.75, 5.00]
QC-09 (1–5)	5.00 [4.00, 5.00]
QC-10 (1–5)	3.00 [3.00, 4.00]
QC-11 (1–5)	4.00 [3.00, 5.00]
QC-12 (1–5)	3.00 [3.00, 4.00]
QC-13 (1–5)	3.00 [3.00, 4.00]
QC-14 (1–5)	3.00 [3.00, 4.00]
QC-15 (1–5)	4.00 [3.00, 4.25]
QC-16 (1–5)	4.00 [3.00, 5.00]
QC-17 (1–5)	3.00 [2.75, 3.00]
QC-18 (1–5)	4.00 [3.00, 4.00]
QC-19 (1–5)	4.00 [3.00, 5.00]
QC-20 (1–5)	4.00 [3.00, 4.25]
QC-21 (1–5)	3.00 [3.00, 4.00]
QC-22 (1–5)	4.00 [3.00, 5.00]

BMI, body mass index; GLP-1, glucagon-like peptide 1; M, months; QC, Questionnaires categorized under Quality of Care. Clinical features as age, HbA1c, BMI, and the number of diabetes medications were expressed in mean ± S.D., and median. The scores of diabetes self-care status, and QC01 to QC-22 were expressed as median [interquartile range].

All patients demonstrated sufficient acceptance of the system during the completion of the RoBoHoN-assisted interview. At the time of the interview, the mean HbA1c level was 7.06 ± 0.69%; two months later, it was 6.86 ± 0.62%. Notably, after two months, participants who underwent RoBoHoN interviews exhibited a 0.19 ± 0.29 percentage point reduction in HbA1c. The distribution of changes in GCS was as follows: A to I: 1 case; I to A: 7 cases; A to A: X cases; I to I: X cases (ChiSq $p < 0.0001$) (Table 3). Fisher's exact test indicated a significant difference in status, with seven instances improving from I to A and one case deteriorating from A to I ($p = 0.02862$).

Table 3. Change in glycemic management status two months after RoBoHoN-mediated interview.

	Adequate GMS after 2 Months	Inadequate GMS after 2 Months
Adequate GMS at baseline	14	1
Inadequate GMS at baseline	7	6

Odds ratio 10.9 (p-value = 0.029 by Fisher exact test). GMS, glycemic management status.

Because the I to A (Adequate to Inadequate) group consisted of only one case, it was combined with the I to I (Inadequate to Inadequate) group, which had six cases. Then, among the three groups of this A to I + I to I, I to A (Inadequate to Inadequate), and A to A (Adequate to Adequate), a Kruskal–Wallis test revealed significant factors such as HbA1c (0M), HbA1c (2M), ΔHbA1c, QC01 (Did RoBoHoN speak clearly?), QC07 (Was RoBoHoN's response natural?), QC15 (Did you feel more pressure when tested during a conversation with a robot phone than when communicating with a healthcare provider?), and QC16 (Did you feel uncomfortable discussing personal matters because of RoBoHoN's childish way of speaking?) (Table 4). Figure 2 shows the distribution of each factor across the four groups of HbA1c at baseline and two months post-intervention and its changes, QC01, QC07, QC15, and QC16.

Table 4. Kruskal–Wallis analyses among 3 groups over two months of change in glycemic management status.

	Adequate to Adequate	Inadequate to Adequate	Inadequate and Adequate to Inadequate	Kruskal–Wallis Analyses among 3 Groups (p-Value)
Number of subjects	14	7	7	
Age, years old	70.50 [62.00, 79.25]	71.00 [64.00, 73.50]	69.00 [65.50, 79.50]	0.958
Gender (M/F)	10/4	6/1	4/3	0.497
HbA1c (0M), %	6.70 [6.45, 7.05]	7.20 [7.10, 7.65]	7.20 [7.15, 7.45]	0.022
HbA1c (2M), %	6.70 [6.32, 6.97]	6.90 [6.80, 7.05]	7.10 [7.10, 7.30]	0.020
ΔHbA1c, %	−0.15 [−0.27, −0.03]	−0.50 [−0.55, −0.30]	0.00 [−0.15, 0.20]	0.016
BMI	24.25 [23.10, 26.65]	26.20 [23.25, 28.95]	22.30 [20.75, 23.85]	0.286
Number of diabetes medications	2.00 [1.00, 2.75]	2.00 [2.00, 3.00]	2.00 [1.00, 2.50]	0.536
Subjects with diabetic complications	4	0	1	0.262
Grade of diabetes care behavior (6–18)	10.00 [7.50, 11.75]	10.00 [9.00, 12.50]	8.00 [8.00, 9.00]	0.264
Understanding of diabetes therapy (3–9)	4.00 [3.00, 4.00]	4.00 [4.00, 5.50]	4.00 [4.00, 4.50]	0.440
Acceptance of diabetes (3–6)	4.00 [4.00, 5.00]	5.00 [4.00, 5.00]	4.00 [3.50, 4.00]	0.104
Effectiveness of Robot interview (2–4)	4.00 [4.00, 4.00]	4.00 [4.00, 4.00]	4.00 [3.50, 4.00]	0.357
QC-01 (1–5)	5.00 [4.00, 5.00]	4.00 [3.50, 4.50]	5.00 [5.00, 5.00]	0.024
QC-02 (1–5)	5.00 [4.00, 5.00]	4.50 [4.00, 5.00]	5.00 [5.00, 5.00]	0.166
QC-03 (1–5)	4.50 [4.00, 5.00]	4.00 [3.00, 4.00]	5.00 [4.50, 5.00]	0.057
QC-04 (1–5)	4.00 [4.00, 4.00]	3.50 [3.00, 4.50]	5.00 [4.50, 5.00]	0.120
QC-05 (1–5)	3.50 [3.00, 4.00]	3.00 [3.00, 3.00]	4.00 [3.50, 4.50]	0.333
QC-06 (1–5)	4.00 [3.00, 4.75]	3.00 [3.00, 4.00]	4.00 [4.00, 5.00]	0.260
QC-07 (1–5)	4.00 [3.00, 4.75]	3.00 [3.00, 4.00]	5.00 [4.00, 5.00]	0.022
QC-08 (1–5)	4.00 [3.00, 5.00]	4.50 [4.00, 5.00]	5.00 [4.00, 5.00]	0.400
QC-09 (1–5)	4.50 [4.00, 5.00]	4.00 [3.50, 5.00]	5.00 [4.50, 5.00]	0.404
QC-10 (1–5)	3.00 [2.25, 4.00]	3.00 [3.00, 3.00]	3.00 [3.00, 5.00]	0.355
QC-11 (1–5)	4.00 [3.00, 4.75]	4.00 [3.00, 4.00]	4.00 [4.00, 5.00]	0.209
QC-12 (1–5)	3.00 [3.00, 3.75]	3.00 [3.00, 3.50]	3.00 [2.50, 4.00]	0.947
QC-13 (1–5)	3.00 [3.00, 3.00]	3.00 [3.00, 3.50]	3.00 [2.50, 4.00]	0.973
QC-14 (1–5)	3.00 [3.00, 3.75]	3.00 [3.00, 4.00]	3.00 [2.50, 4.00]	0.612
QC-15 (1–5)	4.00 [3.00, 4.75]	3.00 [3.00, 4.00]	4.00 [4.00, 5.00]	0.041
QC-16 (1–5)	3.50 [3.00, 4.00]	3.00 [3.00, 4.50]	5.00 [4.50, 5.00]	0.037

Table 4. Cont.

	Adequate to Adequate	Inadequate to Adequate	Inadequate and Adequate to Inadequate	Kruskal–Wallis Analyses among 3 Groups (p-Value)
QC-17 (1–5)	3.00 [2.25, 3.00]	3.00 [3.00, 3.00]	3.00 [2.00, 3.00]	0.506
QC-18 (1–5)	4.00 [3.00, 4.00]	4.00 [3.50, 4.00]	4.00 [4.00, 4.50]	0.357
QC-19 (1–5)	3.50 [3.00, 4.00]	4.00 [3.00, 4.50]	4.00 [3.50, 5.00]	0.735
QC-20 (1–5)	3.00 [3.00, 4.00]	4.00 [3.50, 4.00]	5.00 [4.00, 5.00]	0.138
QC-21 (1–5)	3.00 [3.00, 4.00]	3.00 [3.00, 4.00]	4.00 [3.00, 4.50]	0.740
QC-22 (1–5)	3.50 [3.00, 4.00]	4.00 [4.00, 5.00]	4.00 [4.00, 5.00]	0.126

BMI, body mass index; M, months; QC, questionnaires categorized under Quality of Care. Clinical features as age, HbA1c, BMI, and the number of diabetes medications were expressed in mean ± S.D. The scores of diabetes self-care status, and QC01 to QC-22 were expressed as median [interquartile range].

By further exploring potential contributing factors, we have decided to retain QC01 and QC07 as significant predictors in the model (Table 5).

Table 5. Multiple linear regression to the change in glycemic management status.

Independent Variables	Coefficient	SE	p-Value
HbA1c (0M), %	−0.05219	0.18283	0.7780
QC01	−0.36472	0.15504	0.0280
QC07	−0.35337	0.16189	0.0400
QC15	−0.28527	0.20757	0.1832
QC16	0.03855	0.14311	0.7901

A multivariate regression analysis was conducted with the four categories in the change in glycemic management status (GMS) (Inadequate to Adequate (I to A), Adequate to Adequate (A to A), Inadequate to Inadequate (I to I), Adequate to Inadequate (A to I)) as the dependent variable and assigned scores (I to A—4 points; A to A—3 points; I to I—2 points; A to I—1 point) along with HbA1c, QC01, QC07, QC15, and QC16 as independent variables.

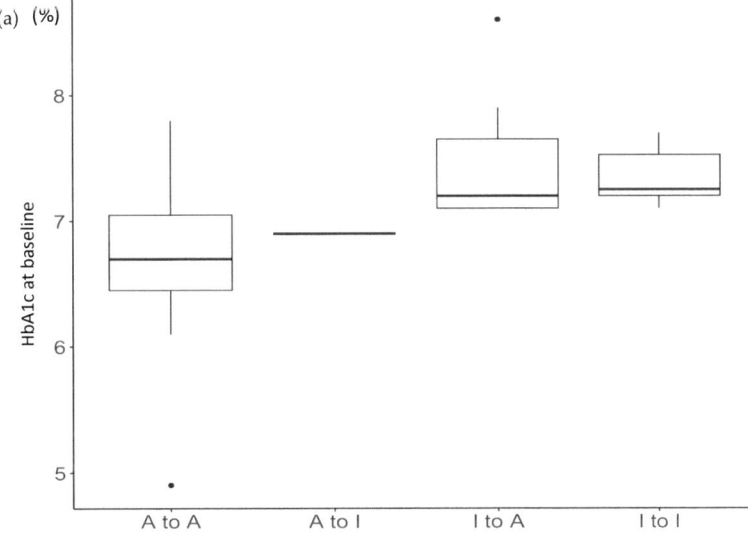

Figure 2. Cont.

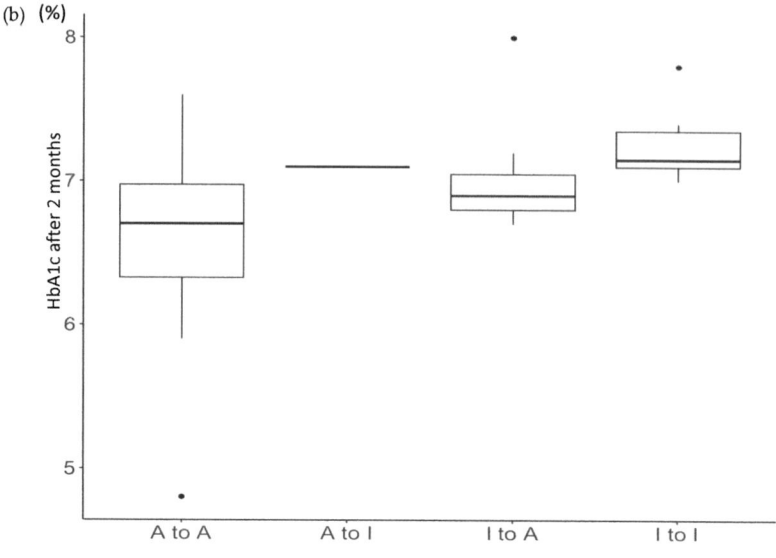

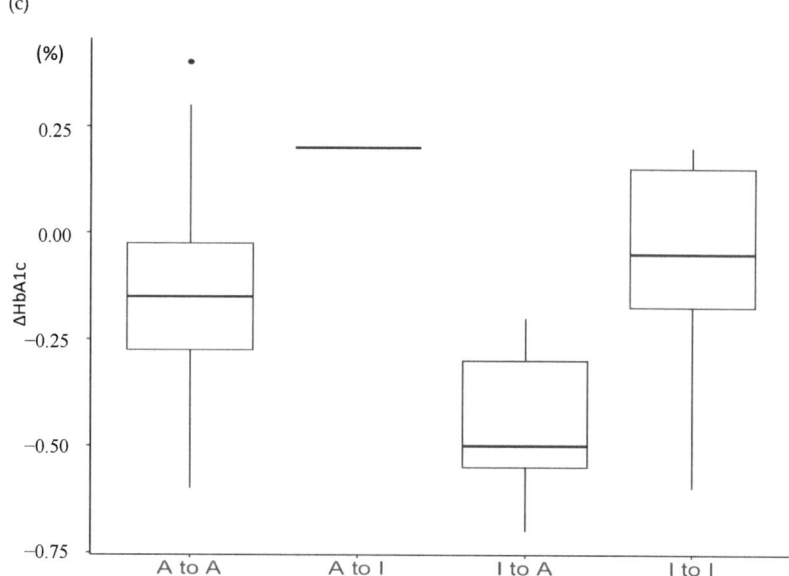

Figure 2. Cont.

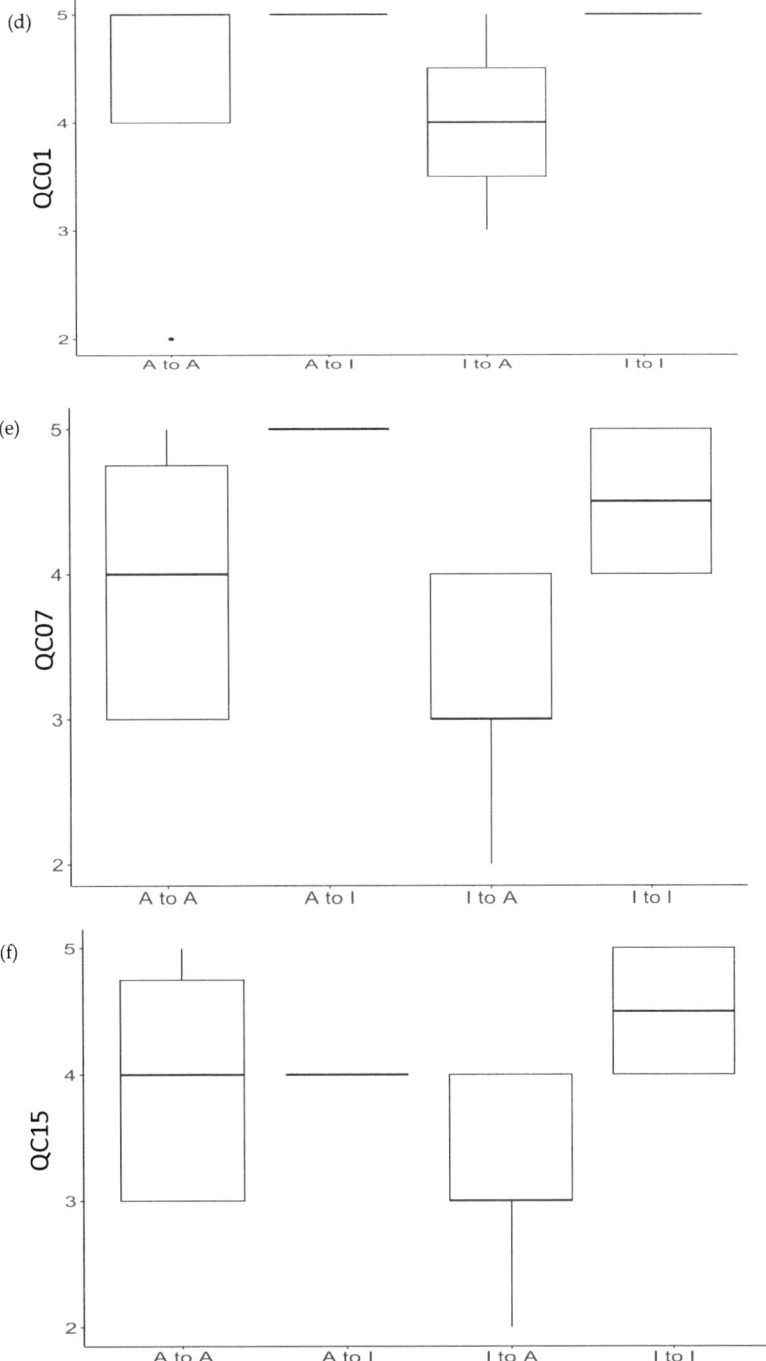

Figure 2. *Cont.*

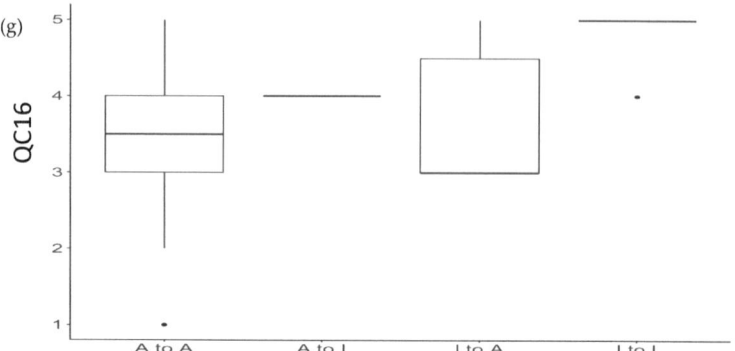

Figure 2. Relationship between the change in glycemic control and the following parameters: (**a**) baseline HbA1c; (**b**) HbA1c two months post-intervention with the RoBoHoN-mediated interview; (**c**) change in HbA1c from baseline to two months post-intervention; (**d**) QC01; (**e**) QC07; (**f**) QC15; (**g**) QC16.

4. Discussion

Despite the absence of any deliberate interventions, instances where the GCS category improved outnumbered those where it deteriorated. When classifying groups based on the presence or absence of improvement, significant differences in response rates were observed for Question 1 (Did RoBoHoN speak clearly?) and Question 7 (Was RoBoHoN's response natural?). Paradoxically, participants who experienced difficulties in understanding the robot's speech and perceived its reactions as unnatural demonstrated improvement in the GCS category.

The participants in this study, with a mean age of 69.5 years, mean BMI of 24.9, and mean HbA1c of 7.06%, closely resemble the demographic characteristics reported in the 2022 JDS JDDM study (mean age of approximately 67.71 years, mean BMI of 24.74, mean HbA1c 7.14%) [15]. This comparison aligns our study population with the broader context of diabetes management in the Japanese population.

The JDS has recently recommended specific glycemic control targets for the elderly, considering factors such as age, functional independence, and medication regimens [11,16]. This approach recognizes that the target HbA1c levels can differ depending on individual circumstances. According to these guidelines, both avoiding hypoglycemia and reducing HbA1c levels were prioritized, with target ranges set for HbA1c, indicating that drastic reductions in HbA1c levels are undesirable. In the cases presented here, none fell below the minimum threshold.

In this study, the robots were strictly limited to data collection without any form of intervention, such as providing medical advice or educational instruction. Although standard practices during routine medical consultations by healthcare professionals often include similar questions, these interactions can become monotonous and may not adequately increase the awareness of individuals managing diabetes. It is presumed that individuals were given the opportunity to reflect on factors such as medication adherence, and eating habits between meals through interactions with robotic inquiries.

Previous research assessing interactions between robots and the elderly has indirectly identified positive impacts [17–20]. Even in interactions with elderly individuals, there was no proactive intervention; instead, the approach remained reactive. In this context, structured inquiries by the robot may have triggered increased awareness among patients, offering a new perspective on the potential benefits of robotic engagement in healthcare environments.

An intriguing aspect of our findings is the responses to questionnaire items QC01 and QC07. Notably, individuals who initially found it challenging to comprehend the robot's speech showed improvement, as did those who initially perceived the robot's responses as unnatural. Equally noteworthy is the lack of significant differences among the three groups regarding retrospective questions (Q18–20) pertaining to reflection. Contrary to conventional expectations, the observed glycemic improvements were not associated with the reflection effects in this study.

Considering QC15 and QC16 together, it is reasonable to hypothesize that individuals who engaged in the interaction with genuine intent to "listen as they would to a human" have experienced improvement. Interestingly, patients who interacted with the robot as if it were a human, despite potential discomfort, seemed to manage the experience smoothly. These findings indicate that instead of prioritizing the superficial naturalness of responses, future system enhancements may benefit more from refining the content and direction of the conversation.

This study has several limitations due to its single-center design and private clinic setting, which inherently limit the generalizability of the findings. However, this study concentrates on examining the potential benefits and practical applications of the robot, acknowledging that although there are limitations, the current dataset offers valuable insights into its effectiveness in a specific healthcare environment.

5. Conclusions

This study indicates that robot-assisted remote diabetes education enhances the support provided by CDCESs to outpatients with T2D, and this approach is applicable to the elderly living in rural regions of Japan. However, it should be tested more extensively.

Author Contributions: K.Y.: patient management. M.I., Y.A. and T.H.: Contributions and comments as expert CDCESs. M.K., F.O., M.S. and S.N.: Contributions as experts in ICT and Robots. K.Y. and N.Y.: clinical reviews and as diabetologists. H.O.: statistical discussions. KY.: data interpretation and writing. All authors have read and agreed to the published version of the manuscript.

Funding: This work was supported by JSPS KAKENHI [Grant Numbers JP18K08505 and JP21K10300]. The sponsor of the study was not involved in the design, data collection, analysis, interpretation of data, report writing, or the decision to submit the paper for publication.

Institutional Review Board Statement: The study was conducted in accordance with the Declaration of Helsinki, and approved by the Ethics Committee of Toyama University Hospital (IRB# R2021083, date of approval as 21 September 2021).

Informed Consent Statement: Informed written consent was obtained from all participants involved in the study. Participants were notified that they could opt out at any time.

Data Availability Statement: The datasets used and analyzed in this study are available from the corresponding author upon reasonable request.

Acknowledgments: The authors thank Enago (Osaka, Japan; www.enago.jp accessed on on 1 January 2024) for editing the English language.

Conflicts of Interest: The authors declare no conflicts of interest.

References

1. Kudoh, R.; Shibayama, T.; Hidaka, K. Nurses' knowledge, attitude, and practice regarding oral management for outpatients with type 2 diabetes: A national survey on certified diabetes educators. *Diabetol. Int.* **2022**, *13*, 407–420. [CrossRef] [PubMed]
2. Powers, M.A.; Bardsley, J.K.; Cypress, M.; Funnell, M.M.; Harms, D.; Hess-Fischl, A.; Hooks, B.; Isaacs, D.; Mandel, E.D.; Maryniuk, M.D.; et al. Diabetes Self-management Education and Support in Adults with Type 2 Diabetes: A Consensus Report of the American Diabetes Association, the Association of Diabetes Care and Education Specialists, the Academy of Nutrition and Dietetics, the American Academy of Family Physicians, the American Academy of PAs, the American Association of Nurse Practitioners, and the American Pharmacists Association. *J. Am. Assoc. Nurse Pract.* **2020**, *33*, 1314–1331. [PubMed]
3. Rodriguez, K.; Ryan, D.; Dickinson, J.K.; Phan, V. Improving Quality Outcomes: The Value of Diabetes Care and Education Specialists. *Clin. Diabetes* **2022**, *40*, 356–365. [CrossRef] [PubMed]
4. Horiguchi, T.; Inagaki, M.; Tasaki, K. The Self-Care Behaviors of Adults With Type 2 Diabetes Within 10 Years After Diagnosis: Relationship Between Self-Care Behaviors, Knowledge and Education. *J. Jpn. Soc. Nurs. Res.* **2021**, *44*, 613–622.

5. Kavookjian, J.; Bzowyckyj, A.S.; DiNardo, M.M.; Kocurek, B.; Kolb, L.E.; Noe, D.; Ryan, D.; Saunders, M.M.; See, M.; Uelmen, S. Current and Emerging Trends in Diabetes Care and Education: 2021 National Practice and Workforce Survey. *Sci. Diabetes Self Manag. Care* **2022**, *48*, 307–323. [CrossRef] [PubMed]
6. Saeedi, P.; Petersohn, I.; Salpea, P.; Malanda, B.; Karuranga, S.; Unwin, N.; Colagiuri, S.; Guariguata, L.; Motala, A.A.; Ogurtsova, K.; et al. Global and regional diabetes prevalence estimates for 2019 and projections for 2030 and 2045: Results from the International Diabetes Federation Diabetes Atlas, 9th ed. *Diabetes Res. Clin. Pract.* **2019**, *157*, 107843. [CrossRef] [PubMed]
7. Iijima, K.; Arai, H.; Akishita, M.; Endo, T.; Ogasawara, K.; Kashihara, N.; Hayashi, Y.K.; Yumura, W.; Yokode, M.; Ouchi, Y. Toward the development of a vibrant, super-aged society: The future of medicine and society in Japan. *Geriatr. Gerontol. Int.* **2021**, *21*, 601–613. [CrossRef] [PubMed]
8. Blanson Henkemans, O.A.; Bierman, B.P.; Janssen, J.; Neerincx, M.A.; Looije, R.; van der Bosch, H.; van der Giessen, J.A. Using a robot to personalise health education for children with diabetes type 1: A pilot study. *Patient Educ. Couns.* **2013**, *92*, 174–181. [CrossRef] [PubMed]
9. Henkemans, O.A.B.; Bierman, B.P.B.; Janssen, J.; Looije, R.; Neerincx, M.A.; van Dooren, M.M.M.; de Vries, J.L.E.; van der Burg, G.J.; Huisman, S.D. Design and evaluation of a personal robot playing a self-management education game with children with diabetes type 1. *Int. J. Hum. Comput. Stud.* **2017**, *106*, 63–76. [CrossRef]
10. Lau, Y.; Chee, D.G.H.; Chow, X.P.; Wong, S.H.; Cheng, L.J.; Lau, S.T. Humanoid robot-assisted interventions among children with diabetes: A systematic scoping review. *Int. J. Nurs. Stud.* **2020**, *111*, 103749. [CrossRef] [PubMed]
11. Araki, E.; Goto, A.; Kondo, T.; Noda, M.; Noto, H.; Origasa, H.; Osawa, H.; Taguchi, A.; Tanizawa, Y.; Tobe, K.; et al. Japanese Clinical Practice Guideline for Diabetes 2019. *Diabetol Int.* **2020**, *11*, 165–223. [CrossRef] [PubMed]
12. American Diabetes Association Professional Practice, C. 2. Diagnosis and Classification of Diabetes: Standards of Care in Diabetes-2024. *Diabetes Care* **2024**, *47*, S20–S42. [CrossRef] [PubMed]
13. Maalouly, E.; Hirano, T.; Yamazaki, R.; Nishio, S.; Ishiguro, H. Encouraging prosocial behavior from older adults through robot teleoperation: A feasibility study. *Front. Comput. Sci.* **2023**, *5*, 1157925. [CrossRef]
14. Available online: https://cocorostore.sharp.co.jp/robohon/ (accessed on 1 January 2024). (In Japanese).
15. Japan Diabetes Clinical Data Management Study Group (JDDM). 2023. Available online: http://jddm.jp/public-information/index-2022/#data_03 (accessed on 1 January 2024).
16. Available online: https://www.jpn-geriat-soc.or.jp/publications/other/diabetes_treatment_guideline.html (accessed on 1 January 2024). (In Japanese)
17. Fan, J.; Ullal, A.; Beuscher, L.; Mion, L.C.; Newhouse, P.; Sarkar, N. Field Testing of Ro-Tri, a Robot-Mediated Triadic Interaction for Older Adults. *Int. J. Soc. Robot.* **2021**, *13*, 1711–1727. [CrossRef] [PubMed]
18. Damholdt, M.F.; Norskov, M.; Yamazaki, R.; Hakli, R.; Hansen, C.V.; Vestergaard, C.; Seibt, J. Attitudinal Change in Elderly Citizens Toward Social Robots: The Role of Personality Traits and Beliefs About Robot Functionality. *Front. Psychol.* **2015**, *6*, 1701. [CrossRef] [PubMed]
19. Nishio, T.; Yoshikawa, Y.; Sakai, K.; Iio, T.; Chiba, M.; Asami, T.; Isoda, Y.; Ishiguro, H. The Effects of Physically Embodied Multiple Conversation Robots on the Elderly. *Front. Robot. AI* **2021**, *8*, 633045. [CrossRef] [PubMed]
20. Fasola, J.; Mataric, M.J. Using Socially Assistive Human–Robot Interaction to Motivate Physical Exercise for Older Adults. *Proc. IEEE* **2012**, *100*, 2512–2526. [CrossRef]

Disclaimer/Publisher's Note: The statements, opinions and data contained in all publications are solely those of the individual author(s) and contributor(s) and not of MDPI and/or the editor(s). MDPI and/or the editor(s) disclaim responsibility for any injury to people or property resulting from any ideas, methods, instructions or products referred to in the content.

Article

Effects of Switching from Degludec to Glargine U300 in Patients with Insulin-Dependent Type 1 Diabetes: A Retrospective Study

Toshitaka Sawamura [1,2,3,4], Shigehiro Karashima [4,*], Azusa Ohbatake [2,4], Takuya Higashitani [2,3,4], Ai Ohmori [1,2], Kei Sawada [1,4], Rika Yamamoto [1,4], Mitsuhiro Kometani [1,2,4], Yuko Katsuda [2,3] and Takashi Yoneda [2,3,4]

1. Department of Internal Medicine, Asanogawa General Hospital, 83 Kosakamachi, Kanazawa 910-8621, Japan; sawaa4211@gmail.com (T.S.); cats2ai@yahoo.co.jp (A.O.); iekadawas0212@yahoo.co.jp (K.S.); kpmmsk@yahoo.co.jp (R.Y.); kometankomekome@yahoo.co.jp (M.K.)
2. Department of Endocrinology and Metabolism, Kanazawa University Graduate School of Medicine, 13-1 Takaramachi, Kanazawa 920-8641, Japan; azusa_k_23@yahoo.co.jp (A.O.); popfunfun@yahoo.co.jp (T.H.); yukatsudayk203@yahoo.co.jp (Y.K.); endocrin@med.kanazawa-u.ac.jp (T.Y.)
3. Department of Diabetes and Endocrinology and Internal Medicine, Fukui Prefectural Hospital, 2-8-1 Yotsui, Fukui 910-8526, Japan
4. Department of Health Promotion and Medicine of the Future, Kanazawa University, 13-1 Takaramachi, Kanazawa 920-8641, Japan
* Correspondence: skarashima@staff.kanazawa-u.ac.jp; Tel.: +81-76-265-2778

Abstract: *Background and Objectives:* Degludec (Deg) and glargine U300 (Gla-300) are insulin analogs with longer and smoother pharmacodynamic action than glargine U100 (Gla-100), a long-acting insulin that has been widely used for many years in type 1 and type 2 diabetes. Both improve glycemic variability (GV) and the frequency of hypoglycemia, unlike Gla-100. However, it is unclear which insulin analog affects GV and hypoglycemia better in patients with insulin-dependent type 1 diabetes. We evaluated the effects of switching from Deg to Gla-300 on the day-to-day GV and the frequency of hypoglycemia in patients with insulin-dependent type 1 diabetes treated with Deg-containing basal-bolus insulin therapy (BBT). *Materials and Methods:* We conducted a retrospective study on 24 patients with insulin-dependent type 1 diabetes whose treatment was switched from Deg-containing BBT to Gla-300-containing BBT. We evaluated the day-to-day GV measured as the standard deviation of fasting blood glucose levels (SD-FBG) calculated by the self-monitoring of blood glucose records, the frequency of hypoglycemia (total, severe, and nocturnal), and blood glucose levels measured as fasting plasma glucose (FPG) levels and hemoglobin A1c (HbA1c). *Results:* The characteristics of the patients included in the analysis with high SD-FBG had frequent hypoglycemic events, despite the use of Deg-containing BBT. For this population, SD-FBG and the frequency of nocturnal hypoglycemia decreased after the switch from Deg to Gla-300. Despite the decrease in the frequency of nocturnal hypoglycemia, the FPG and HbA1c did not worsen by the switch. The change in the SD-FBG had a negative correlation with the SD-FBG at baseline and a positive correlation with serum albumin levels. *Conclusions:* Switching from Deg to Gla-300 improved the SD-FBG and decreased the frequency of nocturnal hypoglycemia in insulin-dependent type 1 diabetes treated with Deg-containing BBT, especially in cases with low serum albumin levels and a high GV.

Keywords: degludec; glargine U300; glycemic variability; hypoglycemia; type 1 diabetes

1. Introduction

The goal of diabetes treatment is to prevent diabetic complications and achieve a long healthy life span. Strict blood glucose control is important to prevent micro- and macro-vascular diabetic complications in patients with diabetes [1,2] However, the Action to Control Cardiovascular Risk in Diabetes (ACCORD) trial, in which the efficiency of the normalization of blood glucose levels via insulin or sulfonylurea was evaluated, revealed that mortality increased through the normalization of hemoglobin A1c (HbA1c) levels via insulin or sulfonylurea [3]. The additional analysis of the ACCORD study revealed that symptomatic, severe hypoglycemia was associated with an increased risk of death [4]. Following this report, more attention has been paid to hypoglycemia. Various studies have shown that overly intensive control induces hypoglycemia and increases the risk of cardiovascular events and mortality [5,6]. Moreover, it has also been noted that hypoglycemia increases the incidence of dementia [7] and the rate of bone fractures [8]. For this reason, avoiding hypoglycemia is one of the most important tasks in diabetes care. In addition to hypoglycemia, the concept of glycemic variability (GV) as a quality of glycemic control has also received attention. GV encompasses various types of variability, including diurnal variability, day-to-day variation, and even seasonal variability. All of these have been reported to correlate with diabetic complications [9]. Among several GV evaluations, the day-to-day GV is a marker that reflects short-to-intermediate GV. A high day-to-day GV is reportedly related to diabetic complications [10,11]. This is explained by the mechanism of exacerbation of oxidative stress due to high day-to-day GV [12]. Moreover, patients with high GV often experience hypoglycemia [13], and hypoglycemia itself could increase the risk of cardiovascular events and mortality [4–6]. Sakamoto et al. reported the factors affecting GV. In the short-term and intermediate GV, beta-cell dysfunction, genetic factors, and insulin resistance have strong contributions. Habitual practice has a major impact on long-term GV. Diet, activity, stress, the effect of medications, and adherence to medications affect all types of GV [9].

As severe beta-cell dysfunction leads to a worsening short-to-intermediate GV, patients with insulin-dependent type 1 diabetes have high day-to-day GV and experience more frequent hypoglycemic events compared with insulin-independent type 2 diabetes [14]. For patients with insulin-dependent type 1 diabetes, subcutaneous multiple insulin injection therapy often used. However, some patients suffer from high day-to-day GV and hypoglycemia, despite the use of subcutaneous multiple insulin injection therapy. Recently, a hybrid closed-loop insulin delivery system could be available in type 1 diabetes. This system uses various combinations of control algorithms, glucose sensors, and insulin pumps. A hybrid closed-loop insulin delivery system could achieve increased time in target, and reductions in HbA1c, hyperglycemia, and hypoglycemia, compared to that with an insulin pump [15]. However, there are some barriers to the use of a hybrid closed-loop insulin delivery system, such as the high financial burden and the difficulty of using it with elderly people and patients with dementia due to the need to handle the machine. Therefore, many patients with insulin-dependent type 1 diabetes are treated with subcutaneous multiple insulin injection therapy.

For the stabilization of blood glucose levels in patients with insulin-dependent type 1 diabetes treated with basal-bolus insulin therapy (BBT), basal insulin has an important role. Insulin glargine U100 (Gla-100) is a long-acting insulin widely used in both type 1 and type 2 diabetes. However, insulin glargine U300 (Gla-300) and insulin degludec (Deg), which have prolonged pharmacodynamic action, can be used recently. Gla-300 achieves a smoother day-to-day GV and a decreased frequency of hypoglycemia compared with Gla-100 in patients with type 1 [16,17] and type 2 [18,19] diabetes. Similarly, Deg also achieves a smoother day-to-day GV and decreases the frequency of hypoglycemia compared with Gla-100 for patients with type 1 [20,21] and type 2 [22,23] diabetes. As Deg has a longer pharmacodynamic action profile than Gla-300, Deg tends to be more commonly used for patients with type 1 diabetes. However, a high day-to-day GV and frequent hypoglycemia in type 1 diabetes can often be observed, despite the use of Deg [14].

The previous reports on type 2 diabetes have shown that Gla-300 can decrease the frequency of hypoglycemia compared with Deg [24,25]. Especially, in patients with low serum albumin levels, Gla-300 could achieve a lower GV and decreased frequency of hypoglycemia than Deg. However, the data about the comparison between these two insulin analogs in type 1 diabetes is limited. We hypothesized that switching from Deg to Gla-300 may improve GV and decrease hypoglycemic events in patients with type

1 diabetes and that there is a group of patients who benefit more from Gla-300 than Deg. The standard deviation of fasting blood glucose levels (SD-FBG), calculated by the self-monitoring of blood glucose (SMBG) records, is used as the marker of day-to-day GV [26]. Here, we evaluated the efficiency of switching from Deg to Gla-300 on the SD-FBG calculated by the SMBG records and the frequency of hypoglycemic events in patients with insulin-dependent type 1 diabetes.

2. Materials and Methods

2.1. Patients

We investigated the outpatients who attended the Endocrinology and Diabetic unit of Fukui Prefectural Hospital and Asanogawa General Hospital from April 2017 to December 2022. The eligible patients were male and female patients including the following: (1) patients with insulin-dependent type 1 diabetes, (2) patients who had been treated with Deg-containing BBT and Deg was switched to Gla-300 for various reasons, and (3) patients who had been instructed to perform SMBG four times/day. Among these patients, we excluded patients as follows: (1) patients who had changed antidiabetic agents or received new nutritional guidance during the observation period, (2) patients who were newly introduced to flash glucose monitoring (FGM) during the observation period, and (3) patients whose HbA1c levels and SMBG records could not be obtained within 1 month and 4 to 6 months after the insulin switch.

2.2. Measurement

The primary endpoint of this study is the change in SD-FBG calculated by SMBG records. SD-FBG was calculated from the 30 records measured before breakfast. The secondary endpoints are the change in the following items: body weight (BW), body mass index (BMI), fasting plasma glucose (FPG), HbA1c, serum creatinine (Cr), estimated glomerular filtration rate (eGFR), serum albumin (Alb), frequency of hypoglycemia, frequency of severe hypoglycemia, frequency of nocturnal hypoglycemia, and a dosage of basal and fasting insulins. Moreover, time above range (TAR) (>180 mg/dL), time in range (TIR) (70 to 180 mg/dL), and time below range (TBL) (<70 mg/dL) were evaluated only in patients with FGM.

The first evaluation points were within 1 month of the insulin switch, and the second evaluation points were 4 to 6 months after the insulin switch. SD-FBG was calculated from the records of SMBG in the previous 30 times. Hypoglycemia is defined as a blood glucose level below 70 mg/dL or having hypoglycemic symptoms. Severe hypoglycemia is defined as a blood glucose level below 54 mg/dL or hypoglycemia that requires treatment assistance from another person. Nocturnal hypoglycemia was defined as hypoglycemia occurring from 00:00 h until the next breakfast.

2.3. Ethics Conduct

This study utilized a retrospective design and was approved by the ethics committee at Fukui Prefectural Hospital (No. 18-69) and Asanogawa General Hospital (No. 216) with a waiver of consent obtained from the committee. All procedures were performed following the 1964 Helsinki Declaration and its later amendments.

2.4. Statistical Analysis

The data are expressed as mean ± SD and were analyzed using the statistical software package EZR version 1.55 (Saitama Medical Center, Jichi Medical University, Saitama, Japan), which is a graphical interface for R (The R Foundation for Statistical Computing, Vienna, Austria) [27]. p-values < 0.05 indicated statistical significance. For the comparisons of the variables, a pairwise t-test was used for normally distributed data, and a Wilcoxson test was used for non-normally distributed data. A correlation analysis was performed using the Pearson test to validate the correlation factors affecting the change in the SD-FBG. No statistical sample size calculations were conducted, as this study is a retrospective design.

3. Results

A total of 27 patients with insulin-dependent type 1 diabetes were switched from Deg-containing BBT to Gla-300-containing BBT from April 2017 to December 2022. Three patients were excluded from the analysis for the following reasons: (1) newly introduced to FGM during the observation period (two patients), and (2) HbA1c level and SMBG records could not be obtained at 4 to 6 months after the insulin switch (one patient). Another 24 patients with insulin-dependent type 1 diabetes whose treatment was switched from Deg-containing BBT to Gla-300-containing BBT were retrospectively analyzed. The clinical characteristics are summarized in Table 1. The patients had a mean age of 56.0 ± 15.2 years, a mean diabetes duration of 14.1 ± 13.6 years, and the percentage of females was 46% (11/24). At the baseline evaluation, the mean levels of HbA1c were $7.8 \pm 0.6\%$, and the mean BMI was 22.1 ± 2.7 kg/m^2, respectively. The mean dosage of fasting insulin and basal insulin were 0.38 ± 0.14 units/kg, and 0.20 ± 0.10 units/kg, respectively. The mean SD-FBG was high, at 58.2 ± 18.2 mg/dL, and the average counts of total hypoglycemia and severe hypoglycemia per month were 7.0 ± 5.6 times/month and 1.0 ± 1.3 times/month, respectively. The percentage of patients with FGM was 58% (14/24).

Table 1. Comparison of parameters before and after the switch from Deg to Gla-300. Data are represented as mean ± SD.

Variable	Baseline	After the Switch	p-Value
Age, years	56.0 ± 15.2		
Male, n (%)	13 (54%)		
Duration of diabetes, years	14.1 ± 13.6		
BW, kg	58.2 ± 9.8	58.3 ± 9.5	0.84
BMI, kg/m^2	22.1 ± 2.7	22.1 ± 2.7	0.80
FPG, mg/dL	135 ± 57.1	142 ± 64.4	0.68
HbA1c, %	7.8 ± 0.6	7.7 ± 0.5	0.69
Cr, mg/dL	0.92 ± 0.47	0.86 ± 0.35	0.27
eGFR, mL/min/1.73 m^2	74.6 ± 29.7	75.0 ± 27.5	0.81
Alb, g/dL	3.9 ± 0.3	3.9 ± 0.3	0.80
Fasting insulin dosage, units	22.2 ± 8.3	22.2 ± 8.1	0.94
Basal insulin dosage, units	12.3 ± 7.1	12.7 ± 6.1	0.27
Fasting insulin dosage, units/kg	0.38 ± 0.14	0.38 ± 0.12	0.81
Basal insulin dosage, units/kg	0.20 ± 0.10	0.21 ± 0.08	0.16
SD-FBG, mg/dL	58.2 ± 18.2	49.7 ± 15.7	0.02
Frequency of total hypoglycemia, times/month	7.0 ± 5.6	6.3 ± 4.6	0.24
Frequency of severe hypoglycemia, times/month	1.0 ± 1.3	1.0 ± 1.1	0.80
Frequency of nocturnal hypoglycemia, times/month	2.5 ± 2.1	1.5 ± 1.3	0.003
TAR, % (n = 14)	38.3 ± 7.4	39.6 ± 11.1	0.58
TIR, % (n = 14)	54.9 ± 7.3	56.4 ± 10.2	0.46
TBR, % (n = 14)	6.8 ± 3.7	4.1 ± 1.6	0.01

BW: body weight; BMI: body mass index; FPG: fasting plasma glucose; HbA1c: Hemoglobin A1c; Cr: creatinine; eGFR: estimated glomerular filtration rate; Alb: albumin; SD-FBG: standard division.

The parameters before and after the insulin switch are also shown in Table 1. The SD-FBG significantly decreased after the switch from Deg to Gla-300 (baseline: 58.2 ± 18.2 mg/dL, after the switch: 49.7 ± 15.7 mg/dL, $p = 0.02$). Moreover, the frequency of nocturnal hypoglycemic events decreased after the insulin switch (baseline: 2.5 ± 2.1 times/month, after the switch: 1.5 ± 1.3 times/month, $p = 0.003$). The frequency of total hypoglycemic events tended to decrease after the insulin switch, but there were no statistical differences. There was no difference in the frequency of severe hypoglycemia before and after the insulin switch. Despite the decrease in nocturnal hypoglycemic events, the HbA1c levels after the insulin switch did not worsen (baseline: 7.8 ± 0.6%, after the switch: 7.7 ± 0.5%, $p = 0.27$). There were no differences in the dosage of fasting insulin (baseline: 22.2 ± 8.3 units, after the switch: 22.2 ± 8.1 units, $p = 0.94$) and basal insulin (baseline: 12.3 ± 7.1 units, after the switch: 12.7 ± 6.1 units, $p = 0.27$) before and after the insulin switch. No changes were observed in other parameters.

In the analysis of the patients with FGM, the TBL decreased after the switch from Deg to Gla-300 (baseline: 6.8 ± 3.7%, after the switch: 4.1 ± 1.6%, $p = 0.01$). Despite the decrease in the TBL, the TAL and TIR after the insulin switch did not worsen.

We described the factors associated with the change in the SD-FBG after the switch from Deg to Gla-300 (Figure 1). The change in the SD-FBG had a negative correlation with the SD-FBG at baseline ($r = -0.52$, $p = 0.002$) and a positive correlation with the Alb ($r = 0.40$, $p = 0.04$). There were no correlations between the change in the SD-FBG and other parameters.

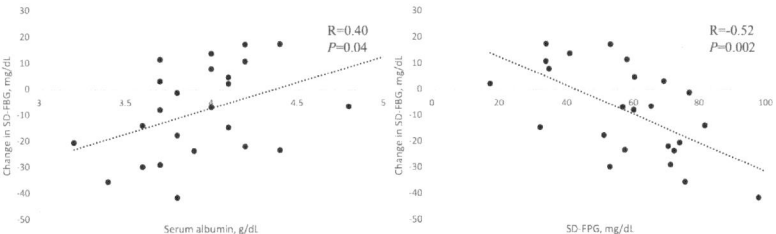

Figure 1. Correlation between the change in SD-FBG and Alb, SD-FBG at baseline.

4. Discussion

Herein, we evaluate the effect of the switch from Deg to Gla-300 on the GV and the frequency of hypoglycemia in patients with insulin-dependent type 1 diabetes treated with Deg-containing BBT. In the first part of the discussion, we describe the unique baseline characteristics of the patients included in this study. In our study, the included patients had a large mean SD-FBG of 58.2 ± 18.2 mg/dL. In a previous Japanese observational study in which the SD-FBG was evaluated in patients with type 1 diabetes, the mean SD-FBG level was 47.5 ± 22.0 mg/dL [28]. In this study, the titer of insulin antibodies was comprehensively measured in patients with type 1 diabetes treated via insulin injection therapy and still had a clinically high GV. In the population who suffered from high GV, the value of the SD-FBG was lower compared with that in our study. This indicated that the population in our research had a markedly lower GV than those under the usual care in Japan. Moreover, the rate of hypoglycemic events was also high compared with that in the previous study. In the SWITCH 1 randomized clinical trial [29], in which the efficiency of Deg compared with Gla-100 was evaluated in patients with type 1 diabetes, the total, nocturnal, and severe hypoglycemia events in patients with Deg were 2.0 times/month, 0.30 times/month, and 0.07 times/month, respectively. Similarly, the BEGIN basal-bolus type 1 trial [20], which is a phase 3, randomized, open-label, treat-to-target, non-inferiority trial of Deg in patients with type 1 diabetes, the total, nocturnal, and severe hypoglycemia in patients with Deg were 3.5 times/month, 0.37 times/month, and 0.02 times/month, respectively. These hypoglycemic rates are lower than those in our study. Therefore, the patients included in our study could be rephrased as patients treated with Deg-containing BBT, but who were not well controlled in terms of GV and hypoglycemia. In this population, switching from Deg to Gla-300 improved the day-to-day GV, expressed by the SD-FBG, and decreased the frequency of nocturnal hypoglycemia. The change in the SD-FBG had a negative correlation with the SD-FBG at baseline and a positive correlation with the Alb.

Deg and Gla-300 improve day-to-day GV, unlike Gla-100, owing to their longer pharmacodynamic effect [16–23]. However, the action mechanisms of these two ultra-long-acting insulin analogs are different. Deg forms a soluble multi-hexameric chain after subcutaneous injection and the zinc moiety of the insulin molecule diffuses slowly from the terminal ends of Deg and gets absorbed into circulation. After the absorption into the circulatory system, almost all Deg binds to albumin and is slowly released from the albumin within the target tissue to achieve a hypoglycemic effect [30]. In contrast, Gla-300 does not bind to albumin when in circulation [31]. The previous report showed that Alb levels fluctuate daily, with high values in the daytime and low values at night [32]. The decrease in the Alb level increases the free insulin levels and could decrease blood glucose levels. Therefore, a high GV and frequent nocturnal hypoglycemia are thought to be improved through switching from Deg to Gla-300.

Kawaguchi et al. reported that a lower GV and a decreased frequency of hypoglycemia are observed in patients with Gla-300 compared with Deg in type 2 diabetes [25]. Their findings indicated that the frequency of nocturnal hypoglycemia in patients with Deg had an association with low serum albumin levels. However, no association was observed between serum albumin levels and the frequency of nocturnal hypoglycemia in patients with Gla-300 [25]. Another report on type 2 diabetes showed that Gla-300 decreased the total and nocturnal hypoglycemia compared with Deg in patients with Alb < 3.8 g/dL [33]. Although these reports differ in that they are based on type 2 diabetes, their results are similar to ours. However, opposite results were also reported on type 2 diabetes. Tibaldi et al. reported that the administration of Deg to patients with type 2 diabetes achieved a greater HbA1c reduction with fewer hypoglycemic events after 6 months from the administration compared with those in patients with Gla-300 [34]. Studies on type 1 diabetes are limited and the results are also controversial. A double-blind crossover euglycemic clump study showed that Gla-300 induced 20% less fluctuation in steady-state glucose infusion rate profiles than that of Deg in a once-daily morning dosing regimen of 0.4 U/kg/day [35]. However, the opposite result was obtained in another double-blind crossover euglycemic clump study [36].

Recently, Miura et al. conducted a multicenter crossover trial on type 1 diabetes in which the efficiency of Deg and Gla-300 on the SD-FBG were evaluated [37]. In this study, 46 patients with insulin-dependent type 1 diabetes were randomly assigned to the Deg-first/Gla-300-s group or the Gla-300-first/Deg-second group and treated with the respective basal insulin for 4-week periods. The primary endpoint of this study was to examine the noninferiority of Deg compared to Gla-300 regarding day-to-day GV evaluated as SD-FBG levels by the SMBG records. This study indicated that the SD-FBG during the Deg treatment period was not inferior to that during the Gla-300 treatment period (mean difference of −6.6 mg/dL, with a 95% CI of −16.1 to 3.0 mg/dL). Among 46 patients included in this study, 32 patients were evaluated using continuous glucose monitoring. In these 32 patients, the TBLs (<70 mg/dL) were shorter during the Gla-300 treatment period, and the TALs (>180 mg/dL) were shorter during the Deg treatment period, respectively. In their conclusion, they identified that these two insulins have comparable glucose-stabilizing effects in patients with insulin-dependent type 1 diabetes. However, there are cases of insulin-dependent type 1 diabetes in which there is a large clinical difference in efficacy between Deg and Gla-300.

There are several important differences between this crossover study [37] and our study. The first was the eligible patients. In our study, patients whose treatments were switched from Deg to Gla-300 for various reasons in real-world medical examinations were included. Therefore, patients with a stable blood glucose control with Deg might not be analyzed. The patients included in our study had a high GV and frequent hypoglycemic events despite the use of Deg-containing BBT. The second difference was the length of observation. Our study has a longer follow-up period compared with the past study by Miura et al. [37]. In general, the longer the observation period, the more it is affected by factors other than simple medication-to-medication differences, including diet and exercise. A past study [37] indicated that treatment with Gla-300 achieved a lower hypoglycemic rate compared to that with Deg. However, there was no difference in the SD-FBG between treatment with Gla-300 and treatment with Deg. Hypoglycemic events could make patients feel hunger and increase appetite. Therefore, the improvement in GV in our study may be affected by the fact that Gla-300 treatment reduced the frequency of hypoglycemia and improved the hypoglycemia-induced increase in appetite. However, no evaluation of appetite and food intake was conducted in our study. The third difference was in the evaluation of factors correlating with the change in the SD-FBG. The previous study [37] did not evaluate the factors influencing the superiority of these ultra-long-acting insulin analogs. In contrast, our study showed that low serum albumin level and a high SD-FBG with Deg-containing BBT are the predictors of the superiority of Gla-300 over Deg in insulin-dependent type 1 diabetes. Knowing these predictor markers may lead to the personalization of treatment in patients with insulin-dependent type 1 diabetes. Although not considered in our current research, the longer pharmacological action of Deg compared with that of Gla-300 should be considered. During the use of Deg, the efficiency and safety of a flexible dosing regimen at fixed intervals with a minimum of 8 h and a maximum of 40 h between each injection was reported [38]. There is no report about a flexible dosing regimen used with Gla-300. However, the pharmacological action of Gla-300 is much shorter than that of Deg. Thus, it is unlikely that the results of using a flexible dosing regimen with Gla-300 will be as favorable as those with Deg. For this reason, Deg is expected to be more useful than Gla-300 in cases where insulin dosing times vary from day to day.

Our study has several limitations. First, this study adopted a retrospective design and was conducted on a small number of patients. In general, larger sample sizes are more likely to yield significant differences when examining differences between two medications. However, even though no significant difference was found in the existing studies with large sample sizes, the present study found a significant difference in the SD-FBG. We believe that the small sample size is a limitation, but also a possible new finding that some groups may benefit from treatment modification. The eligible patients in our study were patients with insulin-dependent type 1 diabetes and whose medications were switched from Deg-containing BBT to Gla-300-containing BBT for various clinical problems and have a high GV and frequent hypoglycemia despite the use of Deg. We would like to inform readers that the switch from Deg to Gla-300 may not be effective in all patients with insulin-dependent type 1 diabetes. The results of this study could be considered applicable to patients with insulin-dependent type 1 diabetes with Deg-containing BBT but who have a high GV or frequent hypoglycemia. The sample size is smaller than existing studies, but the significant difference from Deg to Gla-300, and the fact that the population is different from previous studies, suggest that a treatment change is likely to be effective in certain groups. Therefore, it is desirable to conduct a randomized, prospective study of switching from Deg to Gla-300 in patients with insulin-dependent type 1 diabetes treated with Deg-containing BBT and having high GV or low serum albumin levels, which were shown in our observation study to benefit from the switch from Deg to Gla-300. The next limitation is that the evaluation of GV was made by the records of SMBG which is inferior to that evaluated by CGM. Therefore, it is hoped that in the next study, the assessment of GV will be done using CGM rather than SMBG.

5. Conclusions

Switching from Deg to Gla-300 is effective for improving day-to-day GV and decreasing nocturnal hypoglycemia in patients with insulin-dependent type 1 diabetes and having high day-to-day GV despite the use of Deg-containing BBT. The effectiveness of improving day-to-day GV is greater in cases with low serum albumin levels and large day-to-day GV despite the use of Deg-containing BBT.

Author Contributions: Conceptualization, T.S., and S.K.; methodology, T.S., and S.K.; validation, A.O. (Azusa Ohbatake), M.K., Y.K., and T.Y.; formal analysis, T.S., and S.K.; investigation, T.S., T.H., A.O. (Ai Ohmori), K.S., and R.Y.; data curation, T.S., T.H., A.O. (Ai Ohmori), K.S., and R.Y.; writing—original draft preparation, T.S.; writing—review and editing, S.K., and M.K.; visualization, T.S.; supervision, S.K., A.O. (Azusa Ohbatake), M.K., Y.K., and T.Y.; project administration, S.K., and T.Y. All authors have read and agreed to the published version of the manuscript.

Funding: This research received no external funding.

Institutional Review Board Statement: This study was approved by the ethics committee at Fukui Prefectural Hospital (No. 18-69, 18 March 2019) and Asanogawa General Hospital (No. 216, 26 January 2023) with a waiver of consent obtained from the committee. All procedures were performed following the 1964 Helsinki Declaration and its later amendments.

Informed Consent Statement: Owing to the retrospective study, written informed consent was not necessary. However, we applied an opt-out method to obtain consent for this study by using the poster in the hospital and hospital websites on the internet.

Data Availability Statement: The data are available from the corresponding author upon reasonable request.

Acknowledgments: We thank all the members of staff in the Endocrine and Hypertension group at Kanazawa University Hospital, Fukui Prefectural Hospital, and Asanogawa General Hospital for their support in obtaining clinical data.

Conflicts of Interest: All authors declare that they have no conflicts of interest.

References

1. Diabetes Control and Complications Trial (DCCT) Research Group. The absence of a glycemic threshold for the development of long-term complications: The perspective of the Diabetes Control and Complications Trial. *Diabetes* **1996**, *45*, 1289–1298. [CrossRef]
2. Nathan, D.M.; Cleary, P.A.; Backlund, J.Y.; Genuth, S.M.; Lachin, J.M.; Orchard, T.J.; Raskin, P.; Zinman, B. Intensive diabetes treatment and cardiovascular disease in patients with type 1 diabetes. *N. Engl. J. Med.* **2005**, *353*, 2643–2653.
3. Goff, D.C.; Gerstein, H.C.; Ginsberg, H.N.; Cushman, W.C.; Margolis, K.L.; Byington, R.P.; Buse, J.B.; Genuth, S.; Probstfield, J.L.; Simons-Morton, D.G.; et al. Prevention of cardiovascular disease in persons with type 2 diabetes mellitus: Current knowledge and rationale for the Action to Control Cardiovascular Risk in Diabetes (ACCORD) trial. *Am. J. Cardiol.* **2007**, *99*, 4i–20i. [CrossRef]
4. Bonds, D.E.; Miller, M.E.; Bergenstal, R.M.; Buse, J.B.; Byington, R.P.; Cutler, J.A.; Dudl, R.J.; Ismail-Beigi, F.; Kimel, A.R.; Hoogwerf, B.; et al. The association between symptomatic, severe hypoglycemia and mortality in type 2 diabetes: Retrospective epidemiological analysis of the ACCORD study. *BMJ* **2010**, *340*, b4909. [CrossRef]
5. Goto, A.; Arsh, O.A.; Goto, M.; Terauchi, Y.; Noda, M. Severe hypoglycemia and cardiovascular disease: Systematic review and meta-analysis with bias analysis. *BMJ* **2013**, *347*, f4533. [CrossRef]
6. Hanefeld, M.; Frier, B.M.; Pistrosch, F. Hypoglycemia and cardiovascular Risk: Is There a Major Link? *Diabetes Care* **2016**, *39* (Suppl. 2), 205–209. [CrossRef]
7. Mattishent, K.; Loke, Y.K. Bi-directional interaction between hypoglycemia and cognitive impairment in elderly patients treated with glucose-lowering agents: A systematic review and meta-analysis. *Diabetes Obes. Metab.* **2016**, *18*, 135–141. [CrossRef]
8. Komorita, Y.; Minami, M.; Maeda, Y.; Yoshioka, R.; Ohkuma, T.; Kitazono, T. Prevalence of bone fracture and its association with severe hypoglycemia in Japanese patients with type 1 diabetes. *BMJ Open Diabetes Res. Care.* **2021**, *9*, e002099. [CrossRef] [PubMed]
9. Sakamoto, M. Type 2 Diabetes and Clycemic Variability: Various Parameters in Clinical Practice. *J. Clin. Med. Res.* **2018**, *10*, 737–742. [CrossRef] [PubMed]
10. Hirakawa, Y.; Arima, H.; Zoungas, S.; Ninomiya, T.; Cooper, M.; Hamet, P.; Mancia, G.; Poulter, N.; Harrap, S.; Woodward, M.; et al. Impact of visit-to-visit glycemic variability on the risks of macrovascular and microvascular events and all-cause mortality in type 2 diabetes: The ADVANCE trial. *Diabetes Care* **2014**, *37*, 2359–2365. [CrossRef] [PubMed]
11. Xu, F.; Zhao, L.H.; Su, J.B.; Chen, T.; Wang, X.Q.; Chen, J.F.; Wu, G.; Jin, Y.; Wang, X.H. The relationship between glycemic variability and diabetic peripheral neuropathy in type 2 diabetes with well-controlled HbA1c. *Diabetol. Metab. Syndr.* **2014**, *6*, 139. [CrossRef]
12. Ohara, M.; Fukui, T.; Ouchi, M.; Watanabe, K.; Suzuki, T.; Yamamoto, S.; Yamamoto, T.; Hayashi, T.; Oba, K.; Hirano, T. Relationship between daily and day-to-day glycemic variability and increased oxidative stress in type 2 diabetes. *Diabetes Res. Clin. Pract.* **2016**, *122*, 62–70. [CrossRef]
13. Niskanen, L.; Virkamaki, A.; Hansen, J.B.; Saukkonen, T. Fasting plasma glucose variability as a marker of nocturnal hypoglycemia in diabetes: Evidence from the PREDICTIVE study. *Diabetes Res. Clin. Pract.* **2009**, *86*, e15–e18. [CrossRef] [PubMed]
14. DeVries, J.H.; Bailey, T.S.; Bhargava, A.; Gerety, G.; Gumprecht, J.; Heller, S.; Lane, W.; Wysham, C.H.; Zinman, B.; Bak, B.A.; et al. Day-to-day fasting self-monitored blood glucose variability is associated with risk of hypoglycemia in insulin-treated patients with type 1 and type 2 diabetes: A post hoc analysis of the SWITCH Trials. *Diabetes Obes. Metab.* **2019**, *21*, 622–630. [CrossRef]
15. Garg, S.K.; Weinzimer, S.A.; Tamborlane, W.V.; Buckingham, B.A.; Bode, B.W.; Bailey, T.S.; Brazg, R.L.; Ilany, J.; Slover, R.H.; Anderson, S.M.; et al. Glucose Outcomes with the In-Home Use of a Hybrid Closed-Loop Insulin Delivery System in Adolescent and Adults with Type 1 Diabetes. *Diabetes Technol. Ther.* **2017**, *19*, 155–163. [CrossRef] [PubMed]

16. Home, P.D.; Bergenstal, R.M.; Bolli, G.B.; Ziemen, M.; Rojeski, M.; Espinasse, M.; Riddle, M.C. New Insulin Glargine 300 Units/mL Versus Glargine 100 Units/mL in People with Type 1 Diabetes: A Randomized, Phase 3a, Open-Label Clinical Trial (EDITION 4). *Diabetes Care* 2015, *38*, 2217–2225. [CrossRef] [PubMed]
17. Terauchi, Y.; Koyama, M.; Cheng, X.; Takahashi, Y.; Riddle, M.C.; Bolli, G.B.; Hirose, T. New insulin glargine 300 U/ml versus glargine 100 U/ml in Japanese adults with type 1 diabetes using basal and mealtime insulin: Glucose control and hypoglycemia in a randomized controlled trial (EDITION JP 1). *Diabetes Obes. Metab.* 2016, *18*, 375–383. [CrossRef]
18. Yki-Jarvinen, H.; Bergenstal, R.; Ziemen, M.; Wardecki, M.; Muehlen-Bartmer, I.; Boelle, E.; Riddle, M.C. New Insulin Glargine 300 Units/mL Versus Glargine 100 Units/mL in People with Type 2 Diabetes Using Oral Agents and Basal Insulin: Glucose Control and Hypoglycemia in a 6-Month Randomizes Controlled Trial (EDITION 2). *Diabetes Care* 2014, *7*, 3235–3243. [CrossRef]
19. Terauchi, Y.; Koyama, M.; Cheng, X.; Takahashi, Y.; Riddle, M.C.; Bolli, G.B.; Hirose, T. New insulin glargine 300 U/ml versus glargine 100 U/ml in Japanese people with type 2 diabetes using basal insulin and oral antihyperglycaemic drugs: Glucose control and hypoglycemia in a randomized controlled trial (EDITION JP 2). *Diabetes Metab.* 2016, *18*, 366–374. [CrossRef]
20. Heller, S.; Buse, J.; Fisher, M.; Garg, S.; Marre, M.; Merker, L.; Renard, E.; Russell-Jones, D.; Philotheou, A.; Francisco, A.M.; et al. Insulin degludec, an ultra-longacting basal insulin, versus insulin glargine in basal-bolus treatment with mealtime insulin aspect in type 1 diabetes (BEGIN Basal-Bolus Type 1): A phase 3, randomized, open-label, treat-to-target non-inferiority trial. *Lancet* 2012, *379*, 1489–1497. [CrossRef]
21. Heise, T.; Hermanski, L.; Nosek, L.; Feldman, A.; Rasmussen, S.; Haahr, H. Insulin degludec: Four times lower pharmacodynamic variability than insulin glargine under steady-state conditions in type 1 diabetes. *Diabetes Obes. Metab.* 2012, *14*, 859–864. [CrossRef]
22. Marso, S.P.; McGuire, D.K.; Zinman, B.; Poulter, N.R.; Emerson, S.S.; Pieber, T.R.; Pratley, R.E.; Haahr, P.M.; Lange, M.; Brown-Frandsen, K.; et al. Efficacy and Safety of Degludec versus Glargine in Type 2 Diabetes. *N. Engl. J. Med.* 2017, *377*, 723–732. [CrossRef]
23. Zinman, B.; Philis-Tsimikas, A.; Cariou, B.; Handelsman, Y.; Rodbard, H.W.; Johansen, T.; Endahl, L.; Mathieu, C. Insulin degludec versus insulin glargine in insulin-naïve patients with type 2 diabetes: A 1-year, randomized, treat-to-target trial (BEGIN Once Long). *Diabetes Care* 2012, *35*, 2464–2471. [CrossRef] [PubMed]
24. Rosenstock, J.; Cheng, A.; Ritzel, R.; Bosnyak, Z.; Devisme, C.; Cali, A.M.; Sieber, J.; Stella, P.; Wang, X.; Frías, J.P.; et al. More Similarities Than Differences Testing Insulin Glargine 300 Units/mL Versus Insulin Degludec 100 Units/mL in insulin-Naïve Type 2 Diabetes: The Randomized Head-to-Head BRIGHT Trial. *Diabetes Care* 2018, *41*, 2147–2154. [CrossRef] [PubMed]
25. Kawaguchi, Y.; Sawa, J.; Sakuma, N.; Kumeda, Y. Efficacy and safety of insulin glargine 300 U/mL vs insulin degludec in patients with type 2 diabetes: A randomized, open-label, cross-over study using continuous glucose monitoring profiles. *J. Diabetes Investig.* 2019, *10*, 343–351. [CrossRef] [PubMed]
26. Siegelaar, S.E.; Holleman, F.; Hoekstra, J.B.L.; DeVries, J.H. Glucose variability; Dose It matter? *Endocr. Rev.* 2010, *31*, 171–182. [CrossRef] [PubMed]
27. Kanada, Y. Investigation of the freely-available easy-to use software "EZR" (Easy R) for medical statistics. *Bone Marrow Transpl.* 2013, *48*, 452–458. [CrossRef] [PubMed]
28. Yoneda, C.; Tashima-Horie, K.; Fukushima, S.; Saito, S.; Tanaka, S.; Haruki, T.; Ogino, J.; Suzuki, Y.; Hashimoto, N. Association of monitoring fasting blood glucose variability with insulin antibodies and clinical factors in type 1 diabetes. *Endocr. J.* 2016, *63*, 603–609. [CrossRef]
29. Lane, W.; Bailey, T.S.; Gerety, G.; Gumprecht, J.; Philis-Tsimikas, A.; Hansen, C.T.; Nielsen, T.S.; Warren, M. Effect of insulin Degludec vs Insulin Glargine U100 on Hypoglycemia in Patients with Type 1 Diabetes: The SWITCH 1 Randomized Clinical Trial. *JAMA* 2017, *318*, 33–44. [CrossRef]
30. Jonassen, I.; Havelund, S.; Hoeg-Jensen, T.; Steensgaard, D.B.; Wahlund, P.O.; Ribel, U. Design of the novel protraction mechanism of insulin degludec, an ultra-long-acting basal insulin. *Pharm. Res.* 2012, *29*, 2104–2114. [CrossRef]
31. Heise, T.; Mathieu, C. Impact of the mode of protraction of basal insulin therapies on their pharmacokinetic and pharmacodynamic properties and resulting clinical outcomes. *Diabetes Obes. Metab.* 2017, *19*, 3–12. [CrossRef] [PubMed]
32. Jubiz, W.; Canterbury, J.M.; Reiss, E.; Tyler, F.H. Circadian rhythm in serum parathyroid hormone concentration in human subjects: Correlation with serum calcium, phosphate, albumin, and growth hormone levels. *J. Clin. Investig.* 1972, *51*, 2040–2046. [CrossRef] [PubMed]
33. Kawaguchi, Y.; Sawa, J.; Hamai, C.; Kumeda, Y. Differential Effect of Hypoalbuminemia on Hypoglycemia on Type 2 Diabetes Patients treated with Insulin Glargine 300 U/mL and Insulin Degludec. *Diabetes Ther.* 2019, *10*, 1535–1541. [CrossRef] [PubMed]
34. Tibaldi, J.; Hadley-Brown, M.; Liebl, A.; Haldrup, S.; Sandberg, V.; Wolden, M.L.; Rodbard, H.W. A comparative effectiveness study of degludec and insulin glargine 300 U/mL in insulin-naïve patients with type 2 diabetes. *Diabetes Obes. Metab.* 2019, *21*, 1001–1009. [CrossRef] [PubMed]
35. Bailey, T.S.; Pettus, J.; Roussel, R.; Schmider, W.; Maroccia, M.; Nassr, N.; Klein, O.; Bolli, G.B.; Dahmen, R. Morning administration of 0.4 U/kg/day insulin glargine 300 U/mL provides less fluctuating 24-hour pharmacodynamics and more even pharmacokinetic profiles compared with insulin glargine 100 U/mL in type 1 diabetes. *Diabetes Metab.* 2018, *44*, 15–21. [CrossRef] [PubMed]
36. Heise, T.; Nørskov, M.; Nosek, L.; Kaplan, K.; Famulla, S.; Haahr, H.L. Insulin Degludec: Lower Day-to-Day and Within-day Variability in Pharmacodynamic response Compared with Insulin Glargine 300 U/mK in Type 1 diabetes. *Diabetes Obes. Merab* 2017, *19*, 1032–1039. [CrossRef]

37. Miura, H.; Sakaguchi, K.; Otowa-Suematsu, N.; Yamada, T.; So, A.; Komada, H.; Okada, Y.; Hirota, Y.; Tamori, Y.; Ogawa, W. Effects of Insulin Degludec and Insulin Glargine U300 on Glycemic Stability in Individuals with Type 1 Diabetes. A Multicenter, Randomized Controlled Crossover Study. *Diabetes Obes. Metab.* **2020**, *22*, 2356–2363. [CrossRef]
38. Mathieu, C.; Hollander, P.; Miranda-Palma, B.; Cooper, J.; Franek, E.; Russell-Jones, D.; Larsen, J.; Tamer, S.C.; Bain, S.C.; NN1250-3770 (BEGIN: Flex T1) Trial Investigators. Efficacy and Safety of Insulin Degludec in a Flexible Dosing Regimen vs Insulin Glargine in Patients with Type 1 Diabetes (BEGIN: Flex T1). *J. Clin. Endocrinol. Metab.* **2013**, *98*, 1154–1162. [CrossRef]

Disclaimer/Publisher's Note: The statements, opinions and data contained in all publications are solely those of the individual author(s) and contributor(s) and not of MDPI and/or the editor(s). MDPI and/or the editor(s) disclaim responsibility for any injury to people or property resulting from any ideas, methods, instructions or products referred to in the content.

Article

Effects of Age and Biological Age-Determining Factors on Telomere Length in Type 2 Diabetes Mellitus Patients

Jawaria Ali Tariq [1], KaleemUllah Mandokhail [2], Naheed Sajjad [1], Abrar Hussain [3,*], Humera Javaid [1], Aamir Rasool [4], Hummaira Sadaf [1,5], Sadia Javaid [1] and Abdul Rauf Durrani [6]

1. Department of Biotechnology, Sardar Bahadur Khan Women's University, Quetta 87300, Pakistan; jawaria_ali_tariq@yahoo.com (J.A.T.); drnaheedsajjad@gmail.com (N.S.); chokohollikhhf@gmail.com (H.J.); hummairas1@gmail.com (H.S.); rorquel_roller@yahoo.com (S.J.)
2. Department of Microbiology, University of Balochistan, Quetta 87300, Pakistan; drkaleemullah@gmail.com
3. Department of Biotechnology, Balochistan University of Information Technology, Engineering and Management Sciences, Quetta 87300, Pakistan
4. Institute of Biochemistry, University of Balochistan, Quetta 87300, Pakistan; rasool.amir@gmail.com
5. Maria Sklodowska-Curie National Research Institute of Oncology, Silesian University of Technology, 02-781 Warsaw, Poland
6. Provincial Reference Laboratory (PRL), Fatima Jinnah General and Chest Hospital, Quetta 87300, Pakistan; abdulraufdurrani@gmail.com
* Correspondence: abrarbangash176@hotmail.com

Abstract: *Background and Objectives:* Telomere length (TL) undergoes attrition over time, indicating the process of aging, and is linked to a higher risk of diabetes mellitus type 2 (DM-2). This molecular epidemiological study investigated the correlation between leukocyte TL variations and determinants of molecular aging in 121 Pakistani DM-2 patients. *Materials and Methods:* The ratio of telomere repeats to the SCG copy number was calculated to estimate the TL in each sample through qPCR assays. *Results:* In this study, smaller mean TLs were observed in 48.8% of males (6.35 ± 0.82 kb), 3.3% of underweight patients (5.77 ± 1.14 kb), 61.2% of patients on regular medication (6.50 ± 0.79 kb), 9.1% with very high stress levels (5.94 ± 0.99 kb), 31.4% of smokers (5.83 ± 0.73 kb), 40.5% of patients with low physical activity (6.47 ± 0.69 kb), 47.9% of hypertensive patients (5.93 ± 0.64 kb), 10.7% of patients with DM-2 for more than 15 years, and 3.3% of patients with a delayed onset of DM-2 (6.00 ± 0.93 kb). *Conclusion:* This research indicated a significant negative correlation ($R^2 = 0.143$) between TL and the age of DM-2 patients. This study demonstrated that the correlation of telomere length with age in DM-2 patients was also influenced by various age-determining factors, including hypertension and smoking habits, with significant strong ($R^2 = 0.526$) and moderate ($R^2 = 0.299$) correlations, respectively; sex, obesity, the stress level and age at the onset of diabetes with significant weak correlations ($R^2 = 0.043, 0.041, 0.037,$ and 0.065, respectively), and no significant correlations of medication routine, rate of physical activity, and the durations of DM-2 with age-adjusted telomere length. These results challenge TL as the sole marker of aging, thus highlighting the need for further research to understand underlying factors and mitigate the effect of aging or premature aging on diabetic patients.

Keywords: telomere length; T/S ratio; diabetes mellitus type 2; molecular aging; qPCR assay

1. Introduction

Chronic hyperglycemia indicates type 2 diabetes mellitus (DM-2), a heterogeneous illness manifested by altered insulin production and insulin resistance [1]. DM-2 accounts for >90% of all cases of diabetes mellitus [2]. In addition to the elderly and middle-aged, young people are becoming more and more affected by this condition, especially in non-Caucasian communities [3].

Genetic susceptibility in different populations and the combination of several environmental variables lead to type 2 diabetes mellitus [4]. Diabetes is one of the major health concerns worldwide, especially in Asian countries. Based on recent studies, one in four individuals in the general population in Pakistan has DM-2 [5,6]. Pakistan is a developing nation with a high prevalence of diabetes, as shown by a recent survey through which it was found that 26% of adults in the general population had the disease [5].

Telomeres (DNA and histone protein complexes) shield the chromosome ends from fusion and destruction [7]. Telomere length (TL) undergoes attrition over time, indicating the process (and a potential cause) of aging in human tissues [8]. According to many studies of humans, a short TL assessed in leukocytes is linked to a higher risk of age-related conditions, such as type 2 diabetes [9] and cardiovascular disease [10], along with the individual's lifespan and mortality rate [11]. Leukocyte TL is also linked to environmental exposures (e.g., radiation, smoking cigarettes), health variables (e.g., cholesterol, obesity), and lifestyle factors (e.g., physical activity, dietary habits) [12,13].

Several mechanisms lead to the accelerated molecular aging process in diabetic patients, including telomere shortening, accumulation of advanced glycation end products (AGEs), cellular senescence, inflammation, oxidative stress, and epigenetic modifications. These molecular aging mechanisms can lead to many other age-related disorders, along with diabetes [14,15].

Different studies have explained the effect of diabetes on telomere shortening, some of which suggested an association between diabetes and accelerated aging at the molecular level [15,16]. At the same time, few studies have reported no significant correlation between telomere length and the age of DM-2 patients [17]. A constant rise in the prevalence of DM-2 in Asian countries emphasizes the need for molecular and clinical investigations of this disease, particularly in the elderly population. As no epidemiological data related to Pakistan associating telomere length with the chronological age of DM-2 patients is available to date, the scope of this molecular epidemiological study is to find the telomere length dynamics in leukocytes from diabetic patients and to correlate age-adjusted telomere length variations with various determinants of the molecular aging process in Pakistani patients, with a future perspective to identify biological variables leading to premature molecular aging in DM-2 patients.

2. Materials and Methods

2.1. Experimental Design

The experimental design was planned as suggested by Cawthon [18] and Axelrad et al. [19] for determining telomere lengths by quantitative PCR. Cell populations from different tissues may have different replicative histories, along with the telomere length in those cells. For this research, the telomere length of the leukocytes was measured. For each DNA sample from diabetic patients, the ratio of telomere repeat copy number (T) to SCG copy numbers (S) was calculated to estimate telomere length. The T/S ratio and telomere length are directly proportional because the primer–DNA binding tendency (during initial PCR cycles) and telomere length are directly associated. Thus, both copy numbers (T and S) were measured by comparing the difference in cycle threshold (Ct value) of samples with primers for telomeres and the single copy gene (SCG).

2.2. Inclusion Criteria

Individuals above the age of 39 years with a medical history of type 2 diabetes mellitus were selected for this study. A detailed scrutiny of 350 DM-2 patients was performed through a questionnaire-based survey to minimize the effect of extraneous variables, and 121 patients were selected as a homogenized cohort from the population based on the following:

a. Geographical characteristics (i.e., samples from areas/societies of city away from industrial pollution, and samples from developed residential areas free from a congested population and heavy traffic, also samples from migrants or new residents were avoided);

- b. The patient's medical history (patients having a medical history of a cardiac or any metabolic disorder, any physical disability, genetic disorder, cancer, or any other chronic disease were skipped);
- c. The use of supplements (no samples were collected from the patients taking vitamins or supplements through pills or injections);
- d. Sleep time (samples were collected only from patients with 7–8 h of sleep per day);
- e. Extra physical activity (patients performing excessive exercise and gym workouts were not considered in this study);
- f. Tobacco consumption other than smoking (patients taking smokeless or chewable forms of tobacco were not considered suitable candidates for sampling);
- g. Lifestyle variations (individuals with exhaustive working hours of >6 h/day at the job, individuals below the poverty line, individuals with luxurious lifestyles, individuals with high consumption of fats in diet were also not considered; moreover, patients taking medicines other than metformin were also excluded from this study);
- h. Marital status (only married individuals were considered for further analysis);
- i. Fertility (infertile, menopausal, and post-menopausal patients were not analyzed further).

2.3. Sample Collection

Fresh blood was collected from selected 121 diabetic patients (>39 years of age) voluntarily through the standard venipuncture technique in BD sterile vacutainer blood tubes with EDTA (Becton Dickinson UK Ltd., Oxford, UK; Cat. No. 366643). All the samples were processed for the DNA extraction protocol on the same day of sample collection to maintain uniformity in the procedure.

2.4. DNA Extraction

For the optimization of the DNA extraction protocol and to maintain the integrity and purity of isolated DNA molecules, different methods were scrutinized. The organic method by Shen [20] using a phenol–chloroform solution was found to extract intact and pure DNA molecules, showing a sharp bulky band with no smearing during gel electrophoresis. Therefore, all the blood samples were processed using the organic method for DNA extraction. Extracted DNA was stored in low Tris–EDTA buffer (TE^{-4}, pH = 7.5) at −20 degrees Celsius until further use (not more than three days). Before amplification, the quantity and quality of DNA were measured using a NanoDrop spectrophotometer (Thermo Scientific^{TM}, Waltham, MA USA; Cat. No. ND-2000) by the A260/280 ratio. * All procedures were performed in a biological safety cabinet.

2.5. Oligomers

All oligomers, including standards and primers, were diluted in PCR-grade water to make a stock concentration of 100 pmoles/µL and then kept at −20 °C until further required. Working stocks (10 pmoles/µL) of all the primers were freshly prepared before starting the reaction, and the remaining working primers were stored at 4 °C (not longer than two weeks). Standards and primers (Supplementary Table S1) for the SCG (β-Globin) and telomeres were used, as stated by O'Callaghan and Fenech [21].

2.6. Serial Dilutions

Oligomer standards (both telomeric and SCG standards) were serially diluted in PCR-grade water by the dilution factor (1.68) suggested by O'Callaghan and Fenech [21] to generate the standard curve of Ct values for assay, as shown in Table 1.

Dilutions of the standard oligomers were then added to the PCR tubes with ultra-clear caps for qPCR assays. Additionally, 20 ng of plasmid DNA (pBR-322) was added to each tube of serially diluted standards to maintain the overall mass of the DNA molecule.

Table 1. Serial dilutions of standards.

Telomere Standard (ng/µL)	SCG Standard (ng/µL)
6.10	1.8
3.63	1.07
2.16	0.64
1.29	0.38
0.77	0.23

2.7. Normalization

Varying DNA concentrations may give different average telomere lengths in each sample. To avoid this problem and pipetting errors, DNA normalization was performed by maintaining a consistent concentration of DNA (5 ng/µL) in each sample and diluting extracted DNA samples with PCR-grade water, as practiced by Axelrad et al. [19].

2.8. qPCR Protocol

After diluting all the DNA samples to a final concentration of 5 ng/µL, two master mixes of PCR reagents were prepared: one with telomere forward and reverse primers and the other with the primer pair for the SCG. Aliquots were prepared for no template control and standards plus an extra 5% for pipetting errors (Supplementary Table S2).

All samples were run on a SaCycler-96 (Sacace Biotechnology, Como, Italy) with the SaCycler-96 Real-Time PCR V.7.3 (Sacace Biotechnology, Como, Italy). Both telomere and SCG reactions were run separately using the following program. The reaction program was set as follows: initial denaturation of 10 min followed by 35 cycles of 95 degrees Celsius for 15 s, 60 degrees Celsius for 60 s, and finally a dissociation (or melting) curve.

2.9. Data Analysis

Amplification was observed in standards and samples according to the procedure performed by O'Callaghan and Fenech [21]. The baseline was set (because standards were amplified earlier than the samples), and a standard curve was generated using the C_t readings.

Both telomere and SCG assays were run on each sample. The number of cycles required for fluorescence detection to attain the exponential curve (Ct value) varies between these two assays for a single sample due to the varying DNA copy numbers generated in each assay (Figure 1).

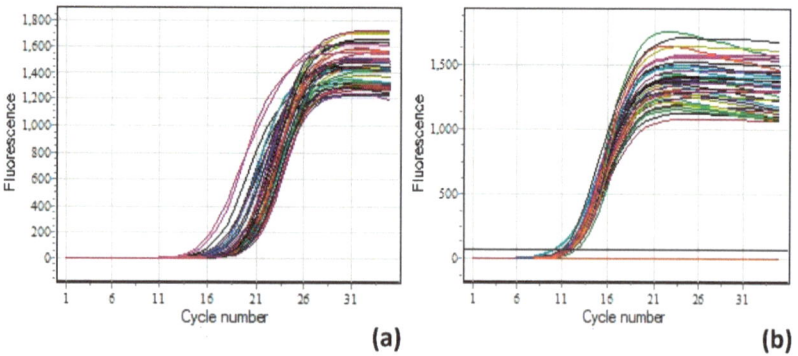

Figure 1. The qPCR assay (colored lines represent each sample). (**a**) Amplification curve of telomeric regions and (**b**) amplification curve of the SCG to evaluate the cycle threshold.

A mean TL of 4270 bp in leukocytes is equal to one T/S ratio unit (qPCR cycles of the telomere standard run over the cycles of the SCG standard run) [22]. In this study, a ratio of 0.98 corresponds to an average of 4185 bp.

After calculating the telomere length for each sample, the effects of different variables (sex, smoking, physical activity, stress level, etc.) on telomere length dynamics in diabetic patients were assessed (Figure 2) using IBM SPSS Statistics 20.

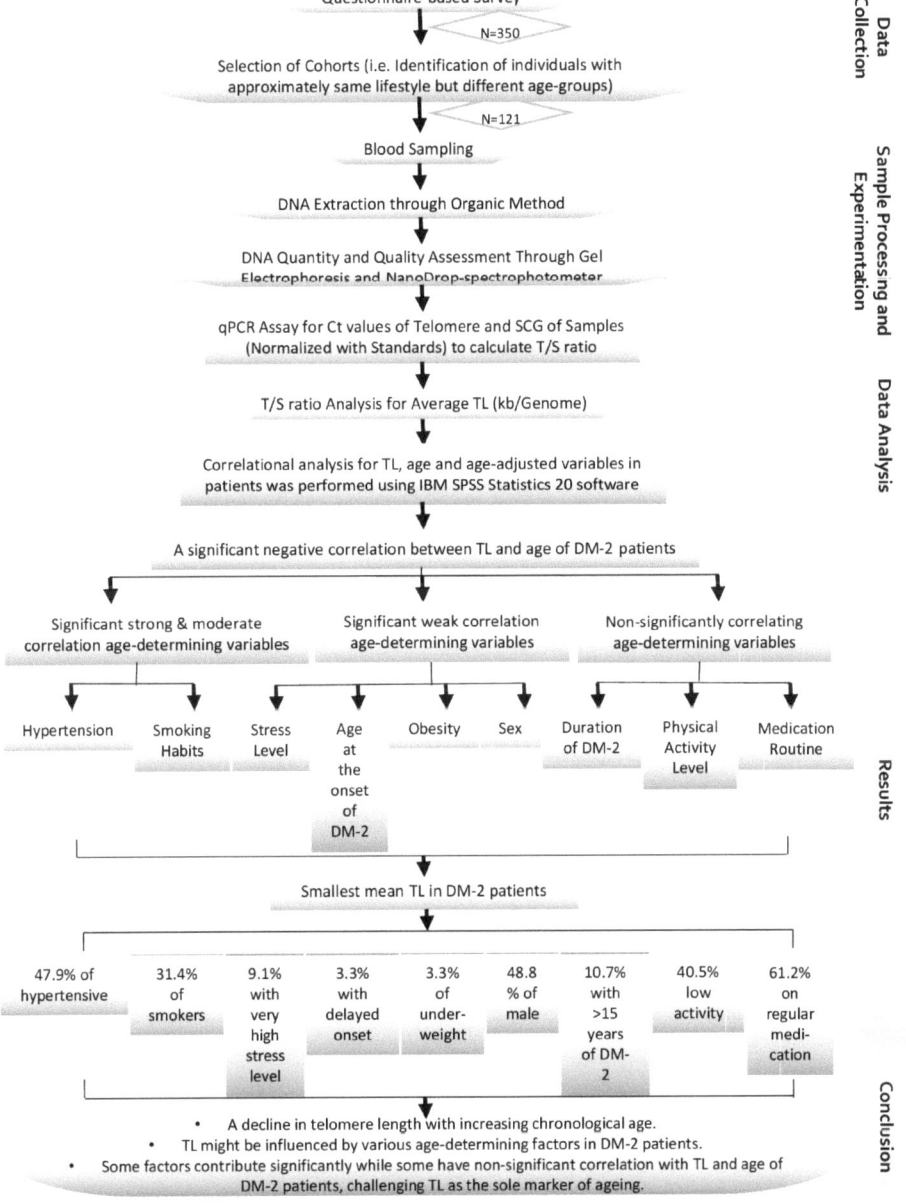

Figure 2. Schematic flowchart of the study.

3. Results

In this study, there is a significant negative correlation (Figure 3) between increasing age and telomere length (because $R^2 = 0.143$, p-value = 0.00) in diabetic patients.

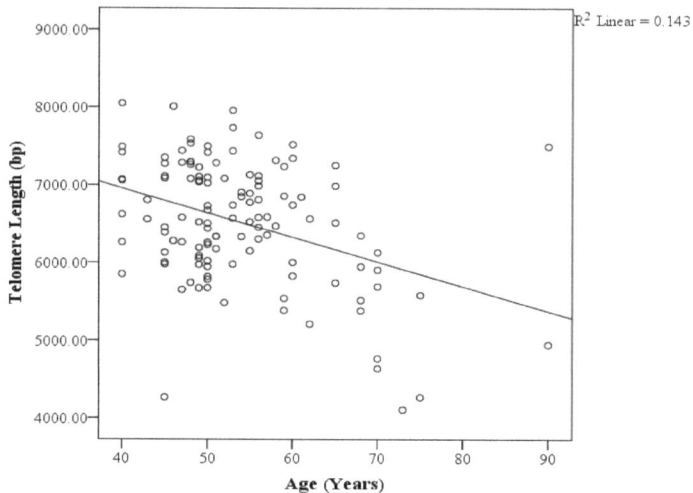

Figure 3. Correlation between telomere length and the age of diabetic patients. A significant moderate negative correlation is presented with a linear regression line.

Along with the significant correlation of increasing age, the correlations of many other age-determining factors (Table 2) with telomere length, including obesity (based on BMI), onset of diabetes, smoking habits, physical activity rate, hypertension, stress level, sex, and medication, were studied in this research.

Table 2. Age-adjusted telomere length by different age-determining variables in type 2 diabetes patients in Pakistan.

Age Determinants	Variables	Mean Telomere Length (kb)	Minimum and Maximum Lengths (kb)	No. of Patients (N)	%N	p Value	R	R²
Sex	Male	6.35 ± 0.82	4.11–7.73	59	48.8	0.02	−0.207 *	0.043
	Female	6.68 ± 0.74	4.26–8.05	62	51.2			
Obesity ***	Underweight	5.77 ± 1.14	4.26–6.90	4	3.3	0.025	0.203 *	0.041
	Normal weight	6.31 ± 0.93	4.11–7.96	36	29.8			
	Pre-obesity	6.63 ± 0.72	4.95–8.05	48	39.7			
	Obesity class I	6.72 ± 0.62	5.65–8.01	20	16.5			
	Obesity class II	6.55 ± 0.60	5.90–7.23	8	6.6			
	Obesity class III	6.76 ± 0.53	6.39–7.64	5	4.1			
Medication	Never	7.00 ± 0.61	6.57–7.44	2	1.7	0.576	−0.051	0.003
	Seldom	6.54 ± 0.75	4.26–7.53	45	37.2			
	Regular	6.50 ± 0.79	4.11–8.05	74	61.2			
Stress Level	Low	6.58 ± 0.65	4.77–8.05	68	56.2	0.036	−0.191 *	0.037
	Moderate	7.18 ± 0.67	6.26–8.01	8	6.6			
	High	6.44 ± 0.89	4.11–7.73	34	28.1			
	Very High	5.94 ± 0.99	4.26–7.11	11	9.1			
Smoking Habits	Primary Smokers	5.83 ± 0.73	4.11–7.10	38	31.4	0.000	0.547 **	0.299
	Secondary Smokers	6.77 ± 0.62	5.38–7.64	43	35.5			
	Non-Smokers	6.91 ± 0.57	5.95–8.05	40	33.1			
Physical Activity	Low Activity	6.47 ± 0.69	5.21–7.64	49	40.5	0.587	−0.050	0.002
	Medium Activity	6.59 ± 0.88	4.11–8.05	45	37.2			
	High Activity	6.53 ± 0.74	4.64–7.73	27	22.3			

Table 2. Cont.

Age Determinants	Variables	Mean Telomere Length (kb)	Minimum and Maximum Lengths (kb)	No. of Patients (N)	%N	p Value	R	R²
Hypertension	Hypertensive	5.93 ± 0.64	4.11–7.06	58	47.9	0.000	0.725 **	0.526
	Non-Hypertensive	7.07 ± 0.44	6.23–8.05	63	52.1			
Disease Onset Age	30–39	6.27 ± 0.99	4.26–8.01	17	14.0	0.005	−0.254 **	0.065
	40–49	6.56 ± 0.68	4.26–7.96	71	58.7			
	50–59	6.62 ± 0.92	4.11–7.73	19	15.7			
	60–69	6.7 ± 0.88	5.39–8.05	10	8.3			
	>70	6.00 ± 0.93	4.77–7.02	4	3.3			
Duration of Diabetes	<1–4	6.59 ± 0.79	4.26–8.05	61	50.4	0.416	−0.075	0.006
	5–9	6.50 ± 0.88	4.11–8.01	24	19.8			
	10–14	6.46 ± 0.76	4.26–7.5	23	19.0			
	15–19	5.92 ± 0.99	4.64–7.29	5	4.1			
	>20	6.64 ± 0.43	6.13–7.35	8	6.6			

* The correlation is significant at the 0.05 level (2-tailed). ** The correlation is significant at the 0.01 level (2-tailed). *** According to the WHO standards (2010). The arithmetic mean (95% CI) was used to calculate the corresponding telomere length in kb, and linear correlation was used to find the significant value (2-tailed), Pearson's correlation coefficient (R), and R-square values.

A typical data set from leukocytes of 59 males and 62 females in the Pakistani population with type 2 diabetes is shown in Table 2. For the females, the leukocytes had a mean telomere length of 6.68 kb/diploid genome (with the longest telomere length in the age group of 40–44-year-old individuals), while in males, leukocytes had a mean TL of 6.35 kb/diploid genome (with the longest telomere length in the age group of 50–54-year-old individuals), as shown in Figure 4.

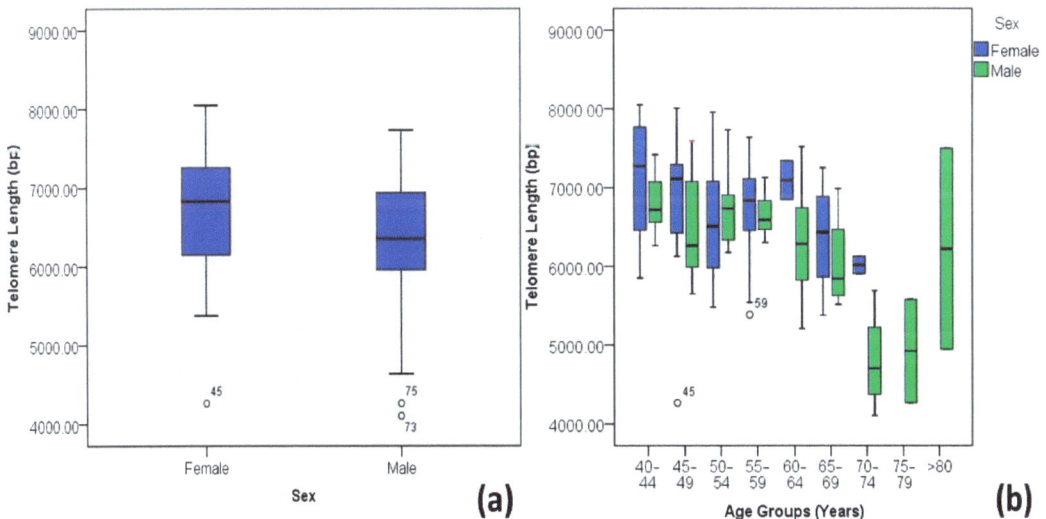

Figure 4. Correlations between sex and age-adjusted telomere length (Outliers are marked by the age of the patients). (**a**) Box plot showing correlations among DM-2 patients and (**b**) plot of age-adjusted correlations.

For obesity (Table 2), the shortest mean telomere length was observed among underweight diabetic patients. The median telomere length tended to increase with increasing BMI (until pre-obesity), while it again declines as the BMI continues to increase (as in obesity class III), as shown in Figure 5.

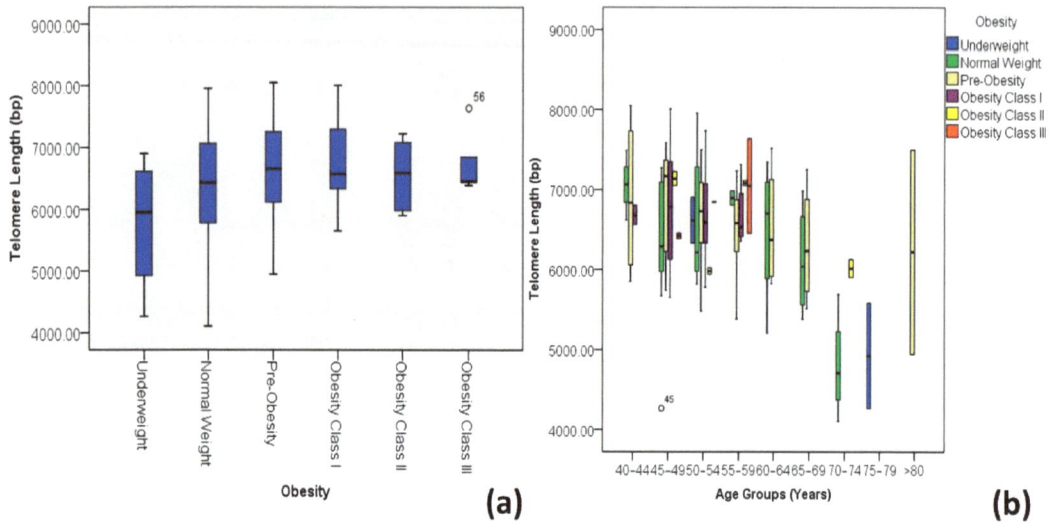

Figure 5. Correlations between obesity and age-adjusted telomere length (Outliers are marked by the age of the patients). (**a**) Box plot showing correlations among DM-2 patients and (**b**) plot of age-adjusted correlations.

Individuals who have never taken medicine for diabetes tend to have longer telomere lengths as compared to those who seldom or regularly take medicines (Figure 6). However, the individuals with the smallest and the longest telomere lengths were on regular medication.

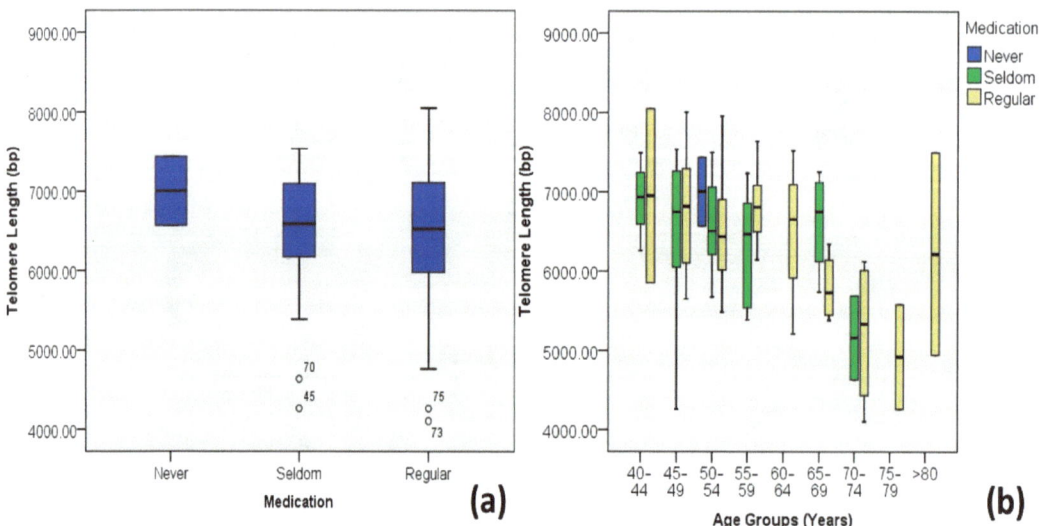

Figure 6. Correlations between routine medication and age-adjusted telomere length (Outliers are marked by the age of the patients). (**a**) Box plot showing correlations among DM-2 patients and (**b**) plot of age-adjusted correlations.

Diabetic individuals with moderate stress levels tend to have longer mean telomeres compared to individuals with high and very high stress levels (Figure 7). On average, primary smokers tend to have smaller telomere lengths compared to secondary smokers and non-smokers (Figure 8).

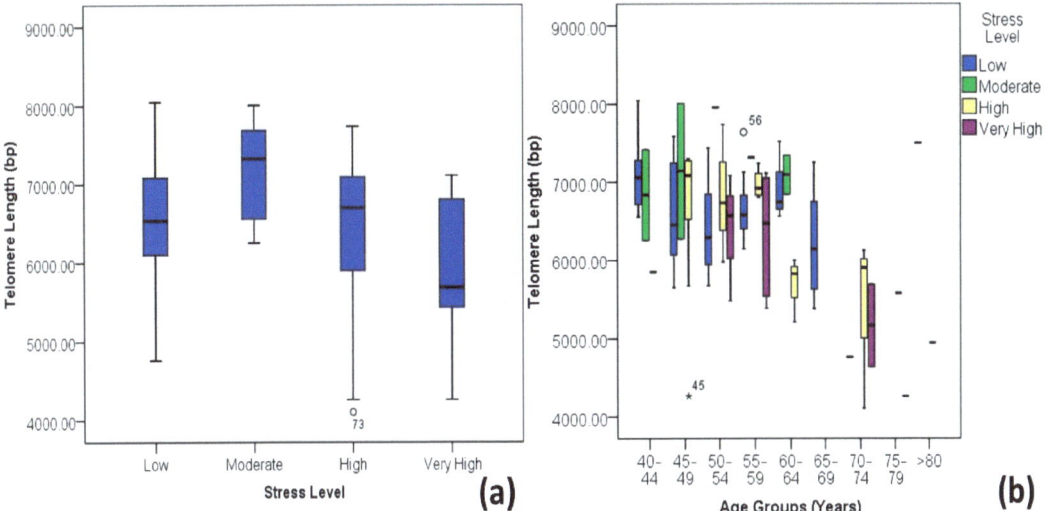

Figure 7. Correlations between stress levels and age-adjusted telomere length (Outliers are marked by the age of the patients). (**a**) Box plot showing correlations among DM-2 patients and (**b**) plot of age-adjusted correlations.

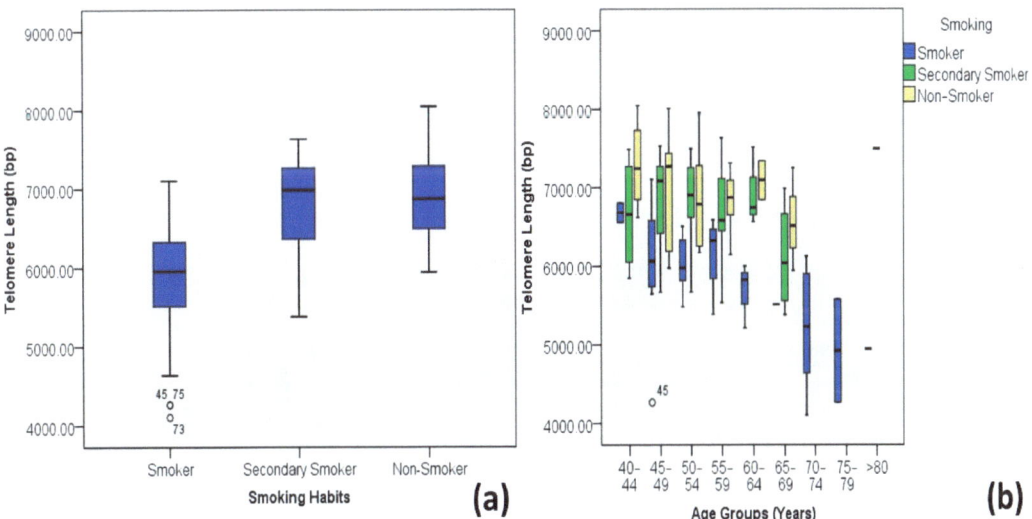

Figure 8. Correlations between smoking habits and age-adjusted telomere length (Outliers are marked by the age of the patients). (**a**) Box plot showing correlations among DM-2 patients and (**b**) plot of age-adjusted correlations.

Individuals with moderate to higher physical activity have longer mean telomere lengths than individuals with a lower activity rate. On the contrary, the individual with the longest telomere length lies in the cohort of individuals with medium physical activity (Figure 9).

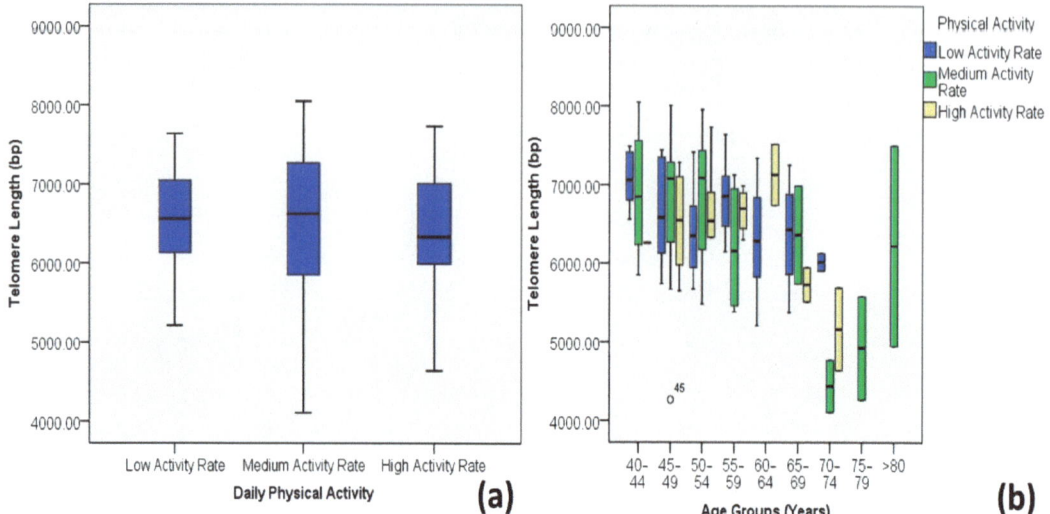

Figure 9. Correlations between physical activity levels and age-adjusted telomere length (Outlier is marked by the age of the patients). (**a**) Box plot showing correlations among DM-2 patients and (**b**) plot of age-adjusted correlations.

There was a clear difference in the range (minimum and maximum telomere lengths) and the mean telomere lengths of hypertensive and non-hypertensive individuals with DM-2. Hypertensive diabetic individuals presented smaller telomere lengths than the non-hypertensive individuals (Figure 10). Most of the individuals with earlier onset of DM-2 tends to have shorter telomere length (Figure 11); however, according to this study, the longer the duration of diabetes, the shorter the telomere length on average (Figure 12).

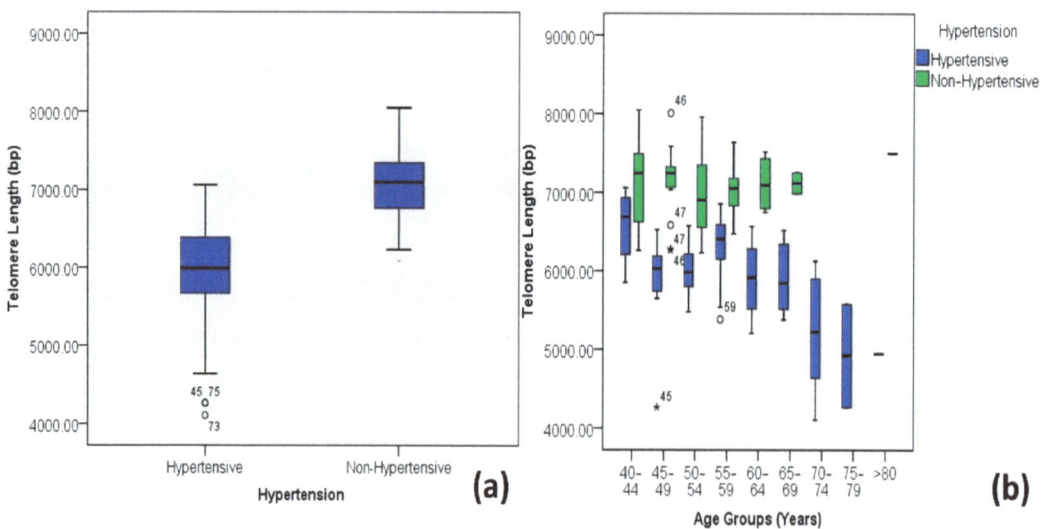

Figure 10. Correlations between hypertension and age-adjusted telomere length (Outliers are marked by the age of the patients). (**a**) Box plot showing correlations among DM-2 patients and (**b**) plot of age-adjusted correlations.

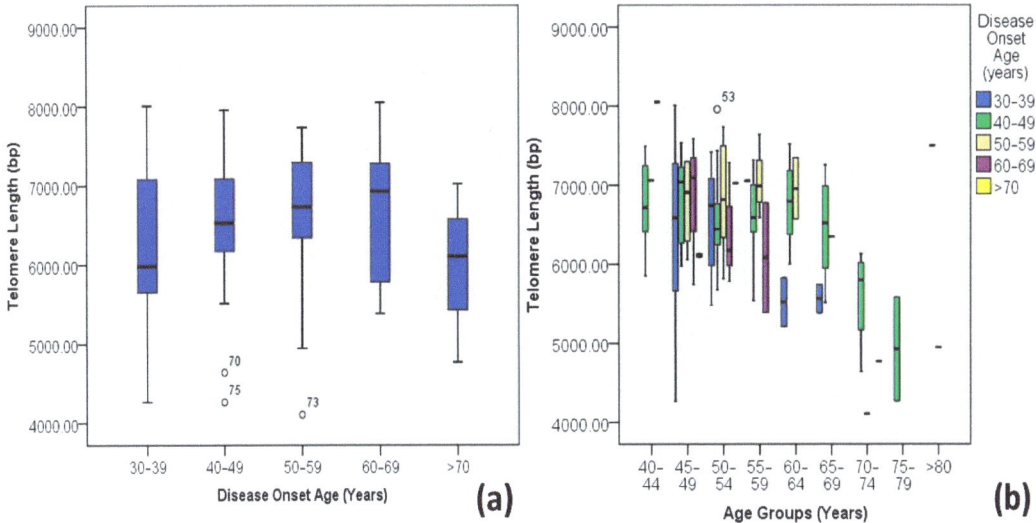

Figure 11. Correlations between the onset of diabetes and age-adjusted telomere length (Outliers are marked by the age of the patients). (**a**) Box plot showing correlations among DM-2 patients and (**b**) plot of age-adjusted correlations.

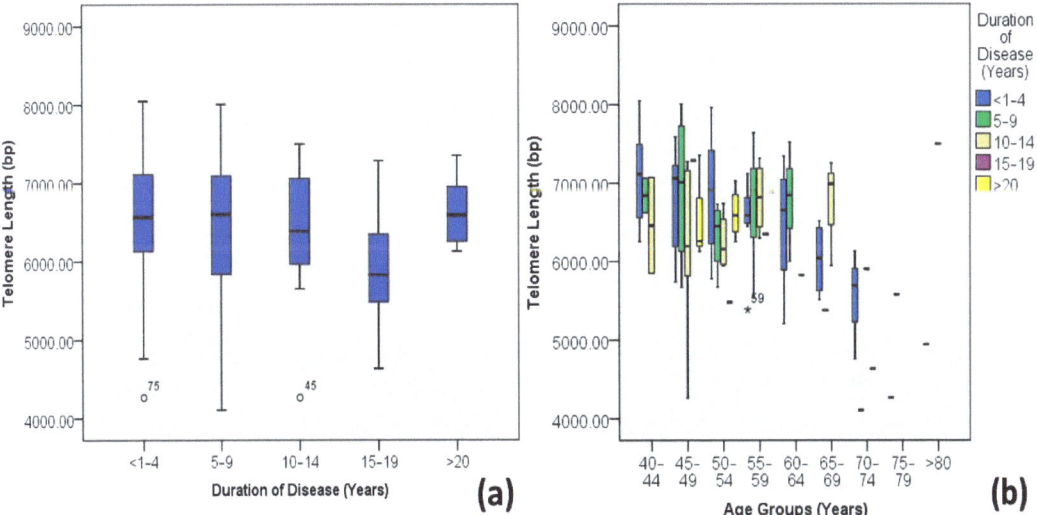

Figure 12. Correlations between the duration of diabetes and age-adjusted telomere length (Outliers are marked by the age of the patients). (**a**) Box plot showing correlations among DM-2 patients and (**b**) plot of age-adjusted correlations.

4. Discussion

Telomere length assays help us to understand the mechanisms of aging because telomere length serves as a unique cellular and molecular marker for studying the aging cell. Many variables are known to influence the telomere length in diabetic patients, including sex, diabetes type, geographical region, BMI, parental age, exercise, oxidative stress, genotype, smoking status, inherited mutations for accelerated aging syndrome, and psychological stress [19]. Although this study presents a few limitations, including the comparatively small sample size in terms of the number of patients (121 patients selected out of 350 patients from the questionnaire-based analysis), the strength of this study lies in the fact that the samples were collected from a homogenous population to avoid the maximum effect of extraneous variables, including geographical characteristics, patient's medical history, use of supplements and dietary habits, sleep time, exhaustive physical activity, tobacco consumption other than smoking, lifestyle variations, age, fertility, and marital status. Therefore, even if additional research with a large sample size will be required to support the results of this study, the sound sampling strategy is the strength of this research, exposing a phenomenon that merits in-depth analysis.

Not only the aging mechanism but also several studies have supported the theory of the association of telomere length with age-related diseases, cancer, and lifestyle changes (e.g., smoking and physical activity). Extensive epidemiological studies on different populations are used to explain or deny the correlations between telomere length and a range of diseases [19]. Many diseases are found to be correlated with a shorter telomere length, including ischemic heart disease [23], lung cancer among smokers [24], Alzheimer's disease [25], high blood pressure [26], diabetes [27], aging [28], and dementia [29]; many of the studies indicate that a few diseases are not significantly correlated with the telomere length, including colorectal cancer [30], age-related macular degeneration [23], and death from infections, cardiac or cerebrovascular disease, or cancer [10].

Diabetes is a serious, chronic condition affecting the lives of individuals and populations worldwide. In 2017, it was projected to have caused four million deaths worldwide, ranking among the top 10 causes of mortality for people [31]. As of 2019, the global prevalence of diabetes was estimated to be 9.3% (about 463 million adults aged 20–79 years). China, India, and Pakistan are expected to have the highest number of diabetic individuals in 2045, with counts of 147, 134, and 37 million, respectively [32].

Different researchers have indicated correlations between a shorter leukocyte telomere length and diabetes and diabetes-associated complications, including impaired glucose tolerance and diabetic macroangiopathy [19]. However, it is yet unknown how diabetes affects TL and molecular age. While some researchers [33,34] have found no age-related telomere length attrition between patients with diabetes mellitus and non-diabetic individuals, others [35,36] have shown an age effect. An association of a shorter telomere length, aging, and diabetes can be alarming for a developing country (Pakistan), which had an increase of 62% in diabetes cases (of 20–79 years individuals) in the past ten years [32]. However, this research indicated a significant moderate negative correlation (Figure 3) between telomere length and the age of diabetic patients, but the mean telomere length tended to decline with an increased duration of diabetes (Figure 12), indicating that although the age of diabetic patients directly influences the telomere length, a longer period of the disease will result in a shorter telomere length. It can be hypothesized that patients with extended periods of DM-2 are exposed to more oxidative stress and related complications of the disease as the chronological age of the patient increases.

The correlation between telomere length and the age of diabetic patients, however, might be accompanied by several other age-determining factors highlighted in this study. Organ dysfunction brought on by aging and diabetes is caused by comparable molecular pathways. The abundance of senescent cells in various tissues increases with age, obesity, and diabetes [37]. Although obesity is associated with shorter telomeres overall [38], studies of older individuals found no relation between telomere length and obesity and no relation with mortality [39]. The present study also indicates a significant but weak correlation ($R^2 = 0.04$) between obesity and telomere length variations among diabetic patients; however, it can be noticed (Figure 5) that the median telomere length increases as the BMI increases but before the obesity limit, i.e., pre-obese individuals; once the BMI crosses the obesity threshold, the median telomere length declines in diabetic individuals.

Studies indicate the association between physical activity levels and telomere length dynamics. Edwards and Loprinzi [40] and Stenbäck et al. [41] independently reported a significant relationship between the physical activity level and telomere length, such that moderate physical activity levels are correlated with a significantly longer peripheral blood mononuclear cell TLs compared to the highest and the lowest quartiles of physical activity. The present study also indicates a weak correlation ($R^2 = 0.002$) between telomere length and the physical activity levels of diabetic patients. However, individuals with moderate physical activity rates tend to have longer telomere lengths on average compared to those with lower or higher activity rates.

Several papers have reported that hypertensive individuals are associated with a shorter telomere length [42–44]. This study also indicates that hypertensive individuals tend to have shorter telomere lengths than non-hypertensive patients, with a significant strong correlation ($R = 0.725$ and $p < 0.001$).

Stressful lifestyles caused by hectic work schedules, family issues, financial burdens, or emotional damage have been correlated with telomere length variations. Many studies suggest that telomere attrition caused by stress could lead to premature aging without adequate recovery [45]. The present study also indicates a significant negative correlation (Pearson's correlation coefficient = -0.191, p-value < 0.05) between telomere length and stress levels in diabetic patients, such that individuals with high or very high stress levels have shorter mean telomere lengths than those having low or moderate stress levels.

The prevention of TL attrition is also associated with healthy life choices, such as not smoking tobacco and drugs or medicines [24,46]. In this study, a highly significant moderate correlation ($R = 0.547$ and p value = 0.000) is observed between telomere length and smoking habits, and a non-significant negligible correlation (Pearson's coefficient = -0.05, p value = 0.576) with regular medication of diabetic individuals. Smokers tend to have shorter telomere lengths as compared to secondary smokers or non-smokers, and individuals who never had taken medicines for diabetes have longer average telomere lengths (7.0 kb) than those who take medicines seldom (6.54 kb) or regularly (6.50 kb).

The clinical and medical applications of the telomere length analysis include prognostic markers and diagnostic indicators [47], cancer research and treatment [48], implications and evaluations of anti-aging therapies [49], and many more. The current study can facilitate future implications to identify factors that have a range of influence (from strong to weak and from significant to non-significant) on the age-adjusted telomere length in DM-2 patients, thus evaluating the premature aging process and its recovery.

5. Conclusions

RT-qPCR is a powerful tool for telomere assays but detects average telomere length, but not on an individual chromosome basis. This method is also extremely sensitive, and so cross-contamination was avoided and the method of analysis was planned appropriately to nullify the chances of erroneous or false results.

The study was conducted on DM-2 patients from Pakistan. The statistical measure of this study suggests a significant negative linear relation between the age of individuals with DM-2 and their telomere length. This finding is important because telomere length is often considered a potential marker of the aging process. However, other factors beyond chronological age may also be influential in determining telomere length, particularly in type 2 diabetes mellitus patients.

Through this research, it is concluded that telomere length may not be the only marker for the aging mechanism as many of the other age-determining factors are worth noting to also contribute to the aging process with significant moderate to strong correlations with telomere length (e.g., smoking habits and hypertension) and significant but weak correlations with telomere length (e.g., sex, obesity, stress level, and age at the onset of diabetes), while some have a non-significant correlation with telomere length (e.g., routine medication, rate of physical activity, and the duration of diabetes) in DM-2 patients. So, this study paved the way to determine the underlying factors that can be further studied to lower the effect of aging or pre-mature aging on diabetic patients.

One more breakthrough of this research includes the further analysis of de-trending cases (individuals with longer telomeres even at older ages and individuals with the shortest telomeres even at younger ages) by asking them for their complete medical checkups and through insights based on their lifestyle. The person with a shorter telomere length was found to be suffering from invasive stage II breast cancer, while the individual with a longer telomere length never had taken any allopathic medicine in his whole life span. So, if properly organized, this assay method (telomere length analysis using qPCR) can also be used as a preliminary diagnostic method for diseases like cancer.

6. Recommendations

Real-time qPCR is an efficient method to measure telomere length, even for large sample sizes requiring a minimal amount of DNA, not only to investigate the correlation with age-related diseases but also for diagnostic purposes to find pre-mature aging in individuals and for other diseases like cancer so that such disorders can be timely treated or cured.

There is a need for further study to find mechanisms of cellular aging, senescence, and disease states where this assay method fails to show a significant correlation of telomere length with age-related factors and diseases. Such studies will help to shift the focus of future aging research from disease-oriented studies to cellular- or molecule-oriented studies.

Although this study finds a correlation between the onset of diabetes and telomere shortening, leading to the theory that diabetes accelerates the molecular aging process, there is a need for further research to fully understand the underlying mechanisms and the extent of the impact of diabetes on telomere length dynamics.

Supplementary Materials: The following supporting information can be downloaded at https://www.mdpi.com/article/10.3390/medicina60050698/s1. Supplemental Table S1: Oligomers (Standards and Primers) used for qPCR analysis of Telomere Length; Supplemental Table S2: Reaction Mix for 20 µL of Final Volume.

Author Contributions: Conceptualization, J.A.T.; Methodology, N.S.; Software, A.R.D.; Validation, J.A.T.; Formal Analysis, H.J., S.J. and A.R.D.; Data Curation, H.J., H.S., S.J. and A.R.D.; Writing—Original Draft Preparation, J.A.T., K.M., A.H. and A.R.; Writing—Review & Editing, A.H.; Supervision, N.S.; Project Administration, N.S. All authors have read and agreed to the published version of the manuscript.

Funding: This research received no external funding.

Institutional Review Board Statement: The design, voluntary consent by participants and religious and ethical perspectives of this research were approved by the Departmental Research Committee (Department of Biotechnology, Sardar Bahadur Khan Women's University, Pakistan) and Board of Advance Studies and Research (Sardar Bahadur Khan Women's University, Pakistan) under the notification number 159/Estt:/ SBKWU/16/1036 dated 27 December 2016.

Informed Consent Statement: Informed consent was obtained from all subjects involved in the study.

Data Availability Statement: The authors can furnish supporting data and datasets for this research via email upon request.

Acknowledgments: We acknowledge the volunteers who consented to provide data and blood samples for this research. We acknowledge the "Higher Education Commission (HEC), Pakistan". We are grateful to Noaman Saeed, CEO of The Biogene Labs and Diagnostic, Islamabad, for the permission to work in his lab.

Conflicts of Interest: The authors declare no conflicts of interest. The funders had no role in the design of the study; in the collection, analyses, or interpretation of data; in the writing of the manuscript; or in the decision to publish the results.

Abbreviations

Body mass index (BMI); cycle threshold value (Ct value); diabetes mellitus type 2 (DM-2); ethylene diamine tetra acetic acid (EDTA); kilobase pair (kb); quantitative real-time polymerase chain reaction (qRT-PCR); single copy gene (SCG); telomere length (TL).

References

1. American Diabetes Association, A. Diagnosis and classification of diabetes mellitus. *Diabetes Care* **2014**, *37* (Suppl. S1), S81–S90. [CrossRef] [PubMed]
2. Banday, M.Z.; Sameer, A.S.; Nissar, S. Pathophysiology of diabetes: An overview. *Avicenna J. Med.* **2020**, *10*, 174–188. [CrossRef] [PubMed]
3. Fazeli Farsani, S.; Van Der Aa, M.; Van Der Vorst, M.; Knibbe, C.; De Boer, A. Global trends in the incidence and prevalence of type 2 diabetes in children and adolescents: A systematic review and evaluation of methodological approaches. *Diabetologia* **2013**, *56*, 1471–1488. [CrossRef]
4. Tremblay, J.; Hamet, P. Environmental and genetic contributions to diabetes. *Metab.-Clin. Exp.* **2019**, *100*, 153952. [CrossRef] [PubMed]
5. Basit, A.; Fawwad, A.; Qureshi, H.; Shera, A. Prevalence of diabetes, pre-diabetes and associated risk factors: Second National Diabetes Survey of Pakistan (NDSP), 2016–2017. *BMJ Open* **2018**, *8*, e020961. [CrossRef] [PubMed]
6. Aamir, A.H.; Ul-Haq, Z.; Mahar, S.A.; Qureshi, F.M.; Ahmad, I.; Jawa, A.; Sheikh, A.; Raza, A.; Fazid, S.; Jadoon, Z. Diabetes Prevalence Survey of Pakistan (DPS-PAK): Prevalence of type 2 diabetes mellitus and prediabetes using HbA1c: A population-based survey from Pakistan. *BMJ Open* **2019**, *9*, e025300. [CrossRef] [PubMed]
7. Blackburn, E.H.; Epel, E.S.; Lin, J. Human telomere biology: A contributory and interactive factor in aging, disease risks, and protection. *Science* **2015**, *350*, 1193–1198. [CrossRef]
8. López-Otín, C.; Blasco, M.A.; Partridge, L.; Serrano, M.; Kroemer, G. The hallmarks of aging. *Cell* **2013**, *153*, 1194–1217. [CrossRef] [PubMed]
9. Willeit, P.; Raschenberger, J.; Heydon, E.E.; Tsimikas, S.; Haun, M.; Mayr, A.; Weger, S.; Witztum, J.L.; Butterworth, A.S.; Willeit, J. Leucocyte telomere length and risk of type 2 diabetes mellitus: New prospective cohort study and literature-based meta-analysis. *PLoS ONE* **2014**, *9*, e112483. [CrossRef]
10. Haycock, P.C.; Heydon, E.E.; Kaptoge, S.; Butterworth, A.S.; Thompson, A.; Willeit, P. Leucocyte telomere length and risk of cardiovascular disease: Systematic review and meta-analysis. *BMJ* **2014**, *349*, g4227. [CrossRef]
11. Arbeev, K.G.; Verhulst, S.; Steenstrup, T.; Kark, J.D.; Bagley, O.; Kooperberg, C.; Reiner, A.P.; Hwang, S.-J.; Levy, D.; Fitzpatrick, A.L. Association of leukocyte telomere length with mortality among adult participants in 3 longitudinal studies. *JAMA Netw. Open* **2020**, *3*, e200023. [CrossRef] [PubMed]
12. Rehkopf, D.H.; Needham, B.L.; Lin, J.; Blackburn, E.H.; Zota, A.R.; Wojcicki, J.M.; Epel, E.S. Leukocyte telomere length in relation to 17 biomarkers of cardiovascular disease risk: A cross-sectional study of US adults. *PLoS Med.* **2016**, *13*, e1002188. [CrossRef] [PubMed]
13. Patel, C.J.; Manrai, A.K.; Corona, E.; Kohane, I.S. Systematic correlation of environmental exposure and physiological and self-reported behaviour factors with leukocyte telomere length. *Int. J. Epidemiol.* **2017**, *46*, 44–56. [CrossRef] [PubMed]
14. Aviv, A.; Hunt, S.C.; Lin, J.; Cao, X.; Kimura, M.; Blackburn, E. Impartial comparative analysis of measurement of leukocyte telomere length/DNA content by Southern blots and qPCR. *Nucleic Acids Res.* **2011**, *39*, e134. [CrossRef] [PubMed]
15. Adaikalakoteswari, A.; Vatish, M.; Lawson, A.; Wood, C.; Sivakumar, K.; McTernan, P.G.; Webster, C.; Anderson, N.; Yajnik, C.S.; Tripathi, G. Low maternal vitamin B12 status is associated with lower cord blood HDL cholesterol in white Caucasians living in the UK. *Nutrients* **2015**, *7*, 2401–2414. [CrossRef] [PubMed]
16. Cheng, F.; Carroll, L.; Joglekar, M.V.; Januszewski, A.S.; Wong, K.K.; Hardikar, A.A.; Jenkins, A.J.; Ma, R.C.W. Diabetes, metabolic disease, and telomere length. *Lancet Diabetes Endocrinol.* **2021**, *9*, 117–126. [CrossRef] [PubMed]
17. Piplani, S.; Alemao, N.N.; Prabhu, M.; Ambar, S.; Chugh, Y.; Chugh, S.K. Correlation of the telomere length with type 2 diabetes mellitus in patients with ischemic heart disease. *Indian Heart J.* **2018**, *70* (Suppl. S3), S173–S176. [CrossRef] [PubMed]
18. Cawthon, R. Telomere measurement by quantitative PCR. *Nucleic Acids Res.* **2002**, *30*, e47. [CrossRef] [PubMed]
19. Axelrad, M.D.; Budagov, T.; Atzmon, G. Telomere length and telomerase activity; a Yin and Yang of cell senescence. *JoVE (J. Vis. Exp.)* **2013**, *22*, e50246.
20. Shen, C.-H. *Diagnostic Molecular Biology*, 2nd ed.; Elsevier: Amsterdam, The Netherlands, 2023.
21. O'Callaghan, N.J.; Fenech, M. A quantitative PCR method for measuring absolute telomere length. *Biol. Proced. Online* **2011**, *13*, 3. [CrossRef]
22. Atzmon, G.; Cho, M.; Cawthon, R.M.; Budagov, T.; Katz, M.; Yang, X.; Siegel, G.; Bergman, A.; Huffman, D.M.; Schechter, C.B. Genetic variation in human telomerase is associated with telomere length in Ashkenazi centenarians. *Proc. Natl. Acad. Sci. USA* **2010**, *107* (Suppl. S3), 1710–1717. [CrossRef]
23. Weischer, M.; Bojesen, S.E.; Cawthon, R.M.; Freiberg, J.J.; Tybjærg-Hansen, A.; Nordestgaard, B.G. Short telomere length, myocardial infarction, ischemic heart disease, and early death. *Arterioscler. Thromb. Vasc. Biol.* **2012**, *32*, 822–829. [CrossRef] [PubMed]

24. Shen, M.; Cawthon, R.; Rothman, N.; Weinstein, S.J.; Virtamo, J.; Hosgood III, H.D.; Hu, W.; Lim, U.; Albanes, D.; Lan, Q. A prospective study of telomere length measured by monochrome multiplex quantitative PCR and risk of lung cancer. *Lung Cancer* **2011**, *73*, 133–137. [CrossRef]
25. Hochstrasser, T.; Marksteiner, J.; Humpel, C. Telomere length is age-dependent and reduced in monocytes of Alzheimer patients. *Exp. Gerontol.* **2012**, *47*, 160–163. [CrossRef] [PubMed]
26. Insel, K.C.; Merkle, C.J.; Hsiao, C.-P.; Vidrine, A.N.; Montgomery, D.W. Biomarkers for cognitive aging part I: Telomere length, blood pressure and cognition among individuals with hypertension. *Biol. Res. Nurs.* **2012**, *14*, 124–132. [CrossRef]
27. Monickaraj, F.; Aravind, S.; Gokulakrishnan, K.; Sathishkumar, C.; Prabu, P.; Prabu, D.; Mohan, V.; Balasubramanyam, M. Accelerated aging as evidenced by increased telomere shortening and mitochondrial DNA depletion in patients with type 2 diabetes. *Mol. Cell. Biochem.* **2012**, *365*, 343–350. [CrossRef] [PubMed]
28. Shammas, M.A. Telomeres, lifestyle, cancer, and aging. *Curr. Opin. Clin. Nutr. Metab. Care* **2011**, *14*, 28–34. [CrossRef]
29. Yaffe, K.; Lindquist, K.; Kluse, M.; Cawthon, R.; Harris, T.; Hsueh, W.-C.; Simonsick, E.M.; Kuller, L.; Li, R.; Ayonayon, H.N. Telomere length and cognitive function in community-dwelling elders: Findings from the Health ABC Study. *Neurobiol. Aging* **2011**, *32*, 2055–2060. [CrossRef] [PubMed]
30. Pauleck, S.; Sinnott, J.A.; Zheng, Y.-L.; Gadalla, S.M.; Viskochil, R.; Haaland, B.; Cawthon, R.M.; Hoffmeister, A.; Hardikar, S. Association of Telomere Length with Colorectal Cancer Risk and Prognosis: A Systematic Review and Meta-Analysis. *Cancers* **2023**, *15*, 1159. [CrossRef]
31. IDF. IDF diabetes atlas. In *Diabetes*; International Diabetes Federation: Brussels, Belgium, 2017; Volume 20, p. 79.
32. Saeedi, P.; Petersohn, I.; Salpea, P.; Malanda, B.; Karuranga, S.; Unwin, N.; Colagiuri, S.; Guariguata, L.; Motala, A.A.; Ogurtsova, K. Global and regional diabetes prevalence estimates for 2019 and projections for 2030 and 2045: Results from the International Diabetes Federation Diabetes Atlas. *Diabetes Res. Clin. Pract.* **2019**, *157*, 107843. [CrossRef]
33. Fyhrquist, F.; Tiitu, A.; Saijonmaa, O.; Forsblom, C.; Groop, P.H.; Group, F.S. Telomere length and progression of diabetic nephropathy in patients with type 1 diabetes. *J. Intern. Med.* **2010**, *267*, 278–286. [CrossRef]
34. Januszewski, A.S.; Sutanto, S.S.; McLennan, S.; O'Neal, D.N.; Keech, A.C.; Twigg, S.M.; Jenkins, A.J. Shorter telomeres in adults with Type 1 diabetes correlate with diabetes duration, but only weakly with vascular function and risk factors. *Diabetes Res. Clin. Pract.* **2016**, *117*, 4–11. [CrossRef]
35. Sharma, R.; Gupta, A.; Thungapathra, M.; Bansal, R. Telomere mean length in patients with diabetic retinopathy. *Sci. Rep.* **2015**, *5*, 18368. [CrossRef] [PubMed]
36. Chaithanya, V.; Kumar, K.; Leela, K.V.; Murugesan, R.; Angelin, M.; Satheesan, A. Impact of telomere attrition on diabetes mellitus and its complications. *Diabetes Epidemiol. Manag.* **2023**, *12*, 100174. [CrossRef]
37. Palmer, A.K.; Gustafson, B.; Kirkland, J.L.; Smith, U. Cellular senescence: At the nexus between ageing and diabetes. *Diabetologia* **2019**, *62*, 1835–1841. [CrossRef] [PubMed]
38. Mundstock, E.; Sarria, E.E.; Zatti, H.; Mattos Louzada, F.; Kich Grun, L.; Herbert Jones, M.; Guma, F.T.; Mazzola, J.; Epifanio, M.; Stein, R.T. Effect of obesity on telomere length: Systematic review and meta-analysis. *Obesity* **2015**, *23*, 2165–2174. [CrossRef]
39. Njajou, O.T.; Cawthon, R.M.; Blackburn, E.H.; Harris, T.B.; Li, R.; Sanders, J.L.; Newman, A.B.; Nalls, M.; Cummings, S.R.; Hsueh, W.-C. Shorter telomeres are associated with obesity and weight gain in the elderly. *Int. J. Obes.* **2012**, *36*, 1176–1179. [CrossRef]
40. Edwards, M.K.; Loprinzi, P.D. Sedentary behavior, physical activity and cardiorespiratory fitness on leukocyte telomere length. *Health Promot. Perspect.* **2017**, *7*, 22–27. [CrossRef]
41. Stenbäck, V.; Mutt, S.J.; Leppäluoto, J.; Gagnon, D.D.; Mäkelä, K.A.; Jokelainen, J.; Keinänen-Kiukaanniemi, S.; Herzig, K.-H. Association of Physical Activity With Telomere Length Among Elderly Adults—The Oulu Cohort 1945. *Front. Physiol.* **2019**, *10*, 444. [CrossRef]
42. Bhupatiraju, C.; Saini, D.; Patkar, S.; Deepak, P.; Das, B.; Padma, T. Association of shorter telomere length with essential hypertension in Indian population. *Am. J. Hum. Biol.* **2012**, *24*, 573–578. [CrossRef]
43. Yeh, J.-K.; Wang, C.-Y. Telomeres and telomerase in cardiovascular diseases. *Genes* **2016**, *7*, 58. [CrossRef]
44. Tellechea, M.L.; Pirola, C.J. The impact of hypertension on leukocyte telomere length: A systematic review and meta-analysis of human studies. *J. Hum. Hypertens.* **2017**, *31*, 99–105. [CrossRef]
45. Parks, C.G.; DeRoo, L.; Miller, D.; McCanlies, E.; Cawthon, R.; Sandler, D. Employment and work schedule are related to telomere length in women. *Occup. Environ. Med.* **2011**, *68*, 582–589. [CrossRef]
46. Xu, Q.; Parks, C.G.; DeRoo, L.A.; Cawthon, R.M.; Sandler, D.P.; Chen, H. Multivitamin use and telomere length in women. *Am. J. Clin. Nutr.* **2009**, *89*, 1857–1863. [CrossRef]
47. Artandi, S.E.; DePinho, R.A. Telomeres and telomerase in cancer. *Carcinogenesis* **2010**, *31*, 9–18. [CrossRef]
48. Zhang, C.; Doherty, J.A.; Burgess, S.; Hung, R.J.; Lindström, S.; Kraft, P.; Gong, J.; Amos, C.I.; Sellers, T.A.; Monteiro, A.N.; et al. Genetic determinants of telomere length and risk of common cancers: A Mendelian randomization study. *Hum. Mol. Genet.* **2015**, *24*, 5356–5366. [CrossRef]
49. Hurvitz, N.; Elkhateeb, N.; Sigawi, T.; Rinsky-Halivni, L.; Ilan, Y. Improving the effectiveness of anti-aging modalities by using the constrained disorder principle-based management algorithms. *Front. Aging* **2022**, *3*, 1044038. [CrossRef]

Disclaimer/Publisher's Note: The statements, opinions and data contained in all publications are solely those of the individual author(s) and contributor(s) and not of MDPI and/or the editor(s). MDPI and/or the editor(s) disclaim responsibility for any injury to people or property resulting from any ideas, methods, instructions or products referred to in the content.

Article

The Relationship between the Ewing Test, Sudoscan Cardiovascular Autonomic Neuropathy Score and Cardiovascular Risk Score Calculated with SCORE2-Diabetes

Andra-Elena Nica [1,2], Emilia Rusu [1,2,*], Carmen Dobjanschi [1,2], Florin Rusu [3], Claudia Sivu [1], Oana Andreea Parlițeanu [4] and Gabriela Radulian [1]

1 Diabetes Department, "Carol Davila" University of Medicine and Pharmacy, 050474 Bucharest, Romania; andra.nica@drd.umfcd.ro (A.-E.N.); carmen.dobjanschi@umfcd.ro (C.D.); claudia.topea@drd.umfcd.ro (C.S.); gabriela.radulian@umfcd.ro (G.R.)
2 "Nicolae Malaxa" Clinica Hospital, 022441 Bucharest, Romania
3 "Doctor Carol Davila" Central Military University Emergency Hospital, 010825 Bucharest, Romania; florinrusumd@yahoo.com
4 "Marius Nasta" Institute of Pneumophysiology, 050159 Bucharest, Romania; oana_andreea@yahoo.com
* Correspondence: emilia.rusu@umfcd.ro; Tel.: +40-7429-59946

Citation: Nica, A.-E.; Rusu, E.; Dobjanschi, C.; Rusu, F.; Sivu, C.; Parlițeanu, O.A.; Radulian, G. The Relationship between the Ewing Test, Sudoscan Cardiovascular Autonomic Neuropathy Score and Cardiovascular Risk Score Calculated with SCORE2-Diabetes. *Medicina* 2024, 60, 828. https://doi.org/10.3390/medicina60050828

Academic Editors: Yuzuru Ohshiro, Kunimasa Yagi and Yasuhiro Maeno

Received: 17 April 2024
Revised: 13 May 2024
Accepted: 15 May 2024
Published: 17 May 2024

Copyright: © 2024 by the authors. Licensee MDPI, Basel, Switzerland. This article is an open access article distributed under the terms and conditions of the Creative Commons Attribution (CC BY) license (https://creativecommons.org/licenses/by/4.0/).

Abstract: *Background and Objectives*: Cardiac autonomic neuropathy (CAN) is a severe complication of diabetes mellitus (DM) strongly linked to a nearly five-fold higher risk of cardiovascular mortality. Patients with Type 2 Diabetes Mellitus (T2DM) are a significant cohort in which these assessments have particular relevance to the increased cardiovascular risk inherent in the condition. *Materials and Methods*: This study aimed to explore the subtle correlation between the Ewing test, Sudoscan-cardiovascular autonomic neuropathy score, and cardiovascular risk calculated using SCORE 2 Diabetes in individuals with T2DM. The methodology involved detailed assessments including Sudoscan tests to evaluate sudomotor function and various cardiovascular reflex tests (CART). The cohort consisted of 211 patients diagnosed with T2DM with overweight or obesity without established ASCVD, aged between 40 to 69 years. *Results*: The prevalence of CAN in our group was 67.2%. In the study group, according SCORE2-Diabetes, four patients (1.9%) were classified with moderate cardiovascular risk, thirty-five (16.6%) with high risk, and one hundred seventy-two (81.5%) with very high cardiovascular risk. *Conclusions*: On multiple linear regression, the SCORE2-Diabetes algorithm remained significantly associated with Sudoscan CAN-score and Sudoscan Nephro-score and Ewing test score. Testing for the diagnosis of CAN in very high-risk patients should be performed because approximately 70% of them associate CAN. Increased cardiovascular risk is associated with sudomotor damage and that Sudoscan is an effective and non-invasive measure of identifying such risk.

Keywords: T2DM; Sudoscan; CAN; CVDs; SCORE2-Diabetes

1. Introduction

Cardiac autonomic neuropathy (CAN) is a severe complication of diabetes mellitus (DM) strongly linked to a nearly five-fold higher risk of cardiovascular mortality [1]. Factors leading to diabetic complications, such as genetic predisposition, environmental signals, insulin resistance, immune dysfunction and inflammation have a similar impact on both microvascular and macrovascular complications [2]. Elucidation of the common mechanism of cell damage in DM put to rest any misconceptions that microvascular disease and macrovascular disease constitute distinct disease entities [3,4].

CAN represents a manageable complication of T2DM, posing an elevated risk of cardiovascular mortality [5]. Patients with CAN are often underdiagnosed due to limited physician awareness of the disease, its frequently late onset with mild or absent symptoms, and the absence of specific diagnostic tests for CAN [6]. Various cross-sectional studies have revealed a wide range of prevalence rates. For instance, studies have shown rates ranging from 16.7% in insulin-dependent diabetic patients participating in the Diabetes Control and Complications Trial (DCCT) cohort to as high as 60% in a community-based sample of type 2 diabetic patients over the age of 65 in Rochester, Minnesota [7,8].

Among the many diagnostic tools available, Ewing and Sudoscan stand out as promising methods for evaluating autonomic neuropathy and its implications for cardiovascular health. Annually, cardiovascular diseases (CVDs) account for approximately 17.9 million deaths, making up 31% of all global fatalities [9]. Public health policies must prioritize preventing cardiovascular issues and managing established risk factors like dyslipidemia, T2DM, and hypertension [10]. Cardiovascular diseases represent the primary contributors to both morbidity and mortality among individuals diagnosed with T2DM [11].

For patients over 40 with T2DM without coronary artery disease (ASCVD) or severe target organ damage (TOD), estimating the 10-year cardiovascular risk using the SCORE2-Diabetes algorithm is recommended. The 2021 ESC Guidelines also suggest using ADVANCE or DIAL models for such risk estimations in diabetic patients [12–14]. The current guidelines recommend using the SCORE2-Diabetes model and to estimate the individual 10-year risk of fatal and non-fatal cardiovascular events [15].

On the other hand, Sudoscan is a useful tool for detecting chronic microvascular complications of T2DM, which are additional cardiovascular risk factors. Its results can provide valuable information for diagnosing and monitoring neuropathy and assessing cardiovascular risk in diabetic patients [8].

As the diabetes epidemic progresses, the increasing prevalence of CAN poses significant life-threatening risks, including arrhythmias and silent myocardial ischemia, particularly impacting the cardiovascular health of patients with T2DM [16].

Managing T2DM demands a comprehensive strategy encompassing lifestyle adjustments and glycemic regulation, alongside mitigating cardiovascular risk factors [17]. Additionally, employing glucose-lowering medications known for their cardiovascular benefits, such as SGLT2 inhibitors [18] and GLP-1 receptor agonists [19], is essential.

This study aimed to explore the subtle correlation between the Ewing test, Sudoscan-cardiovascular autonomic neuropathy score, and cardiovascular risk calculated using SCORE 2 Diabetes in individuals with T2DM. By elucidating this correlation, we aim to shed light on new avenues for comprehensive cardiovascular risk assessment.

2. Materials and Methods

2.1. Study Desgin

This study was cross-sectional, effectuated between June 2019 and June 2020. Ethical approval for the study was obtained from the local Ethics Committee of "Nicolae Malaxa" Clinical Hospital. Informed consent was obtained from all participating patients.

2.2. Study Population

Inclusion Criteria: Individuals diagnosed with Type 2 Diabetes Mellitus (T2DM) and those who were overweight/obese aged between 40 and 69 years were included.

Exclusion Criteria: Patients who did not provide informed consent, had other types of diabetes (type 1 diabetes, latent autoimmune diabetes in adults, maturity-onset diabetes of the young), were outside the age range of 40 to 69 years, pregnant women, those diagnosed with neoplasms within the past five years, had stroke sequelae, a history of myocardial infarction, pelvic limb amputations, pre-existing chronic kidney disease before diabetes diagnosis, and neuropathy from alternative causes (e.g., alcoholism, vitamin B12 deficiency). Type 1 diabetes, latent autoimmune diabetes of adults, and maturity-onset diabetes of the young were diagnosed by performing specific autoantibody tests. The presence of these antibodies led to the exclusion of patients from the study.

Variables: In our article the primary outcome variable was the correlation between the Sudoscan-cardiovascular autonomic neuropathy (CAN) score and the cardiovascular risk calculated using the SCORE2-Diabetes algorithm. This study aimed to investigate the potential association between sudomotor dysfunction, as measured by the Sudoscan-CAN score, and increased cardiovascular risk in individuals with Type 2 Diabetes Mellitus (T2DM).

Examination of patients: Anthropometric indices recorded included height, weight, body mass index (BMI), waist circumference, and waist-to-hip ratio. Blood pressure values in supine and orthostatic positions, heart rate, and smoking status were also documented.

Measurement of biochemical parameters: The following samples were collected from venous plasma after 8 h of fasting: serum glucose, glycated hemoglobin HbA1c, total cholesterol, high-density lipoproteins cholesterol (HDLc), LDL cholesterol, triglycerides, bilirubin, C-reactive protein, serum creatinine, and electrolytes (potassium, magnesium, chloride, sodium, calcium) urea, liver enzymes: aspartate aminotransferase (AST), alanine aminotransferase (ALT), gamma-glutamyl transferase (GGT) and urinary albumin-to-creatinine ratio (ACR).

Diagnosis of CAN: The evaluation included electrocardiograms for assessing the QTc interval, and cardiovascular reflex tests (CART) such as heart rate variability during deep breathing, the Valsalva maneuver, and responses to orthostatic changes. The results of CART were categorized as usual if no abnormal findings were detected. They were considered to indicate mild dysfunction if one out of the five tests was abnormal, moderate dysfunction if two or three of the tests were abnormal, and severe dysfunction if more than three tests were abnormal. All these measurements were performed using the ESP-01-PA Ewing Tester neuropathic measuring and analyzing system with an ECG module. At the same time, the diagnosis of CAN was evaluated by performing a sweat test using Sudoscan assessment. During the test, the patient places their hands and feet on the electrodes. The test takes 3 min to perform, is painless, and requires no subject preparation. Additionally, Sudoscan incorporates built-in algorithms that integrate electrochemical skin conductance with age to generate a score estimating the current risks of CAN (Sudoscan-CAN score) and chronic kidney disease (Sudoscan-Nephro score).

We have also calculated the SCORE2-Diabetes for the patients included in the study. SCORE2-Diabetes is a new algorithm developed, calibrated, and validated to predict 10-year risk of CVD in individuals with type 2 diabetes that enhances identification of individuals at higher risk of developing CVD across Europe [20]. Sex-specific competing risk-adjusted models were used, incorporating traditional risk factors, such as age, smoking, systolic blood pressure, total and HDL-cholesterol alongside diabetes-related variables including age at diabetes onset, glycated hemoglobin (HbA1c), and estimated glomerular filtration rate (eGFR) based on creatinine.

2.3. Statistical Analysis

The statistical analysis of the population was conducted using IBM SPSS v.20. Tests of normality used were Kolmogorov–Smirnov with a Lilliefors significance correction and Shapiro–Wilk statistic. Continuous variables usually distributed were presented as mean ± SD (standard deviation), and non-normal variables were expressed as median (interquartile range [IQR]). In contrast, categorical variables were reported as absolute counts and percentages. The p-value calculation for normally distributed variables was conducted using the ANOVA test, while for non-normally distributed variables, the Kruskal–Wallis test was applied. Statistical significance was determined at a 95% confidence interval. The χ^2 test was utilized for categorical variables—multiple linear regression was used to estimate the independent correlation of the SCORE2-Diabetes risk with results of Sudoscan parameters.

3. Results

The cohort comprised 211 patients diagnosed with T2DM, without established ASCVD, 51.6% being male (n = 109), with a mean age of 58.35 ± 7.18 years and a mean weight of 90.36 ± 17 kg. The prevalence of CAN in our group was 67.2% (n = 142).

In the study group, according SCORE2-Diabetes, four patients (1.9%) were classified with moderate cardiovascular risk, thirty-five (16.6%) with high risk, and one hundred seventy-two (81.5%) with very high cardiovascular risk. Patients' baseline characteristics are presented in Table 1.

Table 1. Anthropometric and biochemical parameters in relation to cardiovascular risk.

Cardiovascular Risk	(5–10%) n = 4		(10–20%) n = 35		(>20%) n = 172		Total		p-Value
	Mean	Std. Deviation	Mean	Std. Deviation	Mean	Std. Deviation	Mean	Std. Deviation	
Age (years)	45	4.92	52	5.27	60	6.59	58	7.18	<0.001
Diabetes duration (years) *	6.50	8	5	5	7.5	8	7	8	0.036
Height (cm)	170.00	8.98	167.14	11.00	166.47	8.71	166.64	9.10	0.701
Weight (kg)	95.00	20.70	88.86	17.42	90.55	16.91	90.36	17.00	0.745
WC (cm)	110.25	21.82	105.49	11.98	106.92	12.94	106.74	12.92	0.722
FPG (mg/dL)	200.50	43.93	162.11	61.97	206.06	91.39	198.67	87.85	0.025
HbA1c (%)	7.26	0.52	7.17	1.19	8.61	2.04	8.34	1.99	<0.001
TC (mg/dL)	212.25	44.63	184.89	50.93	200.75	57.15	198.34	56.09	0.277
HDL-c (mg/dL)	54.10	12.55	47.12	14.40	50.32	13.08	49.86	13.30	0.352
LDL-c (mg/dL)	94.63	9.31	101.85	45.22	109.13	50.53	107.73	49.32	0.668
TGL (mg/dL) *	292.5	612.5	173	169	179	132.5	178	138	0.103
eGFR (mL/min/1.73 m^2)	98.43	17.87	92.74	30.69	74.61	24.27	78.07	26.27	<0.001
GGT (UI/L) *	52.50	62	34	73	38	50	37.5	52	0.05
ACR (mg/g) *	23.13	39.37	12	28.54	31.25	71.33	26.27	49.12	0.78
B12 vitamine (pg/mL) *	889.5	91	427.5	158	350	269	382	364	0.057

Abbreviations: WC = waist circumference, FPG = fasting plasma glucose, HbA1c = glycated hemoglobin, TC = total cholesterol, HDL-c = high-density lipoprotein, LDLc = low-density lipoprotein, TGL = triglycerides, eGFR = estimated glomerular filtration rate, GGT = gamma-glutamyl transferase, ACR = Albumin-to-creatinine ratio, * variables expressed as median, interquartile range [IQR], statistical significance, $p < 0.05$.

Patients with very high cardiovascular risk were older and with longer diabetes duration. However, there were no notable differences in height, weight, or waist circumference across risk categories. Elevated levels of FPG and HbA1c were observed in higher-risk groups. Lipid levels, including TC, HDL-c, TGL, and LDL-c, did not significantly vary among risk groups. Lower eGFR was decreased parallel with increased CVR categories. While GGT showed a marginal difference across risk groups, overall, factors such as age, diabetes duration, FPG, HbA1c, and eGFR were identified as key indicators of cardiovascular risk in diabetic patients (Table 1).

There are significant variations in SBP while lying down among the three risk categories, with higher levels detected in higher-risk groups ($p = 0.041$). There are no notable differences in DBP in the lying position ($p = 0.065$). There are no substantial differences in SBP and DBP while standing across the risk categories ($p > 0.05$). During handgrip exercises, SBP does not significantly differ across risk groups ($p = 0.494$), yet there is a marginal difference in DBP ($p = 0.073$) and a significant difference in heart rate ($p = 0.046$) (Table 2).

The statistical analysis of Sudoscan's parameters in relation to cardiovascular risk reveals significant differences. Specifically, the Sudoscan CAN-score exhibits significant variance among risk groups ($p = 0.002$), indicating a potential correlation between sudomotor dysfunction and cardiac autonomic neuropathy in individuals at higher risk of cardiovascular complications. Similarly, the Sudoscan Nephro-score displays a notable difference across risk categories, with lower scores observed in higher-risk groups ($p = 0.001$), suggesting a potential link between sudomotor dysfunction and renal function in individuals with increased cardiovascular risk. However, scores for Sudoscan parameters related to the feet and hands do not show significant differences across risk categories. These findings suggest that Sudoscan-derived measures may serve as valuable indicators of cardiovascular risk, particularly in assessing cardiac autonomic function and nephropathy in diabetic populations (Table 3).

Table 2. Ewing's parameters in relation to cardiovascular risk.

Cardiovascular Risk	(5–10%) n = 4		(10–20%) n = 35		(>20%) n = 172		Total		p-Value
	Mean	Std. Deviation	Mean	Std. Deviation	Mean	Std. Deviation	Mean	Std. Deviation	
SBP supine position (mmHg)	122.75	21.22	127.29	17.54	134.83	18.13	133.35	18.27	0.041
DBP supine position (mmHg)	88.00	21.86	74.57	9.41	77.10	11.19	76.89	11.24	0.065
HR supine position (bpm)	82.75	10.01	73.23	9.55	76.17	11.00	75.81	10.81	0.147
SBP standing (mmHg)	125.50	10.79	123.69	12.96	129.14	19.91	128.17	18.86	0.286
DBP standing (mmHg)	86.50	15.80	74.86	9.50	76.49	12.31	76.41	12.00	0.181
HR standing (bpm)	84.25	9.22	76.34	10.81	80.00	11.94	79.47	11.77	0.176
SBP Handgrip (mmHg)	137.75	18.12	143.40	16.04	145.72	16.48	145.18	16.41	0.494
DBP Handgrip (mmHg)	89.00	15.87	85.20	14.04	81.10	10.74	81.93	11.53	0.073
HR Handgrip (bpm)	**90.75**	**10.28**	**80.69**	**9.85**	**85.05**	**10.89**	**84.44**	**10.82**	**0.046**
HRVi index	26.00	6.78	19.41	7.01	16.59	7.41	17.25	7.48	0.653
Ewing score *	2	5	3	2	4	3	3	3	0.194

Abbreviations: SBP = systolic blood pressure, DBP = diastolic blood pressure, HR = heart rate, HRVi = heart rate variability index, * variables expressed as median, interquartile range [IQR], statistical significance, $p < 0.05$.

Table 3. Sudoscan's parameters in relation to cardiovascular risk.

Cardiovascular Risk	(5–10%) n = 4		(10–20%) n = 35		(>20%) n = 172		Total		p-Value
	Mean	Std. Deviation	Mean	Std. Deviation	Mean	Std. Deviation	Mean	Std. Deviation	
Sudoscan Nephro-score	**90.00**	**24.59**	**72.34**	**9.29**	**66.06**	**15.39**	**67.56**	**15.19**	**0.001**
Sudoscan CAN-score *	**16.5**	**28**	**31**	**12**	**35**	**13**	**34.5**	**13**	**0.002**
Sudoscan left feet score (uS)	76.00	21.46	77.46	10.78	75.28	14.89	75.65	14.37	0.717
Sudoscan right feet score (uS)	73.50	24.53	76.77	11.28	76.66	13.78	76.62	13.57	0.898
Sudoscan left hand score (uS)	57.75	22.90	71.63	13.19	68.66	15.87	68.95	15.63	0.209
Sudoscan right hand score (uS)	50.75	20.09	69.80	13.23	66.72	16.12	66.93	15.88	0.069

* variables expressed as median, interquartile range [IQR], Statistical significance, $p < 0.05$.

The statistical analysis of the frequency of chronic diabetes complications based on cardiovascular risk indicates significant differences among the various risk categories. For cardiovascular autonomic neuropathy (CAN), the prevalence increases with the degree of cardiovascular risk, with the highest frequency observed in those with a risk greater than 20% (69.80%). The same trend is observed for diabetic polyneuropathy (DPN), chronic kidney disease (CKD), and diabetic retinopathy (DR), where the frequency of complications significantly increases with higher cardiovascular risk. These findings underscore the importance of proper monitoring and management of cardiovascular risk in addressing chronic diabetes complications (Table 4).

Table 4. The frequency of chronic diabetes complications based on cardiovascular risk.

Cardiovascular Risk	5–10% (n = 4)	%	10–20% (n = 35)	%	>20% (n = 172)	%
CAN	2	50.00	20	57.10	120	69.80
DPN	2	50.00	19	54.29	102	59.30
CKD	1	25.00	9	25.71	85	49.42
DR	2	50.00	7	20.00	64	37.21
PAD	0	0.00	0	0.00	34	19.77

Abbreviations: CAN = cardiovascular autonomic neuropathy, CKD = chronic kidney disease, DR = diabetic retinopathy, DPN = diabetic polyneuropathy, PAD = peripheral artery disease.

Correlation of Sudoscan with SCORE 2—Diabetes

The scatterplot shows the relationship between the Sudoscan CAN-score, Sudoscan Nephro-score and the SCORE2-Diabetes. Sudoscan CAN-score showed a significantly positive correlation with SCORE 2-Diabetes and, Sudoscan Nephro-score score showed a significantly negative correlation with SCORE 2-Diabetes (Figure 1a,b).

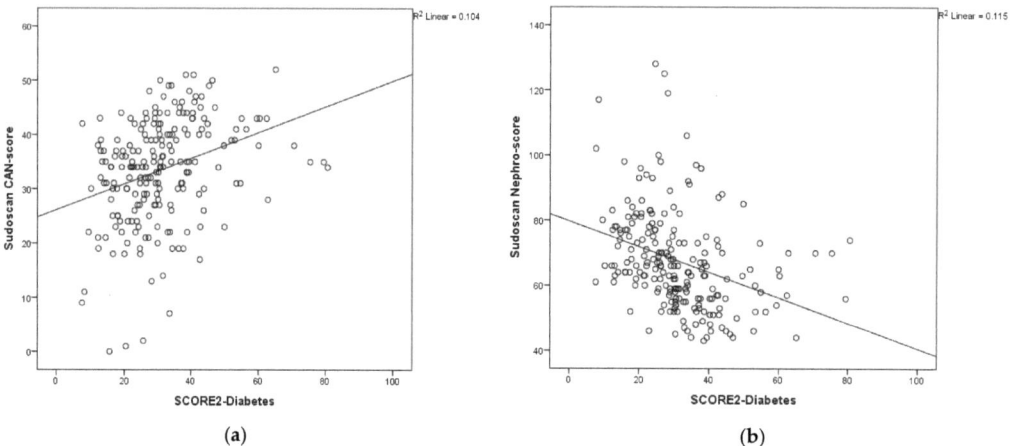

Figure 1. (**a**) Scatterplot showing the relationship between Sudoscan CAN-score on *y*-axis and SCORE 2—Diabetes on *x*-axis. (**b**) Scatterplot showing the relationship between Sudoscan Nephro-score on *y*-axis and SCORE 2—Diabetes on *x*-axis.

The scatterplot shows the relationship between the Sudoscan feets-score, Sudoscan hands-score and the SCORE2-Diabetes. Sudoscan feets-score and Sudoscan hands-score showed a significantly negative correlation with SCORE 2-Diabetes (Figure 2a,b).

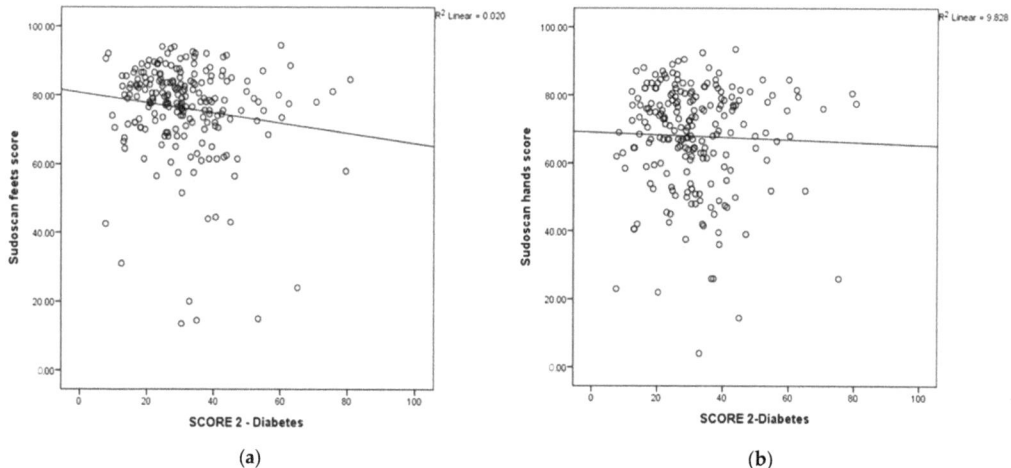

Figure 2. (**a**) Scatterplot showing the relationship between Sudoscan feets-score on *y*-axis and SCORE 2—Diabetes on *x*-axis. (**b**) Scatterplot showing the relationship between Sudoscan hands-score on *y*-axis and SCORE 2—Diabetes on *x*-axis.

The area under the receiver operating characteristic (ROC) curve of the Sudoscan-CAN score to predict very high cardiovascular risk was 0.657 (95%CI: 0.569–0.745) of the total square (Figure 3). The Sudoscan-CAN score cut-off was 39.5, and the test had 34.3% sensitivity and 79% specificity to detect very high cardiovascular risk.

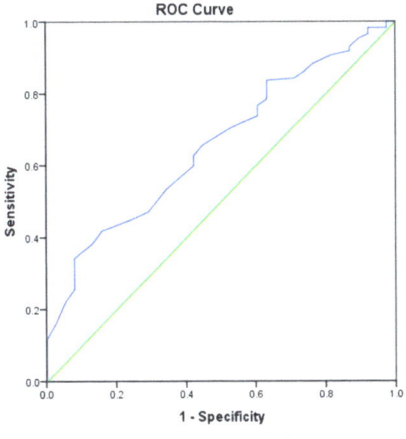

Figure 3. ROC curve of Sudoscan-CAN score in detecting very high cardiovascular risk in patients with T2DM.

The multiple linear regression analysis examining clinical factors associated with SCORE2-Diabetes in patients with T2DM reveals several significant associations. Age and diabetes duration exhibit positive correlations with SCORE2-Diabetes, with standardized β-coefficients of 0.413 and 0.179, respectively, both statistically significant ($p < 0.001$). Similarly, HbA1c, LDL-c Sudoscan feet score, Sudoscan CAN-score, Ewing test score, and SBP in the supine position also demonstrate positive associations with SCORE2-Diabetes, with significant p-values (<0.05). Conversely, eGFR and Sudoscan Nephro-score display negative associations with SCORE2-Diabetes, suggesting that higher eGFR and Sudoscan Nephro-scores are associated with lower SCORE2-Diabetes values. These findings highlight the importance of age, diabetes duration, HbA1c, LDL-c, Sudoscan parameters, eGFR, and SBP in predicting SCORE2-Diabetes in patients with T2DM (Table 5).

Table 5. Clinical factors associated with SCORE2-Diabetes in patients with T2DM using multiple linear regression.

	Standard β-Coefficient	95% CI	p-Value
Age	0.804	[0.585, 1.022]	<0.001
Diabetes duration	0.627	[0.345, 0.908]	<0.001
HbA1c	2.437	[1.617, 3.258]	<0.001
LDL-c	0.048	[0.014, 0.083]	<0.001
eGFR	−0.225	[−0.285, −0.165]	<0.001
Sudoscan Nephro-score	−0.279	[−0.389, −0.169]	<0.001
Sudoscan CAN-score	0.405	[0.233, 0.576]	<0.001
Ewing test score	1.102	[0.327, 1.878]	0.006
SBP supine position	0.169	[0.075, 0.262]	<0.001

Abbreviations: HbA1c = glycated hemoglobin, LDLc = low-density lipoprotein, eGFR = estimated glomerular filtration rate, SBP = systolic blood pressure, statistical significance, $p < 0.05$.

4. Discussion

In this study, we analyzed the correlation between the Sudoscan-CAN score and cardiovascular risk, as calculated using the SCORE2-Diabetes algorithm. The aim was to investigate the potential association between sudomotor dysfunction and increased cardiovascular risk in individuals with Type 2 Diabetes Mellitus (T2DM).

The statistical assessment of Sudoscan parameters in the context of cardiovascular risk showed substantial differences. Notably, the Sudoscan CAN-score varied significantly across risk categories, suggesting an association between sudomotor dysfunction and cardiac autonomic neuropathy in patients at elevated risk for cardiovascular events. Multiple linear regression analysis revealed significant positive correlations between Sudoscan CAN-score and SCORE2-Diabetes in T2DM patients ($p < 0.001$).

The findings of the study highlight the significant cardiovascular risk faced by patients diagnosed with T2DM. The prevalence of CAN was significant within our studied patient cohort, being 67.2%. What is important to mention is that 69.8% of patients with very high cardiovascular risk also associate CAN. With a substantial portion of the cohort (81.5%) classified as having very high cardiovascular risk according to SCORE2-Diabetes, it underscores the critical need for comprehensive risk assessment and management strategies in this population.

In a meta-analysis conducted in 2003, Maser et al. synthesized the evidence base to evaluate the relationship between CAN and the risk of mortality in diabetes. CAN was associated with future risk of mortality both in cases of definite CAN and possible CAN, with a stronger association observed in definite CAN cases [21].

Also, Ewing et al. illustrated a 2.5-year mortality rate of 27.5%, which escalated by 25.5% after 5 years in patients with diabetes and definite CAN [22]. This stands in stark contrast to patients with diabetes and a normal autonomic function test (AFT), who exhibited a mortality rate of only 15% over the same 5-year period. Additionally, CAN also serves as a prognostic indicator for cardiovascular events (CVE) and mortality in the context of intensive glycemic control in type 2 diabetes. This was evidenced by the Action in Diabetes and Vascular Disease: Preterax and Diamicron Modified Release Controlled Evaluation (ADVANCE), Veterans Affairs Diabetes Trial (VADT), and ACCORD Studies [23–25].

Similar investigations, such as the study by T Yuan et al., have highlighted the utility of Sudoscan in formulating a cardiac risk score to evaluate cardiovascular autonomic neuropathy in asymptomatic diabetic patients [26]. Our findings align with these observations, bolstering the clinical relevance of Sudoscan in identifying significant cardiovascular risks among individuals with type 2 diabetes mellitus. While this study focused on creating a predictive model for asymptomatic patients, our research applies Sudoscan measurements directly to established cardiovascular risk scoring systems like SCORE2-Diabetes, providing a practical and efficient tool for assessing risk in a clinical setting. This integration of Sudoscan into routine assessments could facilitate earlier and more personalized interventions, enhancing preventive strategies and patient outcomes in diabetic cardiovascular care.

The research conducted by II Hussein et al. on the assessment of sudomotor function in hypertensive patients with or without T2DM underscores the importance of Sudoscan in evaluating cardiovascular risks [27]. Their findings support our study's conclusions about the predictive relevance of Sudoscan scores for cardiovascular complications in diabetic patients. Both studies highlight the potential of Sudoscan to serve as a vital component of cardiovascular risk management, advocating for its integration into routine clinical practices to enhance early detection and preventive strategies in high-risk patient groups.

Our study introduces the utilization of the Ewing Test and Sudoscan Cardiovascular Autonomic Neuropathy Score as adjunctive tools for assessing cardiovascular risk in T2DM patients. This integration of novel assessment methods provides a more comprehensive evaluation beyond traditional risk scoring algorithms, potentially offering additional insights into cardiovascular health in this population.

Similar investigations, such as those by Vinik et al., have also highlighted the prognostic value of CAN assessments in predicting cardiovascular events in diabetic patients. Our findings parallel these studies, reinforcing the clinical utility of Sudoscan in detecting significant cardiovascular risk. Unlike conventional risk assessment tools, Sudoscan offers a quick and patient-friendly means to assess risk, potentially leading to earlier interventions [26].

The observed correlations between Sudoscan CAN-score and Sudoscan Nephro-score with SCORE2-Diabetes shed light on the interplay between autonomic neuropathy, renal function, and cardiovascular risk in T2DM patients. The positive correlation of Sudoscan CAN-score with cardiovascular risk suggests a potential association between autonomic dysfunction and increased cardiovascular risk. The identification of significant associations between various clinical parameters and SCORE2-Diabetes underscores the multifactorial nature of cardiovascular risk in T2DM patients. This highlights the importance of comprehensive risk stratification strategies that consider not only traditional risk factors but also novel markers such as autonomic function and renal health in guiding therapeutic interventions and optimizing outcomes.

The findings regarding the correlation between Sudoscan feets-score and Sudoscan hands-score with SCORE2-Diabetes suggest a potential role for Sudoscan as a non-invasive tool for assessing peripheral neuropathy in T2DM patients and predicting cardiovascular risk.

The clinical implications of these findings are significant, indicating that measuring sudomotor dysfunction using Sudoscan effectively identifies Type 2 Diabetes Mellitus (T2DM) patients who are at high cardiovascular risk. The inclusion of the Ewing Test and Sudoscan scores as part of cardiovascular risk assessments could serve as a valuable addition to traditional methods, potentially improving early detection and management strategies. This supports the integration of innovative diagnostic tools into routine clinical practice, which could enhance the proactive management of cardiovascular risks in diabetic populations, ultimately improving patient outcomes.

Further research is warranted to elucidate the clinical utility of Sudoscan scores in risk stratification and guiding therapeutic interventions in this population.

Limit

While the study provides valuable insights into the relationship between Ewing Test, Sudoscan scores, and cardiovascular risk in T2DM patients, certain limitations need to be acknowledged. Firstly, these include the relatively small sample size and cross-sectional design. Secondly, all participants included in our research are from Romania, a very-high-risk European region in terms of CV mortality [27]. Also, the study enrolled patients from a single center.

Future prospective studies with larger sample sizes and longitudinal follow-up are needed to validate these findings and explore the utility of novel assessment tools in improving cardiovascular outcomes in T2DM patients.

5. Conclusions

Our study confirms the significant role of sudomotor dysfunction, as measured by Sudoscan, in identifying T2DM patients at high cardiovascular risk. The use of the Ewing Test and Sudoscan scores offers a promising adjunctive tool in the cardiovascular risk assessment of this population. These findings underscore the need for incorporating novel diagnostic tools into standard practice to enhance the early detection and management of cardiovascular risk among diabetic patients.

Author Contributions: Conceptualization, A.-E.N. and E.R.; methodology A.-E.N., C.D. and G.R.; software A.-E.N. and E.R.; validation, A.-E.N., C.D. and G.R.; formal analysis, A.-E.N. and E.R.; investigation, F.R.; resources, A.-E.N., E.R., C.D. and O.A.P.; data curation, C.S. and O.A.P.; writing—original draft preparation, A.-E.N. and E.R.; writing—review and editing, A.-E.N. and E.R.; visualization, F.R.; supervision, G.R.; project administration, A.-E.N.; funding acquisition, A.-E.N. All authors have read and agreed to the published version of the manuscript.

Funding: Publication of this paper was supported by the University of Medicine and Pharmacy Carol Davila, through the institutional program Publish not Perish.

Institutional Review Board Statement: The study was conducted in accordance with the Declaration of Helsinki and approved by the Institutional Review Board of "Nicolae Malaxa" Clinical Hospital, Bucharest, Romania (approval number 2145 on 7 March 2019).

Informed Consent Statement: Informed consent was obtained from all subjects involved in the study.

Data Availability Statement: The data presented in this study are available on request from the corresponding author. The data are not publicly available due to the hospital's privacy policy.

Conflicts of Interest: The authors declare no conflicts of interest.

References

1. Serhiyenko, V.A.; Serhiyenko, A.A. Cardiac autonomic Neuropathy: Risk Factors, Diagnosis and Treatment. *World J. Diabetes* **2018**, *9*, 1. Available online: https://pmc/articles/PMC5763036/ (accessed on 11 April 2024). [CrossRef] [PubMed]
2. Schwartz, S.S.; Epstein, S.; Corkey, B.E.; Grant, S.F.; Gavin, J.R., III; Aguilar, R.B.; Herman, M.E. A Unified Pathophysiological Construct of Diabetes and Its Complications. *Trends Endocrinol. Metab.* **2017**, *28*, 645–655. [CrossRef] [PubMed]
3. Shah, M.S.; Brownlee, M. Molecular and Cellular Mechanisms of Cardiovascular Disorders in Diabetes. *Circ. Res.* **2016**, *118*, 1808–1829. Available online: https://pubmed.ncbi.nlm.nih.gov/27230643/ (accessed on 16 April 2024). [CrossRef] [PubMed]
4. Brownlee, M. The Pathobiology of Diabetic Complications a Unifying Mechanism. *Diabetes* **2005**, *54*, 1615–1625. Available online: http://diabetesjournals.org/diabetes/article-pdf/54/6/1615/381945/zdb00605001615.pdf (accessed on 16 April 2024). [CrossRef] [PubMed]
5. Nica, A.E.; Rusu, E.; Dobjanschi, C.G.; Rusu, F.; Parliteanu, O.A.; Sivu, C.; Radulian, G. The Importance of Evaluating Sudo-motor Function in the Diagnosis of Cardiac Autonomic Neuropathy. *Cureus* **2024**, *16*, 1–10. Available online: https://www.cureus.com/articles/239801-the-importance-of-evaluating-sudomotor-function-in-the-diagnosis-of-cardiac-autonomic-neuropathy (accessed on 17 April 2024).
6. Bönhof, G.J.; Herder, C.; Ziegler, D. Diagnostic Tools, Biomarkers, and Treatments in Diabetic polyneuropathy and Cardiovascular Autonomic Neuropathy. *Curr. Diabetes Rev.* **2022**, *18*, 156–180. [CrossRef] [PubMed]
7. Hicks, C.W.; Wang, D.; Matsushita, K.; Windham, B.G.; Selvin, E. Peripheral Neuropathy and All-Cause and Car-diovascular Mortality in US Adults. *Ann. Intern. Med.* **2021**, *174*, 167–174. [CrossRef] [PubMed]
8. Eleftheriadou, A.; Williams, S.; Nevitt, S.; Brown, E.; Roylance, R.; Wilding, J.P.; Cuthbertson, D.J.; Alam, U. The Prevalence of Cardiac Autonomic Neuropathy in Prediabetes: A Systematic Review. *Diabetologia* **2021**, *64*, 288–303. [CrossRef]
9. Roth, G.A.; Abate, D.; Abate, K.H.; Abay, S.M.; Abbafati, C.; Abbasi, N.; Abbastabar, H.; Abd-Allah, F.; Abdela, J.; Abdelalim, A.; et al. Global, regional, and national age-sex-specific mortality for 282 causes of death in 195 countries and territories, 1980–2017: A systematic analysis for the Global Burden of Disease Study 2017. *Lancet* **2018**, *392*, 1736–1788. [CrossRef]
10. Caussy, C.; Aubin, A.; Loomba, R. The Relationship between Type 2 Diabetes, NAFLD, and Cardiovascular Risk. *Curr. Diabetes Rep.* **2021**, *21*, 15. [CrossRef]
11. Wong, N.D.; Sattar, N. Cardiovascular risk in diabetes mellitus: Epidemiology, assessment and prevention. *Nat. Rev. Cardiol.* **2023**, *20*, 685–695. [CrossRef] [PubMed]
12. Berkelmans, G.F.; Gudbjörnsdottir, S.; Visseren, F.L.; Wild, S.H.; Franzen, S.; Chalmers, J.; Davis, B.R.; Poulter, N.R.; Spijkerman, A.M.; Woodward, M.; et al. Prediction of indi-vidual life-years gained without cardiovascular events from lipid, blood pressure, glucose, and aspirin treatment based on data of more than 500,000 patients with Type 2 diabetes mellitus. *Eur. Heart J.* **2019**, *40*, 2899–2906. [CrossRef]
13. Visseren, F.L.J.; Mach, F.; Smulders, Y.M.; Carballo, D.; Koskinas, K.C.; Bäck, M.; Benetos, A.; Biffi, A.; Boavida, J.M.; Capodanno, D.; et al. 2021 ESC Guidelines on cardio-vascular disease prevention in clinical practice. *Eur. Heart J.* **2021**, *42*, 3227–3337. [CrossRef]
14. Kengne, A.P.; Patel, A.; Marre, M.; Travert, F.; Lievre, M.; Zoungas, S.; Chalmers, J.; Colagiuri, S.; Grobbee, D.E.; Hamet, P.; et al. Contemporary model for cardiovascular risk prediction in people with type 2 diabetes. *Eur. J. Cardiovasc. Prev. Rehabil.* **2011**, *18*, 393–398. [CrossRef]
15. Marx, N.; Federici, M.; Schütt, K.; Müller-Wieland, D.; Ajjan, R.A.; Antunes, M.J.; Christodorescu, R.M.; Crawford, C.; Di Angelantonio, E.; Eliasson, B.; et al. 2023 ESC Guidelines for the management of cardiovascular disease in patients with diabetes. *Eur. Heart J.* **2023**, *44*, 4043–4140.
16. Gavan, D.E.; Gavan, A.; Bondor, C.I.; Florea, B.; Bowling, F.L.; Inceu, G.V.; Colobatiu, L. SUDOSCAN, an Innovative, Simple and Non-Invasive Medical Device for Assessing Sudomotor Function. *Sensors* **2022**, *22*, 7571. [CrossRef]
17. Look AHEAD Research Group; Wing, R.R. Long-term Effects of a Lifestyle Intervention on Weight and Cardiovascular Risk Factors in Individuals with Type 2 Diabetes Mellitus: Four-Year Results of the Look AHEAD Trial. *Arch. Intern. Med.* **2010**, *170*, 1566–1575. Available online: https://jamanetwork.com/journals/jamainternalmedicine/fullarticle/226013 (accessed on 17 March 2024).

18. Palmer, S.C.; Tendal, B.; Mustafa, R.A.; Vandvik, P.O.; Li, S.; Hao, Q.; Tunnicliffe, D.; Ruospo, M.; Natale, P.; Saglimbene, V.; et al. Sodium-Glucose Cotransporter Protein-2 (SGLT-2) Inhibitors and Glucagon-like Peptide-1 (GLP-1) Receptor Agonists for Type 2 Diabetes: Systematic Review and Network Meta-Analysis of Randomised Controlled Trials. *BMJ* **2021**, *372*, m4573. Available online: https://www.bmj.com/content/372/bmj.m4573 (accessed on 17 March 2024). [CrossRef]
19. Marsico, F.; Paolillo, S.; Gargiulo, P.; Bruzzese, D.; Dell'Aversana, S.; Esposito, I.; Renga, F.; Esposito, L.; Marciano, C.; Dellegrottaglie, S.; et al. Effects of glucagon-like pep-tide-1 receptor agonists on major cardiovascular events in patients with Type 2 diabetes mellitus with or without established cardiovascular disease: A meta-analysis of randomized controlled trials. *Eur. Heart J.* **2020**, *41*, 3346–3358. [CrossRef]
20. Pennells, L.; Kaptoge, S.; Østergaard, H.B.; Read, S.H.; Carinci, F.; Franch-Nadal, J.; Petitjean, C.; Taylor, O.; Hageman, S.H.J.; Xu, Z.; et al. SCORE2-Diabetes: 10-year cardiovascular risk estimation in type 2 diabetes in Europe. *Eur. Heart J.* **2023**, *44*, 2544–2556.
21. Maser, R.E.; Mitchell, B.D.; Vinik, A.I.; Freeman, R. The Association between Cardiovascular Autonomic Neuropathy and Mortality in Individuals with Diabetes a Meta-Analysis. *Diabetes Care* **2003**, *26*, 1895–1901. Available online: http://diabetesjournals.org/care/article-pdf/26/6/1895/591627/dc0603001895.pdf (accessed on 16 April 2024). [CrossRef] [PubMed]
22. Ewing, D.J.; Campbell, I.W.; Clarke, B.F. Mortality in Diabetic Autonomic Neuropathy. *Lancet* **1976**, *307*, 601–603. [CrossRef] [PubMed]
23. Pop-Busui, R.; Evans, G.W.; Gerstein, H.C.; Fonseca, V.; Fleg, J.L.; Hoogwerf, B.J.; Genuth, S.; Grimm, R.H.; Corson, M.A.; Prineas, R.; et al. Effects of Cardiac Autonomic Dysfunction on Mortality Risk in the Action to Control Cardiovascular Risk in Diabetes (ACCORD) Trial. *Diabetes Care* **2010**, *33*, 1578–1584. Available online: http://creativecommons (accessed on 16 April 2024). [CrossRef]
24. Pop-Busui, R. Cardiac Autonomic Neuropathy in Diabetes: A Clinical Perspective. *Diabetes Care* **2010**, *33*, 434–441. Available online: https://pubmed.ncbi.nlm.nih.gov/20103559/ (accessed on 16 April 2024). [CrossRef] [PubMed]
25. Zoungas, S.; Arima, H.; Gerstein, H.C.; Holman, R.R.; Woodward, M.; Reaven, P.; Hayward, R.A.; Craven, T.; Coleman, R.L.; Chalmers, J. Effects of intensive glucose control on microvascular outcomes in patients with type 2 diabetes: A meta-analysis of individual partici-pant data from randomised controlled trials. *Lancet Diabetes Endocrinol.* **2017**, *5*, 431–437. [CrossRef]
26. Vinik, A.I.; Ziegler, D. Diabetic Cardiovascular Autonomic Neuropathy. *Circulation* **2007**, *115*, 387–397. Available online: http://www.circulationaha.org (accessed on 16 April 2024). [CrossRef]
27. Luca, S.A.; Bungau, R.M.; Lazar, S.; Potre, O.; Timar, B. To What Extent Does Cardiovascular Risk Classification of Patients with Type 2 Diabetes Differ between European Guidelines from 2023, 2021, and 2019? A Cross-Sectional Study. *Medicina* **2024**, *60*, 334. Available online: https://www.mdpi.com/1648-9144/60/2/334/htm (accessed on 17 April 2024). [CrossRef]

Disclaimer/Publisher's Note: The statements, opinions and data contained in all publications are solely those of the individual author(s) and contributor(s) and not of MDPI and/or the editor(s). MDPI and/or the editor(s) disclaim responsibility for any injury to people or property resulting from any ideas, methods, instructions or products referred to in the content.

Article

Investigation of the Systemic Immune Inflammation (SII) Index as an Indicator of Morbidity and Mortality in Type 2 Diabetic Retinopathy Patients in a 4-Year Follow-Up Period

Nilgun Tan Tabakoglu [1,*] and Mehmet Celik [2]

1. Health Research and Development Center, Faculty of Medicine Hospital, Trakya University, Edirne 22100, Turkey
2. Department of Endocrine and Metabolic Diseases, Faculty of Medicine, Trakya University, Edirne 22100, Turkey; mehmetcelik@trakya.edu.tr
* Correspondence: tabakoglunilgun@gmail.com or nilguntabakoglu@trakya.edu.tr; Tel.: +90-539-454-2793

Abstract: *Background and Objectives*: This study aimed to investigate the relationship between the systemic immune inflammation (SII) index and the development of micro and macro complications and mortality within the first year and the following three years in type 2 diabetic retinopathy patients. *Materials and Methods*: The retrospective study included 523 type 2 diabetic retinopathy patients seen in the endocrinology outpatient clinic of our hospital between January and December 2019. Their demographic and clinical characteristics were analyzed using descriptive statistics. The normal distribution of quantitative data was assessed by the Shapiro–Wilk test. Mann–Whitney U, McNemar–Chi-square, and Cochran's Q tests were used to analyze the SII values and complication rates over time. An ROC analysis determined the sensitivity and specificity of SII. A multiple linear regression analysis examined the relationship between variables and SII, while Spearman's test assessed the correlation between CRP and SII. $p < 0.05$ was accepted as significant. *Results*: The mean age of patients was 63.5 ± 9.3 years, with mean SII values of 821.4 ± 1010.8. Higher SII values were significantly associated with acute–chronic renal failure, peripheral arterial disease, and hospitalization rates in both the first year and the following three years ($p < 0.05$ for all). Significant cut-off values for SII were found for micro- and macrovascular complications and death within the first year ($p < 0.05$ for all). The ROC curve analysis identified an optimal SII cut-off value of >594.0 for predicting near-term (1-year) complications and mortality, with a sensitivity of 73.8% and specificity of 49.4% (area under the ROC curve: 0.629, $p = 0.001$). Multiple linear regression indicated that smoking of at least 20 pack-years had a significant positive effect on SII. The Spearman test showed a weak positive correlation between SII and CRP. *Conclusions*: High SII values predict both early and late acute–chronic renal failure, peripheral arterial disease, and hospitalizations in patients with type 2 diabetic retinopathy. The study also shows that high SII values may predict microvascular and macrovascular complications of type 2 DM and mortality risk in the early period in patients with type 2 diabetic retinopathy. In addition, comorbidities and inflammatory habits, such as long-term smoking, should be considered in the clinical use of SII.

Keywords: systemic immune inflammation; diabetic retinopathy; type 2 diabetes mellitus; microvascular complications; macrovascular complications

Citation: Tabakoglu, N.T.; Celik, M. Investigation of the Systemic Immune Inflammation (SII) Index as an Indicator of Morbidity and Mortality in Type 2 Diabetic Retinopathy Patients in a 4-Year Follow-Up Period. *Medicina* 2024, 60, 855. https://doi.org/10.3390/medicina60060855

Academic Editors: Yuzuru Ohshiro, Kunimasa Yagi and Yasuhiro Maeno

Received: 3 May 2024
Revised: 18 May 2024
Accepted: 21 May 2024
Published: 24 May 2024

Copyright: © 2024 by the authors. Licensee MDPI, Basel, Switzerland. This article is an open access article distributed under the terms and conditions of the Creative Commons Attribution (CC BY) license (https://creativecommons.org/licenses/by/4.0/).

1. Introduction

Diabetes mellitus (DM) is a chronic metabolic disease characterized by hyperglycemia and complications caused by impairments in any or all of the production, release, and effects of insulin, a peptide hormone secreted by the β cells of the islets of Langerhans of the pancreas [1]. It has been reported that 90% of diabetes cases in the world are type 2 DM cases [2]. According to the *Diabetes Atlas* published periodically by the International Diabetes Federation, the number of people diagnosed with diabetes globally by the end of 2021 was recorded as 537 million. Even more remarkably, this number is estimated to reach 783 million by 2045 [3]. Chronic complications of type 2 DM are categorized under two main headings as micro- and macrovascular complications: nephropathy, retinopathy, and neuropathy are defined as microvascular complications, while coronary artery disease (CAD), peripheral arterial disease (PAD), and cerebrovascular disease (CVD) are defined as macrovascular complications [4,5]. Chronic hyperglycemia is a critical factor in the formation of these complications. The duration of diabetes and comorbidities are other factors affecting the development of complications. Diabetic retinopathy (DR) is the most common microvascular complication and is a precursor of other microvascular complications. One out of every three patients with type 2 DM develops DR. It is also the most common cause of preventable new cases of blindness in people aged 20–74 years [6].

Fundus examination is critical to detect the first signs of DR. Initial signs include damage to the endothelium of the retinal vasculature, microaneurysms, hemorrhage, and exudate development. With DR progression, capillary occlusion, neovascularization, and subsequent retinal detachment may develop. Retinal lesions are detected with an ophthalmoscope. Retinal lesions are divided into two types according to their intensity: non-proliferative diabetic retinopathy (NPDR) and proliferative diabetic retinopathy (PDR). Fundus fluorescein angiography (FFA) and optical coherence tomography (OCT) are used for diagnosis and follow-up. These are costly, invasive, and specialized methods. At the same time, each of the methods used to monitor the development of other micro and macro complications in patients with type 2 DM increases the financial burden. Therefore, it is important to have additional biomarkers for the diagnosis and follow-up of both DR and other complications in type 2 DM patients with DR in many peripheral regions.

Hyperglycemia-induced glycosylated proteins damage cells, causing dysfunction and initiating the inflammatory process by inducing the release of tumor necrosis factor-alpha (TNF-α), free radicals, interleukin-6 (IL-6), and C-reactive protein (CRP). This is recognized as one of the main mechanisms of vascular damage, which is the cause of the chronic complications of diabetes [2].

During the inflammatory response, platelets mediate the activation of circulating leukocytes and their adhesion to the endothelial surface, leading to both changes in the number of circulating leukocytes and endothelial damage [7].

The determination of systemic immune inflammation (SII) cut-off values in patients with DR may be a warning for the prevention of other complications and mortality.

SII derived from laboratory parameters has recently come to the forefront in determining the prognosis of many diseases. SII, which can be calculated simply by using platelet, lymphocyte, and neutrophil values from complete blood count (CBC) values, was first developed in 2014 to predict survival rates of individuals with hepatocellular carcinoma [8,9].

There is increasing evidence that there may be an association between SII and metabolic derangements and their components [10]. Moreover, considering that macro- and microvascular complications of type 2 DM have chronic inflammation and metabolic derangements as common risk factors, it is a valid assumption that people with high SII levels have a higher risk of these complications.

Numerous studies have examined the relationship between type 2 DM, its complications, and SII. However, no study has addressed the association of SII with duration in predicting macrovascular and other microvascular complications and mortality rates in type 2 diabetic retinopathy patients.

This study was conducted to fill the existing literature gap and examine the relationship between SII levels, complications, and mortality rates in type 2 diabetic retinopathy patients during the first year of follow-up and the subsequent three years after the index was calculated.

2. Materials and Methods

This study was designed as a single-center and retrospective study. Between January and December 2019, 523 patients over 18 years old who were admitted to the endocrinology outpatient clinic of our hospital with a DR diagnosis of type 2 DM were included in the study. Patients diagnosed with type 1 DM, patients with acute infectious disease, sepsis, chronic inflammatory and rheumatologic diseases, active malignancies, other endocrinologic disorders, patients under 18 years of age, pregnant women, and breastfeeding women were excluded. For this study, TÜTF-GOBAEK 2023/341 approval was received from the ethics committee of our Faculty of Medicine, dated 29 September 2023. DR progression was defined as increased macular edema, decreased visual acuity, and the conversion of NPDR to PDR.

Demographic information, laboratory data at the time of admission, medical history, and examination findings of the patients included in the study were obtained from the patient file records of the hospital. The demographic information of the patients was determined as age, gender, alcohol, and smoking. The diagnosis of DR was confirmed according to the EURETINA guidelines, and type 2 DM was confirmed according to the American Diabetes Association 2019 criteria [11,12]. The medical history of the patients included the duration of DM, a diagnosis and duration of hypertension, a diagnosis of heart failure, a diagnosis of cerebrovascular disease, insulin use, and antitriglyceridemic and antihyperlipidemic drug use. Laboratory parameters in the study were measured in the Biochemistry Department of our hospital using an automated analyzer. The SII value was obtained using the following: platelet count at admission x neutrophil count/lymphocyte count ratio [13,14]. Macro and micro complications and clinical findings of the patients during follow-up were obtained through outpatient clinic records, phone calls, and an e-pulse system [15].

Descriptive statistics regarding the demographic and clinical characteristics of patients were shown as mean, standard deviation, number, and %. The suitability of quantitative data for normal distribution was examined with the Shapiro–Wilk test. The Mann–Whitney U test was used to compare SII values between groups. The McNemar–Chi-square test was used in the comparison of complication rates between the first year and the following three years. The Related-Samples Cochran's Q test was used in the comparison of complication rates between the baseline, first year, and the following three years. Using the Receiver Operating Characteristic (ROC) analysis method and the area under the curve (AUC), the cut-off points and the sensitivity and specificity values were calculated for SII. The relationship between SII and the variables hypertension, at least 20 pack-years of smoking, antihyperlipidemic/antitriglyceridemic drug use, DM insulin usage, and gender was examined using multiple linear regression analysis. The relationship between CRP and SII was examined using the Spearman test. $p < 0.05$ value was accepted as statistical significance. The SPSS 20.0 package program (IBM SPSS Statistics for Windows, Version 20.0. Armonk, NY, USA: IBM Corp.) was used to analyze the data.

3. Results

The demographic and clinical characteristics of the patients are shown in Table 1. The average age of the patients was 63.5 ± 9.3 years, and 40.3% were women. The average SII values were found to be 821.4 ± 1010.8.

The rate of polyneuropathy was 39.4% in the first year and 43.3% in the following three years. Retinopathy progression was detected as 20.3% in the first year and 28.8% in the following three years. While the rate of microvascular complications was 74.6% in the first year, this rate decreased to 61.8% in following three years ($p < 0.001$). While the rate of macrovascular complications was 45.9% in the first year, this rate decreased to 30.4% in following three years ($p < 0.001$). Similarly, a significant decrease was observed in the rates of micro- and macrovascular complications, microvascular complications and death, macrovascular complications and death, and micro- and macrovascular complications and death combinations in following three years ($p < 0.001$; Table 2).

Table 1. Demographic and clinical characteristics of the patients.

	N = 523
Age, *years*	63.5 ± 9.3
Gender, *female*	211 (40.3)
Duration of DM, *years*	
<5 years	18 (3.4)
5–10 years	103 (19.7)
>10 years	402 (76.9)
Insulin use status in DM, *yes*	362 (69.2)
Hypertension, *yes*	489 (93.5)
Duration of hypertension, *years*	
≤5 years	37 (7.6)
>5 years	452 (92.4)
Stroke, *yes*	76 (14.5)
Heart failure, *yes*	145 (27.7)
Lung pathology, *yes*	25 (4.8)
Smoking status, ≥20 *packets/year*	265 (50.7)
Continued smoking in the last 10 Years, *yes*	133 (25.4)
Antihyperlipidemic/antitriglyceridemic drug use, *yes*	364 (70.3)
SII	821.4 ± 1010.8
C-Reactive Protein, *mg/dL*	1.9 ± 6.4

Mean ± Standard deviation, n (%), SII: Systemic immune inflammation index.

Table 2. Patients' first year and following three years' health outcomes and complication values.

	The First-Year	The Following Three Years	*p*
Polyneuropathy, *yes*	206 (39.4)	179 (43.3)	**<0.001**
Retinopathy progression, *yes*	106 (20.3)	119 (28.8)	**<0.001**
Coronary artery disease, *yes*	154 (29.4)	88 (21.3)	0.555
Acute-chronic renal failure, *yes*	252 (48.2)	207 (50.1)	**<0.001**
Peripheral artery disease, *yes*	138 (26.4)	96 (23.2)	0.362
Hospitalization for any reason, *yes*	363 (69.4)	310 (75.1)	**<0.001**
Hospitalization for DM, *yes*	39 (7.5)	34 (8.2)	0.291
Surgery for Retinopathy, *yes*	522 (99.8)	386 (93.5)	**<0.001**
Death, *yes*	108 (20.7)	29 (7.0)	**<0.001**
Microvascular Complications, *yes*	390 (74.6)	323 (61.8)	**<0.001**
Macrovascular Complications, *yes*	240 (45.9)	159 (30.4)	**<0.001**
Micro + Macrovascular complications, *yes*	201 (38.4)	142 (27.2)	**<0.001**
Microvascular Complications + Death, *yes*	91 (17.4)	27 (5.2)	**<0.001**
Macrovascular Complications + Death, *yes*	73 (14.0)	23 (4.4)	**<0.001**
Micro + Macrovascular complications + Death, *yes*	61 (11.7)	22 (4.2)	**<0.001**

The comparison of patients' SII values in the first year and the following three years according to the presence of micro and macro complications, hospitalization for any reason, and hospitalization for DM parameters is shown in Table 3. Accordingly, the SII values of those with acute–chronic renal failure, peripheral artery disease, and hospitalization were found to be significantly higher both in the first year and following three years ($p < 0.05$ for all). No significant difference was found in the SII values in terms of Polyneuropathy, CAD, Retinopathy Progression, and hospitalization due to DM ($p > 0.05$ for all).

The comparison of SII values in the first year and the following three years according to micro- and macrovascular complications and death is shown in Table 4. Accordingly, in the first year, SII values were found to be significantly higher in those with microvascular and macrovascular complications, in those who died, and in combinations of these parameters ($p < 0.05$ for all). In following three years, no significant difference was found in the SII values in terms of micro- and macrovascular complications and death ($p > 0.05$ for all).

The results of the ROC analysis of SII values regarding micro- and macrovascular complications and death are shown in Table 5. Accordingly, the area under the curve (AUC) values in the first year were 0.608 ($p < 0.001$) for microvascular complications, 0.570 ($p = 0.005$) for macrovascular complications, 0.618 ($p < 0.001$) for death, 0.589 ($p < 0.001$) for micro- and macrovascular complications, and 0.629 ($p = 0.001$) for micro- and macrovascular complications and death. The area under the curve values in the following three years were not significant ($p > 0.05$ for all).

The results of the ROC analysis of the SII values regarding micro- and macrovascular complications and death in the first year are shown graphically in Figure 1. It was found that the SII values reached the highest diagnostic accuracy rate in the first year in the micro- and macrovascular complications and death combination, and at the cut-off point calculated as >594.0, the sensitivity value was 73.8%, and the specificity value was 49.4%.

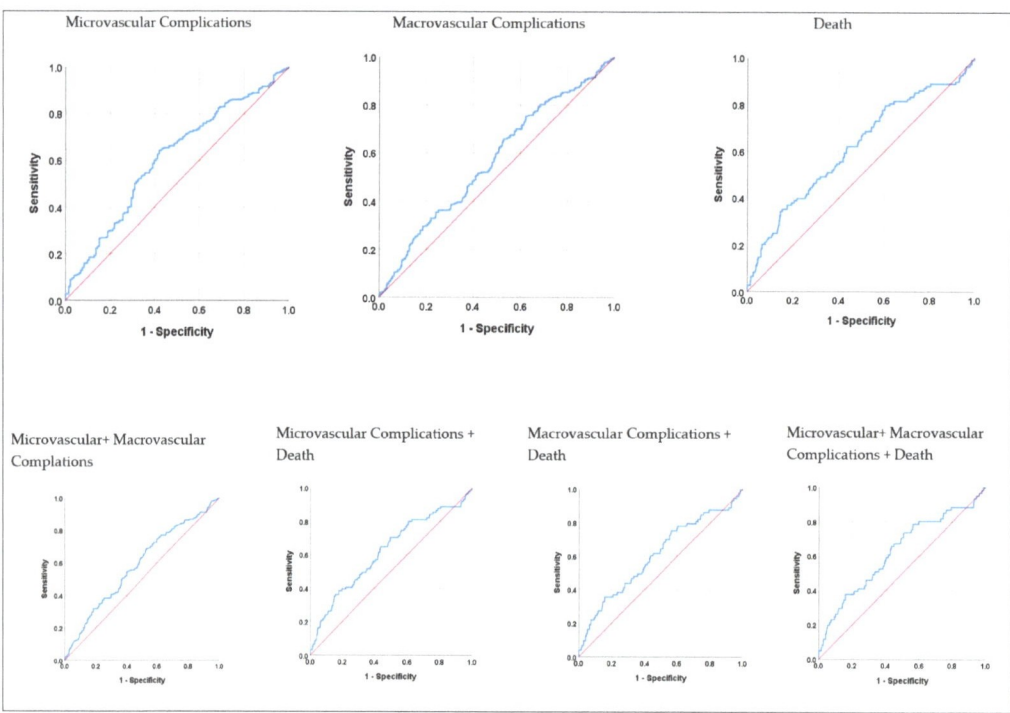

Figure 1. ROC analysis results of SII values for microvascular and macrovascular complications and death in the first year. Blue line: Represents the performance of the SII, showing the balance between sensitivity and specificity; Red line: Represents a random classifier with no discrimination ability and an AUC of 0.5.

Table 3. Comparison of SII values according to micro and macro complications, hospitalization for any reason, and hospitalization for DM parameters.

		Polyneuropathy	Coronary Artery Disease	Retinopathy Progression	Acute-Chronic Renal Failure	Peripheral Artery Disease	Hospitalization for Any Reason	Hospitalization for DM
The first-year	No	803 ± 1151	772.4 ± 692	824.7 ± 1050.9	706.5 ± 566.9	756.4 ± 668.5	670.4 ± 490.3	833.1 ± 1041.8
	Yes	849.6 ± 747	938.6 ± 1521.2	808.2 ± 838.8	944.9 ± 1322.7	1002.6 ± 1610.9	887.9 ± 1163.2	675.4 ± 463.8
	p	0.076	0.124	0.997	0.001	0.006	0.018	0.474
The following three years	No	700.4 ± 562	712.4 ± 584.1	739.8 ± 609.7	680.7 ± 553.5	711.9 ± 609.5	609.7 ± 361.8	744.7 ± 610.6
	Yes	770.5 ± 633.2	798.7 ± 629	708.6 ± 555.8	780.6 ± 629.5	793.1 ± 539	771 ± 648.9	576.2 ± 331.7
	p	0.119	0.085	0.831	0.016	0.012	0.016	0.139

Mean ± Standard deviation.

Table 4. Comparison of SII values according to microvascular and macrovascular complications and death.

		Microvascular Complications	Macrovascular Complications	Death	Micro + Macrovascular Complications	Microvascular Complications + Death	Macrovascular Complications + Death	Micro + Macrovascular complications + Death
The first-year	No	640.5 ± 471.1	736.5 ± 639.8	729.9 ± 593.1	725.7 ± 1122.6	735.6 ± 596.4	747.4 ± 606.5	746.7 ± 602.6
	Yes	883.0 ± 1131.6	921.4 ± 1315.2	1172.9 ± 1861.7	820.1 ± 1422.2	1228.1 ± 2004.8	1277.4 ± 2206.4	1386.8 ± 2393.7
	p	<0.001	0.006	<0.001	0.001	<0.001	0.003	0.001
The following three years	No	912.2 ± 1423.9	831.4 ± 1145.8	723.2 ± 582.6	821.8 ± 1122.6	818.4 ± 1023.3	815.4 ± 1020.1	814.3 ± 1019.4
	Yes	765.1 ± 628.4	798.2 ± 598.3	844.4 ± 737.7	820.1 ± 620.9	874.2 ± 755.4	951.7 ± 786.9	982.1 ± 791.4
	p	0.772	0.195	0.661	0.081	0.718	0.306	0.175

Mean ± Standard deviation.

Table 5. ROC analysis results of SII values regarding microvascular and macrovascular complications and death.

		AUC	p	Cut-Off	Sensitivity (%)	Specificity (%)
The first-year	Microvascular Complications	0.608	**<0.001**	>540.5	64.1	57.9
	Macrovascular Complications	0.570	**0.005**	>467.6	75.4	37.8
	Death	0.618	**<0.001**	>994.0	35.2	84.8
	Micro + Macrovascular complications	0.589	**<0.001**	>531.1	68.7	47.2
	Microvascular Complications + Death	0.622	**<0.001**	>634.2	64.8	56.3
	Macrovascular Complications + Death	0.607	**0.004**	>1019.2	35.6	84.2
	Micro + Macrovascular complications + Death	0.629	**0.001**	>594.0	73.8	49.4
The following three years	Microvascular Complications	0.508	0.776	>1019	84.5	23.5
	Macrovascular Complications	0.536	0.192	>467.6	74.2	34.3
	Death	0.524	0.711	>838.5	37.9	77.8
	Micro + Macrovascular complications	0.550	0.080	>657.7	52.8	58.3
	Microvascular Complications + Death	0.521	0.757	>838.5	40.7	74.2
	Macrovascular Complications + Death	0.563	0.371	>852.2	43.5	75.0
	Micro + Macrovascular complications + Death	0.585	0.223	>852.2	45.5	75.1

A multiple linear regression analysis was conducted to investigate the effects of various factors on SII. The model included hypertension, smoking (at least 20 pack-years), antihyperlipidemic and antitriglyceridemic drug use, DM insulin usage, and gender as predictors. The overall model was not statistically significant ($F(7, 510) = 1.065$, $p = 0.385$), indicating that these factors collectively did not explain a significant portion of the variance in SII ($R^2 = 0.014$, Adjusted $R^2 = 0.001$).

Among the predictors, only at least 20 pack-years of smoking showed a statistically significant positive effect on SII (B = 180.053, $p = 0.034$, Table 6). The other factors, including hypertension, antihyperlipidemic and antitriglyceridemic medication use, DM insulin usage, and gender, did not show a significant relationship with SII ($p > 0.05$ for all, Table 6).

Table 6. Multiple linear regression analysis results for predictors of Systemic Immune Inflammation (SII).

Variable	B	p-Value
Hypertension	−132.996	0.292
At least 20 pack-years of smoking	180.053	**0.034**
Antihyperlipidemic and antitriglyceridemic drug use	2.806	0.967
DM insulin usage	11.876	0.861
Gender	−22.822	0.758

The Spearman correlation analysis showed a weak positive correlation between SII and CRP ($r = 0.266$; $p < 0.001$; Table 7).

Table 7. Spearman correlation analysis results of SII and CRP in type 2 diabetic retinopathy patients.

		SII	CRP
SII	r	1.000	0.266
	p		<0.001
CRP	r	0.266	1.000
	p	<0.001	

r: Spearman's correlation coefficient. Values range between +1 and −1: +1 indicates a perfect positive correlation; −1 indicates a perfect negative correlation; and 0 indicates no correlation. p: Statistical significance value: $p < 0.05$ indicates that the correlation is significant at the 95% confidence level; $p < 0.01$ indicates significance at the 99% confidence level. CRP: C-Reactive protein, mg/dL; N: Sample size, which is 523 in this study.

4. Discussion

This study investigated the relationship between the predictive value of SII on other microvascular complications of type 2 DM (nephropathy and neuropathy), macrovascular complications (coronary artery disease, peripheral artery disease, and cerebrovascular disease), and mortality rates in patients diagnosed with DR, which is one of the most common microvascular complications of type 2 DM, in early and late periods (the first year and the following three years). The findings of our study show that high SII values were associated with increased mortality and the development of other microvascular and macrovascular complications and were statistically significant in predicting the likelihood of mortality and the development of complications, especially in the first year following the calculation of the SII value.

Type 2 DM is a chronic disease characterized by persistent hyperglycemia. Due to hyperglycemia, glucose reacts non-enzymatically with lipids and proteins, releasing advanced glycation end products (AGEs) and causing oxidative stress. The resulting oxidative stress increases the levels of reactive oxygen species (ROS), leading to impaired blood flow in small blood vessels and impaired tissue nutrition, leading to microvascular complications. In large blood vessels, it facilitates the development of atherosclerotic plaques, which narrow the arteries and restrict blood flow, leading to macrovascular complications [16,17]. Sustained hyperglycemia triggers the immune response, leading to the synthesis of pro-inflammatory cytokines, C-reactive protein, tumor necrosis factor (TNF)-α, interleukin-6 (IL-6), and chemokines [18]. Chronic inflammation disrupts the structure of small vessels and contributes to the development of microvascular complications. In large vessels, they enter atherosclerotic plaques and destabilize them. They cause plaque ruptures [19]. Endothelial dysfunction develops. Another condition that contributes to endothelial dysfunction is adipokines (adiponectin, leptin, and resistin). Adiponectin is anti-inflammatory, whereas leptin and resistin are inflammatory. Adiponectin increases nitric oxide (NO) levels. However, in type 2 DM, adiponectin is decreased, which leads to a decrease in the NO level and an increase in the endothelin-1 level, which disrupts the balance in small vessels in favor of vasoconstriction and leads to the development of microvascular complications. In large vessels, endothelial dysfunction plays an important role in the emergence of macrovascular complications by facilitating vasoconstriction, inflammation, and oxidative stress [20].

DR is a microvascular complication of diabetes in the eye with an increasing prevalence. Once it occurs, it is irreversible. It can progress to vision loss. Current treatments aim to halt its progression. Chronic minimal inflammation plays a role in its pathogenesis. An animal model study in diabetic mice showed that the progression of DR was slowed when inflammatory pathways were blocked [21]. Another study examined the levels of inflammatory proteins in patients with DR using proteomic techniques and found an increase in these levels [22]. These studies demonstrate the role of chronic minimal inflammation in the development and progression of DR and suggest that the inhibition of inflammation may prevent the progression of the disease.

Many inflammatory cells are involved in developing and maintaining chronic minimal inflammation. The most prominent feature of this inflammatory process is the increased permeability of the vascular wall secondary to inflammation, allowing inflammatory cells to pass from the vessels to the tissues. Neutrophils are the first cells to arrive when inflammation occurs. Under chronic inflammation, neutrophils are constantly activated, leading to the formation and accumulation of neutrophil extracellular traps (NETs). This leads to vascular occlusion, tissue damage, and the exacerbation of inflammation, increasing damage at the site of inflammation. In a study comparing the neutrophil levels of patients with type 2 DM and healthy individuals, it was found that the neutrophil levels of the patient group were significantly higher [22]. Significant increases in the number of neutrophil, monocyte–macrophage, and platelet cells have been recorded in type 2 DM patients during the inflammation process. In a systematic meta-analysis study examining the changes in hematologic parameters of patients with type 2 DM and type 1 DM and a control group of healthy individuals, it was shown that patients with type 2 DM had significantly higher absolute neutrophil, absolute monocyte, absolute lymphocyte, and absolute basophil counts and relative neutrophil and basophil counts [23]. In addition, another study reported that hematologic parameters, such as the neutrophil-to-lymphocyte ratio (NLR) and monocyte-to-platelet ratio, were higher in patients with diabetes and that these increases could be considered a sign of inflammatory response [24]. In another study, the neutrophil levels of patients with type 2 DM and healthy individuals were compared, and it was found that the neutrophil levels of the patient group were significantly higher [25]. In another study, it was concluded that NET levels were elevated in type 2 DM patients and that NETs play a role in the pathogenesis of both DM and microvascular complications, especially diabetic nephropathy and cardiac disease associated with DM [26]. One of the other cells involved in chronic minimal inflammation is monocyte–macrophages, which maintain and exacerbate the inflammatory process by secreting pro-inflammatory cytokines (IL-1β, TNF-α, IL-6, IL-8, MCP-1, and IL-1) [27]. Another cell is platelets, which both form thrombosis with their own activation and increase their activation by binding to leukocyte subcellular groups [28]. An increase in neutrophil, monocyte–macrophage, and platelet cell numbers indicates inflammation. These results are compatible with our study because the neutrophil, monocyte, and platelet values were found to be high in patients with DR in our study (SII: 821.4 ± 1010.8; Table 1).

SII, one of the most frequently used inflammatory markers in recent years, is less affected by physiopathologic changes. This makes SII more reliable and more sensitive in showing existing inflammation [29]. Other advantages are that it is cost-effective and can be calculated with hemogram results that can be performed even in peripheral areas. In a study of 500 type 2 DM patients with and without DR, SII was found to be an inflammatory marker that can be used for early diagnosis of DR [30]. In our study, our patients had a diagnosis of DR, but we found that SII had no predictive value in DR progression in the first year and the following three years of follow-up (retinopathy progression $p > 0.05$; Table 3). This may be because the SII values of all patients already diagnosed with DR were higher than the SII of only patients with type 2 DM due to inflammation. In a study in which 100 uncomplicated type 2 DM patients were included, and many biomarkers were examined, the mean SII value was found to be 470.91 [31]. In our study, all of our patients were type 2 DM patients with DR diagnosis, and the mean value for SII was found to be 821.4. Due to the existing microvascular complications, the mean SII value in our study was almost twice as high compared to uncomplicated type 2 DM in the other study. For this reason, chronic minimal inflammation may have continued in the following years but may not have reached very high levels with its progression. Another reason may be that 108 patients died within one year after the calculated SII values, and 29 patients died within the following three years (Table 2). As a result, we do not know how many of these deceased patients had DR progression. In a study of 300 type 2 DM patients on whether SII has a predictive value for diabetic nephropathy and cardiovascular diseases in patients with type 2 DM, predictive SII values were found for cardiovascular diseases and diabetic nephropathy [31]. In a study investigating the relationship between SII and diabetic neuropathy in 1460 hospitalized type 2 DM patients in China, it was found that SII levels were significantly higher in patients with diabetic neuropathy [32]. In another study in which 584 patients with diabetic nephropathy (DN) due to type 2 DM, type 2 DM patients without DN, and a control group were investigated for the relationship between SII and DN, it was found that high SII levels were associated with DN [33]. The results of our study are consistent with the literature. SII was found to be statistically significant in predicting type 2 diabetic retinopathy patients to be diagnosed with acute or chronic renal failure due to diabetic nephropathy in the first year and the following three years, the development of peripheral arterial disease, and the hospitalization of patients for any reason (Table 3). It was not statistically significant in predicting coronary artery disease, hospitalization due to diabetes, and the development of polyneuropathy both in the first year and the following three years (Table 3). This may be due to the fact that the basal inflammation was high in those who died, and chronic minimal inflammation was ongoing due to DR, and therefore, the patient's immune systems were active, or other microvascular complications may have started due to the development of DR but were not detected because they were not yet diagnosed. In addition, it has been observed that type 2 diabetic retinopathy patients are more likely to be hospitalized due to complications. As patients who develop micro- and macrovascular complications are mostly patients with a poor glycemic index, who do not pay attention to their diet and do not follow the controls, it is natural that they are frequently hospitalized due to complications that develop as a result. Another reason may be the need for a larger sample group, which is also one of the limitations of our study.

One of the most important results of this study is that SII values have significant predictive power for parameters, such as micro- and macrovascular complications and death and combinations of these parametrics in the early period, but this predictive power decreases with increasing duration. In the first year, the significant association between high SII values and these clinical outcomes ($p < 0.05$; Table 4) suggests that SII may be a valuable biomarker for short-term risk assessment. However, the lack of a significant difference supporting this association over the following three years ($p > 0.05$; Table 4) suggests that the long-term clinical predictive value of SII is limited. This finding emphasizes the need for further investigation of the dynamic changes in inflammatory markers over time and their impact on long-term clinical outcomes.

The focus of our study was to examine the relationship between the SII values of patients with type 2 diabetic retinopathy and the development of micro- and macrovascular complications and death in the first year and the following three years. When the SII values of the patients were analyzed by an ROC analysis for microvascular complications, macrovascular complications, death, micro- and macrovascular complications, microvascular complications and death, macrovascular complications and death, micro- and macrovascular complications and death, and micro- and macrovascular complications and death (composite endpoint) for the first year and following three years, a statistically significant optimal SII value was confirmed for all parameters in the first year ($p < 0.05$; Table 5). For following three years, the SII values determined were not statistically significant ($p > 0.05$; Table 5). This may be due to the relationship between the existing inflammation and duration. In general, our SII values were found to be high because type 2 DM, DR, and its complications are based on chronic minimal inflammation, and SII is an inflammatory marker. This is consistent with the literature. The ROC curve analysis confirmed that the optimal SII value for the composite endpoint, >594.0, predicted the composite endpoint with a higher sensitivity of 73.8%, specificity of 49.4%, and AUC= 0.629 compared to other parameters ($p = 0.001$; Table 5; Figure 1).

We applied multiple linear regression analysis to determine whether the variables in our study affected the SII values. We found that only "at least 20 pack-years of smoking" had a statistically significant positive effect on SII (B = 180.053, $p = 0.034$). This indicates that individuals who have smoked for at least 20 pack-years have higher levels of systemic immune inflammation. This finding is consistent with previous studies linking heavy smoking to increased inflammatory responses [34]. In contrast, (B = −132.996, p = 0.292), the use of antihyperlipidemic and antitriglyceridemic drugs (B = 2.806, p = 0.967), use of DM insulin (B = 11.876, p = 0.861), and gender (B = −22.822, p = 0.758) did not show statistically significant relationships with SII. These results suggest that these factors do not have a significant impact on systemic immune inflammation in this study population. This may be due to various reasons, such as the sample size not being large enough to detect subtle effects, or these variables affect SII through mechanisms not captured in this analysis. Further research is needed to better understand and manage conditions associated with inflammation by exploring other potential determinants and mechanisms affecting SII.

Another inflammation biomarker, CRP, is known to rise in response to inflammation in the body and is widely used. In a 2015 meta-analysis study examining the relationship between CRP and diabetic retinopathy, it was concluded that CRP could be used as a biomarker to determine the severity of diabetic retinopathy [35]. Given that SII is an inflammatory index, we examined the relationship between CRP values and SII values in our patients using the Spearman test. Consistent with the above literature, we found a weak positive relationship between the two inflammatory markers (r = 0.266; p < 0.001; Table 7). However, we did not find any studies in the literature regarding CRP's ability to predict early and late-term mortality and morbidity in patients with type 2 diabetic retinopathy.

This study found that the SII index is an effective predictor of morbidity and mortality in patients with type 2 diabetic retinopathy. High SII values were significantly associated with acute and chronic renal failure, peripheral arterial disease and hospitalization rates. These findings suggest that the SII index can be used to predict microvascular and macrovascular complications and mortality risk in patients with type 2 diabetic retinopathy, especially within 1 year of calculating SII values. However, combining SII with other biomarkers and artificial intelligence methods may improve diagnostic accuracy. Integrating various biomarkers and AI methods has shown potential to improve diagnostic processes in clinical applications, as demonstrated in studies involving hemogram-based decision tree models to differentiate COVID-19 patients [36]. Similarly, the use of additional biomarkers and AI in combination with SII may improve the long-term prediction of complications and mortality risk in type 2 diabetic retinopathy patients.

Our study had several limitations, the most important of which was that it was a retrospective and single-center study. Another one is the small number of participants. Moreover, the specificity values of the SII are low; therefore, they should be carefully interpreted in their clinical usage.

5. Conclusions

Our research is significant in determining the predictive cut-off value of the SII for microvascular and macrovascular complications, mortality, and composite endpoints in the first year among patients with type 2 diabetic retinopathy. Additionally, both in the first year and the following three years, it is valuable for identifying SII values that can predict the development of acute or chronic renal failure due to diabetic nephropathy, peripheral arterial disease, and hospital admissions for any reason among these patients. This capability enables the early diagnosis of potential complications in patients with type 2 diabetic retinopathy, leading to timely treatment. Thus, patients at risk can be identified through simple hemogram parameters used to calculate the SII, potentially preventing or halting the progression of diabetic complications. This not only helps prevent a decline in patients' quality of life but also results in significant healthcare cost savings. Moreover, predicting the risk of death within the first year could also enhance patient survival rates.

In addition, in the clinical use of SII, patients' histories of comorbidities, as well as their inflammatory habits, such as long-term smoking, should also be taken into consideration.

Author Contributions: Conceptualization, N.T.T. and M.C.; methodology, N.T.T.; software, N.T.T.; validation, N.T.T. and M.C.; formal analysis, N.T.T.; investigation, N.T.T.; resources, N.T.T.; data curation, N.T.T.; writing—original draft preparation, N.T.T. and M.C.; writing—review and editing,

N.T.T. and M.C.; visualization, N.T.T.; supervision, N.T.T. and M.C. All authors have read and agreed to the published version of the manuscript.

Funding: This research received no funding.

Institutional Review Board Statement: Ethics committee approval for this study was obtained from the Ethics Committee of the Faculty of Medicine of our university with the decision number and protocol code TÜTF-GOBAEK 2023/341 dated 29 September 2023, and all procedures in this study were performed under the Declaration of Helsinki and its subsequent amendments.

Informed Consent Statement: This study was conducted retrospectively as a file review.

Data Availability Statement: Our study data contain personal information of patients and, therefore, are not available for sharing due to the 'Personal Data Protection Law' and ethical reasons.

Acknowledgments: We would like to thank Necdet Sut and Hande Guçlu for their assistance in conducting this study.

Conflicts of Interest: The authors declare no conflicts of interest.

References

1. National Diabetes Consensus Group. Diabetes Mellitus Tanı, Sınıflama ve İzlem İlkeleri. In *Diyabet Tanı ve Tedavi Rehberi 2019*, 9th ed.; Balcı, M.K., Ed.; Turkish Diabetes Foundation: İstanbul, Turkey, 2020; pp. 16–17.
2. Bhattacharyya, S.; Jain, N.; Verma, H.; Sharma, K. A Cross-sectional Study to Assess Neutrophil Lymphocyte Ratio as a Predictor of Microvascular Complications in Type 2 Diabetes Mellitus Patients. *J. Clin. Diagn. Res.* **2021**, *15*, 59. [CrossRef]
3. Kocaeli, A.A.; Gül, Ö.Ö. Diabetes Mellitusun Epidemiyolojisi. In *Diabetes Mellitusun Tanı, Tedavi ve İzlemi*, 1st ed.; İmamoğlu, Ş., Ersoy, C.Ö., Eds.; Bursa Uludağ University: Bursa, Turkey, 2022; pp. 43–57.
4. De Ferranti, S.D.; Boer, I.H.; Vivian, F.; Fox, C.S.; Golden, S.H.; Lavie, C.J.; Magge, S.N.; Marx, N.; McGuire, D.K.; Orchard, T.J.; et al. Type 1 diabetes mellitus and cardiovascular disease: A scientific statement from the American Heart Association and American Diabetes Association. *Circulation* **2014**, *130*, 1110–1130. [CrossRef]
5. Jameson, J.; Fauci, A.S.; Kasper, D.L.; Hauser, S.L.; Longo, D.L.; Loscalzo, J. *Harrison's Principles of Internal Medicine*, 20th ed.; McGraw-Hill Education: New York, NY, USA, 2018; pp. 2850–2889.
6. Lonardo, A. Liver Fibrosis: More than meets the eye. *Ann. Hepatol.* **2024**, *29*, 101479. [CrossRef] [PubMed]
7. Bian, X.; He, J.; Zhang, R.; Yuan, S.; Dou, K. The Combined Effect of Systemic Immune-Inflammation Index and Type 2 Diabetes Mellitus on the Prognosis of Patients Undergoing Percutaneous Coronary Intervention: A Large-Scale Cohort Study. *J. Inflamm. Res.* **2023**, *16*, 6415–6429. [CrossRef] [PubMed]
8. Guo, W.; Song, Y.; Sun, Y.; Du, H.; Cai, Y.; You, Q.; Fu, H.; Shao, L. Systemic immune-inflammation index is associated with diabetic kidney disease in Type 2 diabetes mellitus patients: Evidence from NHANES 2011–2018. *Front. Endocrinol.* **2022**, *13*, 1071465. [CrossRef] [PubMed]
9. Qin, Z.; Li, H.; Wang, L.; Geng, J.; Yang, Q.; Su, B.; Liao, R. Systemic immune-inflammation index is associated with increased urinary albumin excretion: A population-based study. *Front. Immunol.* **2022**, *13*, 863640. [CrossRef] [PubMed]
10. Wang, P.; Guo, X.; Zhou, Y.; Li, Z.; Yu, S.; Sun, Y.; Hua, Y. Monocyte-to-high-density lipoprotein ratio and systemic inflammation response index are associated with the risk of metabolic disorders and cardiovascular diseases in general rural population. *Front. Endocrinol.* **2022**, *13*, 944991. [CrossRef]
11. Schmidt-Erfurth, U.; Garcia-Arumi, J.; Gerendas, B.S.; Midena, E.; Sivaprasad, S.; Tadayoni, R.; Wolf, S.; Loewensteing, A. Guidelines for the management of retinal vein occlusion by the European Society of Retina Specialists (EURETINA). *Ophthalmologica* **2019**, *242*, 123–162. [CrossRef] [PubMed]
12. Buse, J.B.; Wexler, D.J.; Tsapas, A.; Rossing, P.; Mingrone, G.; Mathieu, C.; D'Alessio, D.A.; Davies, M.J. 2019 update to: Management of hyperglycemia in type 2 diabetes, 2018. A consensus report by the American Diabetes Association (ADA) and the European Association for the Study of Diabetes (EASD). *Diabetes Care* **2020**, *43*, 487–493. [CrossRef]
13. Wang, R.-H.; Wen, W.X.; Jiang, Z.-P.; Du, Z.-P.; Ma, Z.-H.; Lu, A.-L.; Li, H.-P.; Yuan, F.; Wu, S.B.; Guo, J.W.; et al. The clinical value of neutrophil-to-lymphocyte ratio (NLR), systemic immune-inflammation index (SII), platelet-to-lymphocyte ratio (PLR) and systemic inflammation response index (SIRI) for predicting the occurrence and severity of pneumonia in patients with intracerebral hemorrhage. *Front. Immunol.* **2023**, *14*, 1115031. [CrossRef]
14. Ozkan, U.; Gurdogan, M. TyG index as a predictor of spontaneous coronary artery dissection in young women. *Postgrad. Med.* **2023**, *135*, 669–675. [CrossRef] [PubMed]
15. Birinci, Ş.A. Digital Opportunity for Patients to Manage Their Health: Turkey National Personal Health Record System (The e-Nabız). *Balkan Med. J.* **2023**, *40*, 215–221. [CrossRef] [PubMed]
16. Archundia Herrera, M.C.; Subhan, F.B.; Chan, C.B. Dietary patterns and cardiovascular disease risk in people with type 2 diabetes. *Curr. Obes. Rep.* **2017**, *6*, 405–413. [CrossRef] [PubMed]
17. Caussy, C.; Aubin, A.; Loomba, R. The relationship between type 2 diabetes, NAFLD, and cardiovascular risk. *Curr. Diab. Rep.* **2021**, *21*, 15. [CrossRef] [PubMed]

18. Devaraj, S.; Dasu, M.R.; Jialal, I. Diabetes is a proinflammatory state: A translational perspective. *Expert Rev. Endocrinol. Metab.* **2010**, *5*, 19–28. [CrossRef] [PubMed]
19. Nguyen, D.V.; Shaw, L.C.; Grant, M.B. Inflammation in the pathogenesis of microvascular complications in diabetes. *Front. Endocrinol.* **2012**, *3*, 170. [CrossRef] [PubMed]
20. Zitouni, K.; Steyn, M.; Earle, K.A. Residual renal and cardiovascular disease risk in conventionally-treated patients with type 2 diabetes: The potential of non-traditional biomarkers. *Minerva Med.* **2018**, *109*, 103–115. [CrossRef] [PubMed]
21. Portillo, J.-A.C.; Yu, J.-S.; Vos, S.; Bapputty, R.; Corcino, Y.L.; Hubal, A.; Daw, J.; Arora, S.; Sun, W.; Lu, Z.-L.; et al. Disruption of retinal inflammation and the development of diabetic retinopathy in mice by a CD40-derived peptide or mutation of CD40 in Müller cells. *Diabetologia* **2022**, *65*, 2157–2171. [CrossRef]
22. Youngblood, H.; Robinson, R.; Sharma, A.; Sharma, S. Proteomic biomarkers of retinal inflammation in diabetic retinopathy. *Int. J. Mol. Sci.* **2019**, *20*, 4755. [CrossRef]
23. Bambo, G.M.; Asmelash, D.; Alemayehu, E.; Gedefie, A.; Duguma, T.; Kebede, S.S. Changes in selected hematological parameters in patients with type 1 and type 2 diabetes: A systematic review and meta-analysis. *Front. Med.* **2024**, *11*, 1294290. [CrossRef]
24. Zhang, Y.; Liu, H. Correlation between insulin resistance and the rate of neutrophils-lymphocytes, monocytes-lymphocytes, platelets-lymphocytes in type 2 diabetic patients. *BMC Endocr. Disord.* **2024**, *24*, 42. [CrossRef] [PubMed]
25. Giovenzana, A.; Carnovale, D.; Phillips, B.; Petrelli, A.; Giannoukakis, N. Neutrophils and their role in the aetiopathogenesis of type 1 and type 2 diabetes. *Diabetes Metab. Res. Rev.* **2022**, *38*, e3483. [CrossRef] [PubMed]
26. Njeim, R.; Azar, W.S.; Fares, A.H.; Azar, S.T.; Kassouf, H.K.; Eid, A.A. NETosis contributes to the pathogenesis of diabetes and its complications. *J. Mol. Endocrinol.* **2020**, *65*, R65–R76. [CrossRef] [PubMed]
27. Nirenjen, S.; Narayanan, J.; Tamilanban, T.; Subramaniyan, V.; Chitra, V.; Fuloria, N.K.; Wong, L.S.; Ramachawolran, G.; Sekar, M.; Gupta, G.; et al. Exploring the contribution of pro-inflammatory cytokines to impaired wound healing in diabetes. *Front. Immunol.* **2023**, *14*, 1216321. [CrossRef] [PubMed]
28. Klisic, A.; Scepanovic, A.; Kotur-Stevuljevic, J.; Ninic, A. Novel leukocyte and thrombocyte indexes in patients with prediabetes and type 2 diabetes mellitus. *Eur. Rev. Med. Pharmacol. Sci.* **2022**, *26*, 2775–2781. [CrossRef] [PubMed]
29. Dziedzic, E.A.; Gasior, J.S.; Tuzimek, A.; Paleczny, J.; Junka, A.; Dabrowski, M.; Jankowski, P. Investigation of the associations of novel inflammatory biomarkers-Systemic Inflammatory Index (SII) and Systemic Inflammatory Response Index (SIRI)-With the severity of coronary artery disease and acute coronary syndrome occurrence. *Int. J. Mol. Sci.* **2022**, *23*, 9553. [CrossRef] [PubMed]
30. Wang, S.; Pan, X.; Jia, B.; Chen, S. Exploring the Correlation Between the Systemic Immune Inflammation Index (SII), Systemic Inflammatory Response Index (SIRI), and Type 2 Diabetic Retinopathy. *Diabetes Metab. Syndr. Obes.* **2023**, *16*, 3827–3836. [CrossRef] [PubMed]
31. Suvarna, R.; Biswas, M.; Shenoy, R.P.; Prabhu, M.M. Association of clinical variables as a predictor marker in type 2 diabetes mellitus and diabetic complications. *J. Biomed.* **2023**, *43*, 335–340. [CrossRef]
32. Li, J.; Zhang, X.; Zhang, Y.; Dan, X.; Wu, X.; Yang, Y.; Chen, X.; Li, S.; Xu, Y.; Wan, Q.; et al. Increased Systemic Immune-Inflammation Index Was Associated with Type 2 Diabetic Peripheral Neuropathy: A Cross-Sectional Study in the Chinese Population. *J. Inflamm. Res.* **2023**, *16*, 6039–6053. [CrossRef]
33. Duman, T.T.; Ozkul, F.N.; Balci, B. Could Systemic Inflammatory Index Predict Diabetic Kidney Injury in Type 2 Diabetes Mellitus? *Diagnostics* **2023**, *13*, 2063. [CrossRef]
34. Saint-André, V.; Charbit, B.; Biton, A.; Rouilly, V.; Possémé, C.; Bertrand, A.; Rotival, M.; Bergstedt, J.; Patin, E.; Albert, M.L.; et al. The Milieu Intérieur Consortium. Smoking changes adaptive immunity with persistent effects. *Nature* **2024**, *626*, 827–835. [CrossRef] [PubMed]
35. Song, J.; Chen, S.; Liu, X.; Duan, H.; Kong, J.; Li, Z. Relationship between C-reactive protein level and diabetic retinopathy: A systematic review and meta-analysis. *PLoS ONE* **2015**, *10*, e0144406. [CrossRef] [PubMed]
36. Dobrijević, D.; Andrijević, L.; Antić, J.; Rakić, G.; Pastor, K. Hemogram-based decision tree models for discriminating COVID-19 from RSV in infants. *J. Clin. Lab. Anal.* **2023**, *37*, e24862. [CrossRef] [PubMed]

Disclaimer/Publisher's Note: The statements, opinions and data contained in all publications are solely those of the individual author(s) and contributor(s) and not of MDPI and/or the editor(s). MDPI and/or the editor(s) disclaim responsibility for any injury to people or property resulting from any ideas, methods, instructions or products referred to in the content.

Article

BCG Vaccination Suppresses Glucose Intolerance Progression in High-Fat-Diet-Fed C57BL/6 Mice

Haruna Arakawa [1] and Masashi Inafuku [1,2,*]

[1] Faculty of Agriculture, University of the Ryukyus, Senbaru 1, Nishihara 903-0213, Japan; hana20001714@yahoo.co.jp
[2] The United Graduate School of Agricultural Sciences, Kagoshima University, Kagoshima 890-0065, Japan
* Correspondence: h098648@agr.u-ryukyu.ac.jp; Tel.: +81-98-895-8978; Fax: +81-98-895-8734

Abstract: *Background and Objectives*: *Mycobacterium bovis* Bacillus Calmette–Guérin (BCG) vaccine administration has been suggested to prevent glucose metabolism abnormalities and fatty liver in genetically obese *ob/ob* mice; however, it is not clear whether the beneficial effects of BCG are also observed in the progression of glucose intolerance induced by a high-fat diet (HFD). Therefore, the effects of BCG vaccination on changes in glucose tolerance and insulin response were investigated in HFD-fed C57BL/6 mice. *Materials and Methods*: We used the BCG Tokyo 172 strain to determine effects on abnormalities in glucose metabolism. For vaccination, five-week-old male mice were injected intraperitoneally with BCG and maintained on a HFD for three weeks. The mice were regularly subjected to intraperitoneal glucose tolerance and insulin tolerance tests (IGTTs and ITTs). These tests were also performed in mice transplanted with bone marrow cells from BCG-vaccinated donor mice. *Results*: Significant effects of BCG vaccination on blood glucose levels in the IGTTs and ITTs were observed from week 12 of the experiment. BCG vaccination significantly improved changes in fasting glucose and insulin levels, insulin resistance indexes, and glucagon-to-insulin ratios in conjunction with the HFD at the end of the experiment. Significant inhibitory effects in the IGTTs and ITTs on glucose intolerance were also observed with transplantation with bone marrow cells derived from BCG-vaccinated donor mice. *Conclusions*: BCG vaccination significantly delayed glucose intolerance progression, suggesting a beneficial effect of BCG on the pathogenesis of type 2 diabetes. It has also been suggested that the effects of BCG vaccination may be at least partially due to an immune memory (trained immunity) for hematopoietic stem and progenitor cells of the bone marrow.

Keywords: BCG; glucose intolerance; insulin resistance; trained immunity; nonalcoholic fatty liver disease

1. Introduction

Obesity, especially visceral obesity, contributes to the pathogenesis of metabolic syndrome, a cluster of metabolic abnormalities including hyperlipidemia, hypertension, and insulin resistance (IR) [1]. It is well known that IR is the primary indicator of type 2 diabetes mellitus (T2DM). T2DM pathogenesis is also considered to be linked to the innate and adaptive immune systems, which are recognized as important etiological components in the development of IR [2–4]. Per the International Diabetes Federation, T2DM is the most common type of diabetes (accounting for approximately 90% of all cases). In 2021, more than one in ten adults had diabetes mellitus globally, and the number of people with this disease will continually increase in the future [5]. The increasing prevalence of this condition makes it a public health problem of paramount importance. T2DM imposes a significant personal and public health burden in terms of the number of people affected, complications, and expenses incurred by national health and social care systems [6]. Therefore, the discovery and development of new treatments that regulate glucose and metabolic homeostasis are urgently needed.

An imbalance in energy homeostasis is a hallmark of T2DM [7]. Altered immune surveillance and impaired host defenses have been observed in patients suffering from obesity and T2DM, which may predispose patients to infection caused by germs such as *Mycobacterium tuberculosis* (Mtb) [8,9]. Animal and human studies have also indicated an increased susceptibility to Mtb infection in type 1 diabetes mellitus (T1DM), which is commonly known as juvenile-onset diabetes and characterized by an absolute deficiency in insulin production by the autoimmune destruction of islet β-cells [10,11]. To protect against Mtb infection and its progression to tuberculosis, an attenuated strain of *M. bovis* was used to develop the Bacillus Calmette–Guérin (BCG) vaccine over 100 years ago. The nonspecific effects of BCG were first used for bladder cancer treatment over 40 years ago [12]. Thereafter, the off-target effects of BCG have been shown to protect against infectious and noninfectious diseases, including T1DM [13–18]. It has been reported that repeated BCG vaccinations in long-term diabetics can restore blood sugars to near normal by resetting the immune system and by increasing glucose utilization through a metabolic shift to aerobic glycolysis, a high-glucose-utilization state. [18]. However, to the best of our knowledge, only a few studies have examined the effects of BCG vaccination on T2DM. It has also been reported that in leptin receptor-deficient *db/db* mice, multiple BCG injections significantly decreased blood glucose levels and increased glucose uptake in bone marrow cells [19]. Our previous study demonstrated that a single intravenous administration of BCG significantly decreased serum insulin levels and the insulin resistance index in a homeostatic model assessment for insulin resistance (HOMA-IR) in leptin-deficient *ob/ob* mice [20]. However, to the best of our knowledge, no information is available on the effect of BCG vaccination on diet-induced glucose intolerance. In the current study, we investigate the effects of prior BCG vaccination on the progression of glucose intolerance in high-fat and chow-diet-fed mice.

2. Materials and Methods

2.1. Animals, Diet, and Microorganisms

All animal experiments were approved by the Animal Care and Use Committee of the University of the Ryukyus (approval numbers: A2022003 and A2022007) and conducted per their guidelines. Male C57BL/6JmsSlc (CD45.2; referred to as B6-Ly5.2) mice were purchased from Japan SLC Inc. (Shizuoka, Japan). C57BL/6-Ly5.1 (CD45.1; referred to as B6-Ly5.1) mice were maintained in our animal laboratory. The mice were randomly housed in environmentally enriched cages (5 animals per cage) under a controlled environment (at 24 °C ± 1 °C in a 12 h day/night cycle with lights on from 07:00 to 19:00). After one week of adaptation, the mice were randomly divided into experimental groups (n = 10/group) for each experiment. All animals had free access to food and water during the experiment.

Products for a commercial chow diet (12 kcal% fat, CE-2 diet) and a high-fat diet (HFD, 30 kcal% fat, Quick fat diet) were purchased from CLEA Japan, Inc. (Tokyo, Japan). The BCG Tokyo 172 strain was purchased from the Japan BCG Laboratory (Tokyo, Japan) and suspended at 5×10^8 colony-forming units (CFU)/mL in phosphate-buffered saline (PBS) before use.

2.2. BCG Vaccination in HFD-Fed Mice

For vaccination, five-week-old-B6-Ly5.2 male mice were intraperitoneally (i.p.) injected with BCG (5×10^7 CFU/100 µL) for the BCG group, and with vehicle PBS for the chow and control groups (Figure 1A). All mice were fed a chow diet for three weeks, after which the chow group continued on the chow diet, and the control and BCG groups transferred to a HFD for the rest of the experiment. After the thirty-week feeding period, the mice were sacrificed after 12 h of starvation by exsanguination from the heart under isoflurane anesthesia to minimize suffering.

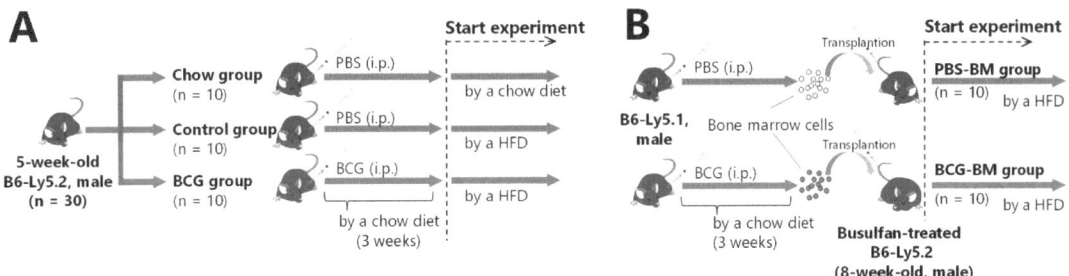

Figure 1. Schematic diagram of the experimental design. (**A**) To assess the effect of BCG vaccination on glucose intolerance progression in HFD-fed mice. (**B**) To assess the effect of bone marrow transplantation from BCG-vaccinated mice on glucose intolerance progression in HFD-fed mice. PBS-BM group were transplanted with bone marrow from PBS-treated mice; BCG-BM group were transplanted with bone marrow from BCG-vaccinated mice.

2.3. Bone Marrow Transplantation from BCG-Vaccinated Mice

Donor B6-Ly5.1 male mice were i.p. injected with vehicle PBS or BCG (5×10^7 CFU/100 µL) and then fed a chow diet for three weeks (Figure 1B). Their bone marrow samples were obtained by flushing their femurs and tibias with Eagle's minimum essential medium, followed by resuspension in PBS for transplantation. Recipient 8-week-old B6-Ly5.2 male mice were treated with i.p. busulfan injection for five days (20 mg/kg body weight/day) and transplanted intravenously with 1×10^7 nucleated cells from the donor's bone marrow 24 h after the last busulfan injection. In this experiment, mice transplanted with bone marrow cells derived from PBS-treated mice served as the PBS-BM group, and those from BCG-vaccinated mice served as the BCG-BM group. After bone marrow transplantation, all mice were fed an HFD.

2.4. Intraperitoneal Glucose Tolerance Test and Insulin Tolerance Test

At 8, 12, and 24 weeks after the start of the experiment, intraperitoneal glucose tolerance tests (IGTTs) were performed to assess whether the mice exhibited alterations in peripheral glucose regulation. The mice fasted for 12 h and were injected i.p. with D-glucose (1 g/kg body weight). Their blood glucose levels were measured before and at 30, 60, 90, and 120 min post glucose injection. For the insulin tolerance test (ITT), mice fasted for 4 h and were injected with human recombinant insulin (1 U/kg body weight; Wako Pure Chemical Industries, Ltd., Osaka, Japan) i.p., and blood glucose levels were measured before and at 30, 60, 90, and 120 min post injection. For both tests, blood was obtained from the tail vein, and glucose levels were measured using a glucometer (Free Style Precision Neo, Abbott Laboratories, Green Oaks, IL, USA).

2.5. Measurement of Biochemical Parameters in Serum

Serum triglyceride (TG), total cholesterol and glucose levels, and the activities of hepatopathy indicators, as well as the activities of alanine aminotransferase (ALT) and aspartate aminotransferase (AST), were measured using a commercial enzymatic kit (Wako Pure Chemical Industries, Ltd., Osaka, Japan). Serum insulin and glucagon levels were measured using enzyme-linked immunosorbent assay kits purchased from Morinaga Institute of Biological Science, Inc. (Kanagawa, Japan), and Wako Pure Chemical Industries, Ltd., respectively. The homeostatic indexes for the quantification of insulin resistance and beta cell function (HOMA-IR and HOMA-β) were calculated as previously described [21].

2.6. Measurement of Hepatic Lipid Levels

Hepatic lipids were extracted and purified using a previously reported method [22]. We determined hepatic TG levels using commercial enzymatic kits (Wako Pure Chemical Industries).

2.7. Histopathological Examination

The pancreas was excised and immediately fixed in 10% neutral formalin solution. Formalin-fixed samples were embedded in paraffin and cut into 4 μm thick sections. Paraffinized tissue sections were stained with hematoxylin and eosin (H&E) per a standard protocol for microscopic evaluation. The sizes of islets were calculated from digital images using Image J software (version 1.54i, NIH, Bethesda, MA, USA).

2.8. Flow Cytometry

The monoclonal antibodies (mAb) used in this study included APC/Cyanine7-conjugated anti-mouse CD45.1 mAb (clone A20), PerCP-cyanine5.5-conjugated anti-mouse CD45.1 mAb (clone 104), and non-labeled anti-mouse CD16/CD32 mAb (clone 2.4G2), bought from BioLegend Inc. (San Diego, CA, USA), Thermo Fisher Scientific Inc. (Waltham, MA, USA), and BD Biosciences (Milpitas, CA, USA), respectively. Before staining with the labeled mAb, isolated splenocyte was preincubated with anti-CD32/CD16 mAb (2.4G2, BD Biosciences) to prevent the nonspecific Fc-receptor-mediated binding of mAbs. The stained cells were analyzed on a FACSCanto II flow cytometer with the FACSDiva software program (version 5.0, BD Biosciences).

2.9. Statistical Analyses

All data are expressed as the mean ± SEM. The statistical significance of the difference between the two experimental groups was determined using the Student's t-test. To determine the significance of the differences among mean values in the three experimental groups, the differences among the mean values were inspected using the Tukey–Kramer multiple comparison test. The threshold for statistical significance was set at $p < 0.05$.

3. Results

3.1. Effect of BCG Vaccination on Glucose Intolerance in HFD-Fed Mice

To assess the effect of BCG vaccination on the progression of glucose dysmetabolism in HFD-fed mice, we performed IGTTs and ITTs to measure the ability of mice to retain circulatory glucose levels over time after administering glucose and insulin, and calculated the area under the curve (AUC) from these results. Changes in blood glucose levels and the AUC in the GTT at week 8 of the experimental period did not differ significantly among all experimental groups (Figure 2A). Although no significant differences were detected in fasting glucose levels, the blood glucose levels in the HFD-fed control group at all measurement time points after glucose administration and the AUC were significantly increased compared with those in the chow group at week 12. However, when comparing the mice in the HFD-fed groups, blood glucose levels at 90 and 120 min after glucose administration and the AUC in the BCG group were significantly lower than those in the control group. The IGTT performed at week 24 revealed that HFD feeding led to marked hyperglycemia; however, BCG vaccination significantly reduced these abnormal levels and suppressed the increase in the AUC. As shown in Figure 2B, blood glucose levels after 4 h of fasting in the chow and BCG groups were significantly lower than those in the control group, although there were no significant differences in the mean dietary intake among all experimental groups. Significant inhibitory effects of BCG vaccination on blood glucose levels after insulin administration were observed only at week 12 of the experimental period in this study.

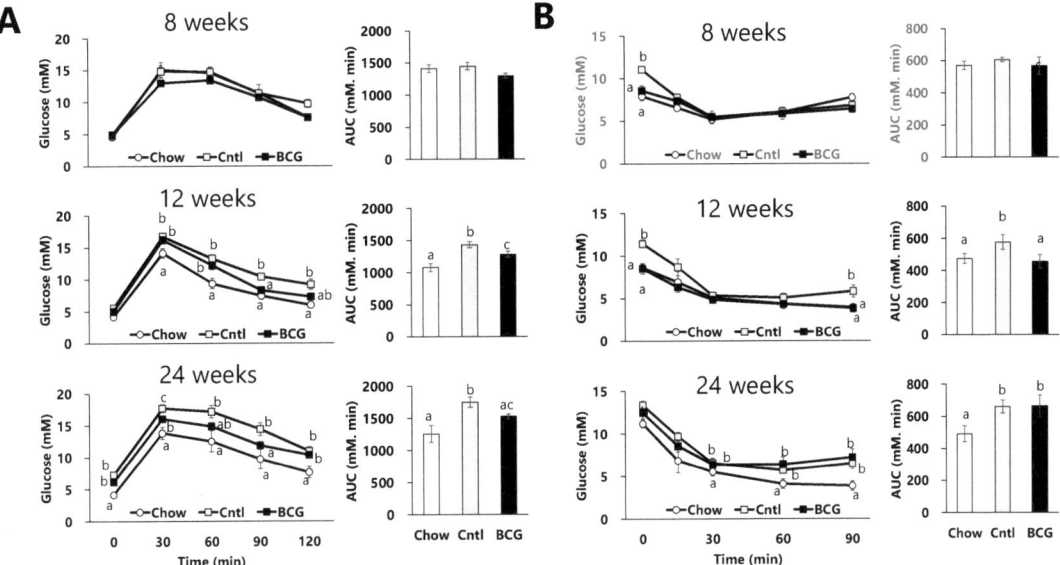

Figure 2. Effect of BCG vaccination on the progression of glucose intolerance in high-fat-diet-fed mice. (**A**) Effect of BCG vaccination on the intraperitoneal glucose tolerance test (IGTT). (**B**) Effect of BCG vaccination on the insulin tolerance test (ITT). Chow; chow group, Cntl; control group, BCG; BCG group. Data are shown as the mean ± SEM. Different letters indicate significant differences among the experimental groups using Tukey–Kramer multiple comparison test ($p < 0.05$).

3.2. Effect of BCG Vaccination on Growth Parameters, Blood Parameters, and Hepatic Lipid Content

We assessed the effects of BCG on growth and serum parameters (Table 1). HFD consumption resulted in significant increments in final body weight, liver weight, serum TG level, and hepatic TG content. BCG vaccination tended to inhibit these increments in liver weight and hepatic TG content. The serum cholesterol level and hepatopathy indicators did not differ significantly among all experimental groups.

Table 1. Growth and serum parameters in chow- and high-fat-diet-fed mice at the end of the experimental period.

Parameters	Chow	Cntl	BCG
Growth			
Final body weight (g)	27.5 ± 0.6 [a]	39.7 ± 1.8 [b]	36.6 ± 1.5 [b]
Liver weight (g)	0.95 ± 0.03 [a]	1.70 ± 0.19 [b]	1.30 ± 0.08 [ab]
Serum			
Triglyceride (mg/dL)	54.4 ± 4.9 [a]	69.7 ± 3.3 [b]	60.3 ± 6.1 [b]
Total cholesterol (mg/dL)	160 ± 12	188 ± 8.6	161 ± 9.5
AST (IU/L)	29.3 ± 5.2	26.9 ± 5.6	21.8 ± 5.9
ALT (IU/L)	4.41 ± 1.24	4.02 ± 0.50	4.47 ± 0.44
Hepatic			
Triglyceride (mg/g liver)	31.3 ± 0.6 [a]	57.4 ± 6.6 [b]	40.6 ± 5.6 [b]

Chow: chow group; Cntl: control group; BCG: BCG group; AST: aspartate aminotransferase; ALT: alanine aminotransferase. Data are shown as the mean ± SEM. Different letters indicate significant differences among the experimental groups using Tukey–Kramer multiple comparison test ($p < 0.05$).

3.3. Effect of BCG Vaccination on Glucose Metabolism Parameters

Fasting serum glucose, insulin, and glucagon levels in the control group were significantly higher than those in the chow group (Figure 3A–C). A significant decrease in the glucagon-to-insulin ratio and significant increments in HOMA-IR and HOMA-β in the control group were observed compared with those in the chow group (Figure 3D,E). BCG vaccination significantly decreased fasting serum glucose levels and insulin levels compared with the control group (Figure 3A,B). Although no significant effects of BCG vaccination on glucagon levels were observed (Figure 3C), the HFD-induced decrease in the glucagon-to-insulin ratio was significantly alleviated by BCG vaccination (Figure 3D). HOMA-IR in the BCG group was significantly decreased compared with that in the control group; meanwhile, HOMA-β did not differ significantly between the control and BCG groups (Figure 3E,F).

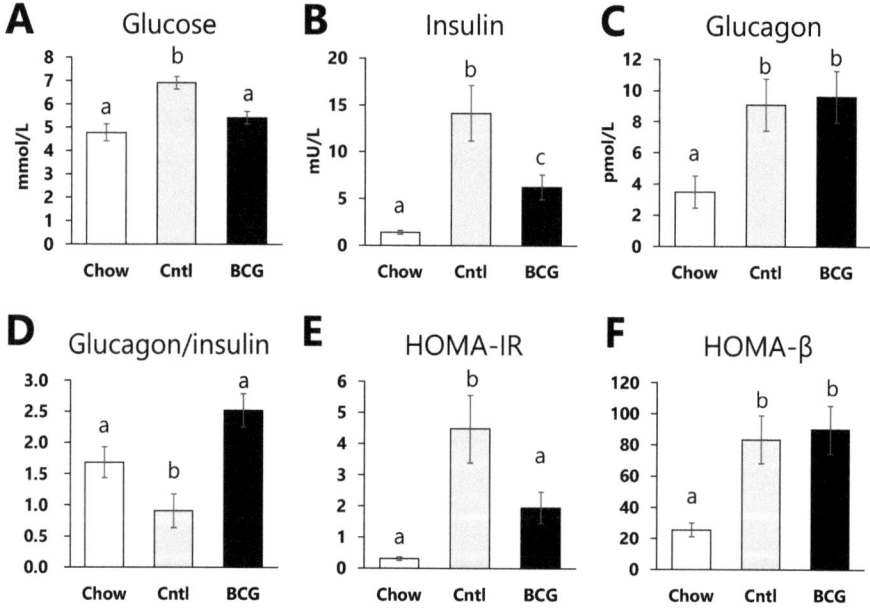

Figure 3. Effect of BCG vaccination on serum parameters related to glucose metabolism in high-fat-diet-fed mice at the end of the experiment. (**A**) Serum glucose level. (**B**) Serum insulin level. (**C**) Serum glucagon level. (**D**) Glucagon-to-insulin ratio. (**E**) Homeostatic model assessment for insulin resistance (HOMA-IR). (**F**) Homeostatic model assessment for beta cell function (HOMA-β). Chow: chow group; Cntl: control group; BCG, BCG group. Data are presented as the mean ± SEM. Different letters indicate significant differences among the experimental groups using Tukey–Kramer multiple comparison test ($p < 0.05$).

3.4. Effect of BCG Vaccination on Pancreatic Islet Size

As shown in Figure 4, the sizes of the pancreatic islets in the control group were larger than in the chow group. These significant increments in islet size were inhibited by BCG-vaccinated mice.

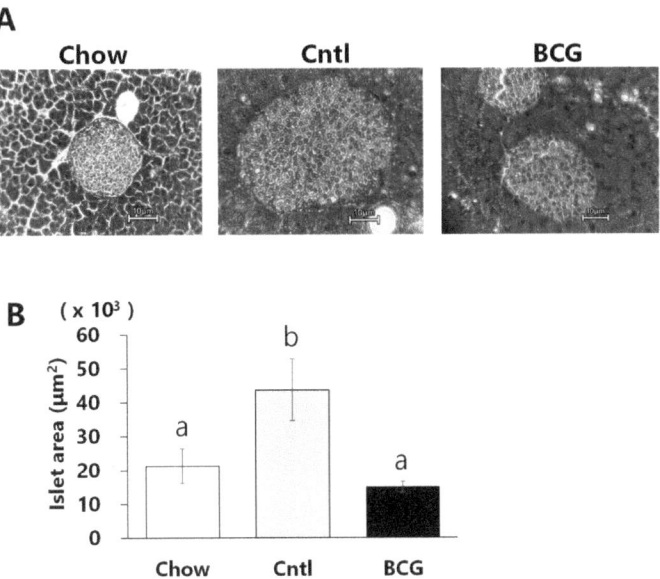

Figure 4. Effect of BCG vaccination on pancreatic islet size in high-fat-diet-fed mice at the end of the experiment. (**A**) Representative H&E histology of pancreatic samples from mice in each experimental group (scale bar = 100 μm). (**B**) The size of pancreatic islets. Chow: chow group; Cntl, control group; BCG, BCG group. Data are shown as the mean ± SEM. Different letters indicate significant differences among the experimental groups using Tukey–Kramer multiple comparison test ($p < 0.05$).

3.5. Effect of Bone Marrow Transplantation from BCG-Vaccinated Mice on Glucose Intolerance Progression

To assess whether immunomodulation induced by BCG vaccination affects the progression of glucose intolerance in HFD-fed mice, recipient B6-Ly5.2 mice were transplanted with bone marrow from donor B6-Ly5.1 mice in which BCG vaccination performed three weeks earlier had caused immune changes. At week 8, the results of the IGTTs and ITTs did not differ significantly between the PBS- and BCG-BM groups (Figure 5A,B). Blood glucose levels and AUCs were significantly decreased in the BCG-BM group compared with those in the PBS-BM group at week 12, although BCG vaccination did not affect the ITT results. Maximal blood glucose levels after 30 min of glucose administration were significantly decreased in the BCG-BM group compared with those in the PBS-BM group at week 24. Although blood glucose levels after insulin administration in the BCG-BM group tended to be lower than those in the PBS-BM group, the AUC of the BCG-BM group was significantly decreased compared to the PBS-BM group. More than 90% of the immune cells of recipient mice used in these studies were of donor origin (Figure 5C).

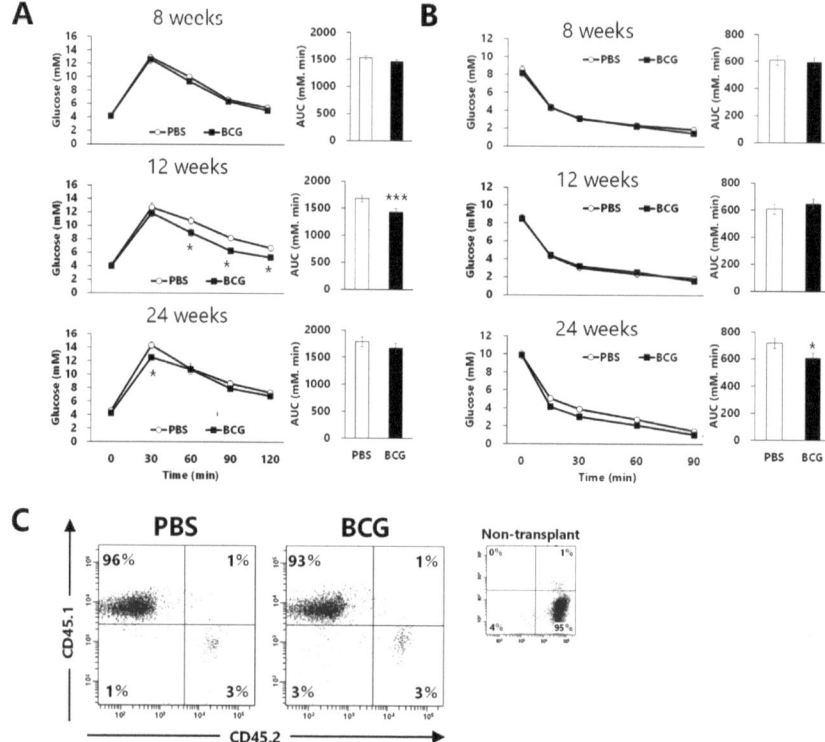

Figure 5. Effect of bone marrow transplantation from BCG-vaccinated mice on the progression of glucose intolerance in high-fat-diet-fed mice. (**A**) Effect of BCG vaccination on intraperitoneal glucose tolerance test (IGTT). (**B**) Effect of BCG vaccination on insulin tolerance test (ITT). (**C**) Representative flow cytometry plot of CD45.1 (Ly5.1) and CD45.2 (Ly5.2). PBS: PBS-BM group that were transplanted with bone marrow from PBS-treated mice; BCG: BCG-BM group that were transplanted with bone marrow from BCG-vaccinated mice. Data are shown as the mean ± SEM. The asterisk shows significant differences as compared with the PBS-BM group using the Student's t-test (* $p < 0.05$, *** $p < 0.001$).

4. Discussion

BCG is a microorganism that was developed as a vaccine for tuberculosis 100 years ago, and its off-target effects have been found to range from cancer treatment to protection against infectious and noninfectious diseases [12–18]. Non-obese diabetic (NOD) mice are well-studied spontaneous models of autoimmune diabetes, but they mimic only some features of T1DM [17,23]. Three decades of research have indicated that BCG administration permanently cures diabetes when administered to NOD mice [23,24]. Many studies have shown that BCG has therapeutic promise for T1DM in humans [17,25–27]; meanwhile, others have suggested that BCG is not useful for T1DM [28–30]. T1DM onset is typically associated with the rapid loss of pancreas function from the T-cell autoimmune attack on the insulin-secreting cells of the islets of Langerhans. The impact of BCG on human blood sugars in T1DM appears to be driven by immune and immune–metabolic effects [31]. The transfer of immune cells from BCG-vaccinated NOD mice prevented the occurrence of overt diabetes in the recipients, while the transfer from untreated donors did not [23]. It has been demonstrated that BCG can reset the immune system on the cellular level by inducing suppressive regulatory T cells and killing the autoreactive cytotoxic T cells that attack insulin-secreting cells [32,33]. It is also known that aerobic glycolysis is suppressed in individuals with T1DM, and oxidative phosphorylation, which is a metabolic pathway involving low glucose utilization, high ketone production, and high Krebs cycle utilization, is predominant instead. Recent studies have shown that BCG treatment switches the systemic metabolism from overactivated oxidative phosphorylation to accelerated aerobic glycolysis, suggesting that this leads to the lowering of blood sugar levels [17,18]. This suggests that the BCG induction of aerobic glycolysis has broader applicability to other forms of hyperglycemia, including T2DM [17].

Our previous study suggested that intravenous BCG administration may be therapeutic in relation to the inhibition of fat accumulation and reducing fasting insulin levels and the insulin resistance index, HOMA-IR, in an obese T2DM model of leptin-deficient *ob/ob* mice [20]. Shpilsky et al. recently revealed a significant reduction in blood sugar levels and body weight gain by four BCG injections in the rear footpads of *db/db* mice in a model of T2DM, although a single injection failed to have a clinical outcome [19]. Herein, we examined the beneficial effects of single and prior BCG vaccination on HFD-induced glucose intolerance in wild-type C57BL/6 mice. Although significant effects of HFD intake on changes in blood sugar and the AUC in both the IGTTs and ITTs were observed from week 12 onward, blood glucose levels after 4 h of fasting were significantly lower in both the chow and BCG groups than in the control group at week 8 (Figure 2). These data suggest that prior BCG vaccination not only lowers blood glucose levels, but also inhibits glucose dysmetabolism progression in HFD-fed mice.

At the end of the experimental period, fasting blood glucose, insulin levels, and HOMA-IR in BCG-vaccinated mice were significantly lower than in the control mice (Figure 3A,B,E). Fasting insulin levels and HOMA-IR are one approach to measuring IR [34]. T2DM is characterized by hyperglycemia, typically due to the interaction of IR and impaired beta cell function [35]. We observed that BCG vaccination does significantly not affect HOMA-β (Figure 3F). Pancreatic β-cell function is commonly estimated using HOMA-β, which in individuals with T2DM increases between years 4 and 3 before diagnosis, and then decreases until diagnosis [36]. Glucose intolerance in HFD-fed mice with reduced insulin sensitivity is suggested to impair β-cell function in pancreatic islets, resulting in excessive β-cell proliferation and increased islet size [37,38]. It has also been reported that the dietary intake of an HFD to induce IR in rodents results in an increase in pancreatic islet size [39,40]. The H&E staining showed that pancreatic islets were herein larger in the control group than in the chow group, and this increase was significantly diminished by BCG vaccination (Figure 4). These data suggest that prior BCG vaccination alleviates the development of β-cell dysfunction in HFD-fed mice and consequently improves insulin tolerance.

The glucagon-to-insulin ratio was ameliorated in the BCG group compared with that in the control group, although the serum glucagon levels were largely comparable between both experimental groups (Figure 3C,D). Recent studies have indicated that the lower the glucagon-to-insulin ratio, the greater the likelihood of suffering from nonalcoholic fatty liver disease (NAFLD), cardiovascular disease, and metabolic syndrome in patients with T2DM [41,42]. Intravenous BCG administration to *ob/ob* mice has been shown to improve the hepatic fat accumulation state and serum levels of high-molecular-weight adiponectin, which is considered a more sensitive marker of metabolic dysfunction [20,43]. In the present study, no significant differences were observed; however, inhibitory tendencies were observed in liver weight ($p = 0.07$) and hepatic TG content ($p = 0.06$) after BCG vaccination (Table 1). Therefore, these results suggest that pre-vaccination with BCG not only delays the progression of glucose dysmetabolism, but also has beneficial effects on various diseases associated with lipid metabolism abnormalities, such as metabolic syndrome, in HFD-fed mice.

Although innate immune cells are usually considered to be able to respond de novo to stimuli but not to form immunological memories, which have previously been considered only a part of adaptive immunity, it has been found that even organisms without an adaptive immune response can protect themselves against reinfection with pathogens [44].

"Trained immunity", a term first coined in 2011, refers to the immunological memory responses of innate immune cells in response to past pathogen infections [45]. This phenomenon was clarified in humans, and it was found that BCG vaccination can lead to the epigenetic reprograming of monocytes, resulting in enhanced proinflammatory responses to secondary invasion with unrelated pathogens [46]. It has also been reported that BCG vaccination induces persistent epigenetic, transcriptional, and functional changes in hematopoietic stem and progenitor cells in human bone marrow [47]. Several studies have suggested that BCG-induced trained immunity may be partly related to the heterologous beneficial off-target effects of BCG vaccination [48–50]. Therefore, we assessed whether BCG-induced trained immunity modulates the progression of glucose intolerance in HFD-fed mice. Bone marrow cells were harvested from donor B6-Ly5.1 mice vaccinated with BCG or administered with PBS three weeks previously and transplanted into busulfan-conditioned mice. The progression of HFD-induced glucose intolerance was inhibited in recipient B6-Ly5.2 mice with> 90% myeloid cells derived from BCG-vaccinated mice compared with those with myeloid cells not affected by BCG (Figure 5). These data suggest that the inhibitory effects of BCG vaccination on glucose intolerance in HFD-fed mice may be at least partially mediated by trained immunity. Faustman and colleagues reported that in vitro and in vivo BCG treatments can improve the baseline glucose transport of monocytes, which is deficient in T1DM individuals [51], and also reported that the glucose uptake of peripheral monocytes isolated from T2DM subjects and the bone marrow cells of T2DM models of *db/db* mice were augmented by in vitro BCG treatment [19]. Therefore, it is suggested that such changes in glucose metabolism in bone marrow cells and their derived myeloid cells are relevant to the results of this study. The main limitation of our study is that it did not identify the molecular mechanisms and key functional cells for the inhibitory effects of BCG vaccination on glucose metabolism abnormalities, and these remain to be addressed in future studies.

5. Conclusions

Our study aimed to investigate, for the first time, the effects of prior BCG vaccination on glucose intolerance progression induced by HFD intake in normal mice. BCG vaccination significantly delayed the progression of glucose intolerance and tended to inhibit hepatic lipid accumulation and reductions in the glucagon-to-insulin ratio, suggesting a beneficial effect of BCG on the development of T2DM and NAFLD, which are frequently associated with features of metabolic syndrome. Furthermore, it has been suggested that the effects of BCG vaccination may be at least partially due to immune memory and trained immunity for hematopoietic stem and progenitor cells of bone marrow. In this study, we used the BCG Tokyo 172 strain to determine effects on abnormalities in glucose metabolism. However, it has been suggested that the BCG dose, BCG strain, and timing of BCG administration are important for achieving efficacy in human T1DM [17]. Further studies are required to understand the effects of differences in dose administration, timing, and strains of the BCG vaccine, together with the identification of the molecular mechanisms and key functional cells of the beneficial effects of BCG vaccination.

Author Contributions: Conceptualization, M.I.; methodology, H.A. and M.I.; validation, H.A. and M.I.; formal analysis, H.A. and M.I.; investigation, H.A. and M.I.; data curation, H.A. and M.I.; writing—original draft preparation, H.A. and M.I.; writing—review and editing, H.A. and M.I.; visualization, H.A.; supervision, M.I.; project administration, M.I.; funding acquisition, M.I. All authors have read and agreed to the published version of the manuscript.

Funding: This work was supported by JSPS KAKENHI Grant Number 22K11878.

Institutional Review Board Statement: All animal experiments were approved by the Animal Care and Use Committee of the University of the Ryukyus on 17 May, 2022 (approval numbers: A2022003 and A2022007) and conducted in accordance with their guidelines.

Informed Consent Statement: Not applicable.

Data Availability Statement: The data presented in this study are available in this article.

Acknowledgments: We would like to thank Japan BCG Laboratory and Goro Matsuzaki for preparing the BCG suspension.

Conflicts of Interest: The authors declare no conflicts of interest.

References

1. Formiguera, X.; Canton, A. Obesity: Epidemiology and clinical aspects. *Best Pract. Res. Clin. Gastroenterol.* **2004**, *18*, 1125–1146. [CrossRef] [PubMed]
2. Patel, P.S.; Buras, E.D.; Balasubramanyam, A. The Role of the Immune System in Obesity and Insulin Resistance. *J. Obes.* **2013**, 616193. [CrossRef] [PubMed]
3. Pedicino, D.; Francesca, A.; Alessandro, V.; Trotta, F.; Liuzzo, G. *Type 2 Diabetes, Immunity and Cardiovascular Risk: A Complex Relationship*; InTech: Houston, TX, USA, 2012.
4. Zhou, T.; Hu, Z.; Yang, S.; Sun, L.; Yu, Z.; Wang, G. Role of Adaptive and Innate Immunity in Type 2 Diabetes Mellitus. *J. Diabetes Res.* **2018**, 7457269. [CrossRef] [PubMed]
5. Sun, H.; Saeedi, P.; Karuranga, S.; Pinkepank, M.; Ogurtsova, K.; Duncan, B.B.; Stein, C.; Basit, A.; Chan, J.C.N.; Mbanya, J.C.; et al. IDF Diabetes Atlas: Global, regional and country-level diabetes prevalence estimates for 2021 and projections for 2045. *Diabetes Res. Clin. Pract.* **2022**, *183*, 109119. [CrossRef]
6. Sinclair, A.; Saeedi, P.; Kaundal, A.; Karuranga, S.; Malanda, B.; Williams, R. Diabetes and global ageing among 65-99-year-old adults: Findings from the International Diabetes Federation Diabetes Atlas, 9(th) edition. *Diabetes Res. Clin. Pract.* **2020**, *162*, 108078. [CrossRef] [PubMed]
7. Gao, A.W.; Cantó, C.; Houtkooper, R.H. Mitochondrial response to nutrient availability and its role in metabolic disease. *EMBO Mol. Med.* **2014**, *6*, 580–589. [CrossRef] [PubMed]
8. Lachmandas, E.; Van Den Heuvel, C.N.A.M.; Damen, M.S.M.A.; Cleophas, M.C.P.; Netea, M.G.; Van Crevel, R. Diabetes Mellitus and Increased Tuberculosis Susceptibility: The Role of Short-Chain Fatty Acids. *J. Diabetes Res.* **2016**, 6014631. [CrossRef]
9. Radhakrishnan, R.K.; Thandi, R.S.; Tripathi, D.; Paidipally, P.; McAllister, M.K.; Mulik, S.; Samten, B.; Vankayalapati, R. BCG vaccination reduces the mortality of Mycobacterium tuberculosis–infected type 2 diabetes mellitus mice. *JCI Insight* **2020**, *5*, e133788. [CrossRef] [PubMed]
10. Jiang, Y.; Zhang, W.; Wei, M.; Yin, D.; Tang, Y.; Jia, W.; Wang, C.; Guo, J.; Li, A.; Gong, Y. Associations between type 1 diabetes and pulmonary tuberculosis: A bidirectional mendelian randomization study. *Diabetol. Amp; Metab. Syndr.* **2024**, *16*, 60. [CrossRef]
11. Sugawara, I.; Mizuno, S. Higher Susceptibility of Type 1 Diabetic Rats to Mycobacterium tuberculosis Infection. *Tohoku J. Exp. Med.* **2008**, *216*, 363–370. [CrossRef]
12. Jamshidi, P.; Danaei, B.; Mohammadzadeh, B.; Arbabi, M.; Nayebzade, A.; Sechi, L.A.; Nasiri, M.J. BCG Vaccination and the Risk of Type 1 Diabetes Mellitus: A Systematic Review and Meta-Analysis. *Pathogens* **2023**, *12*, 581. [CrossRef] [PubMed]
13. Arts, R.J.W.; Moorlag, S.; Novakovic, B.; Li, Y.; Wang, S.Y.; Oosting, M.; Kumar, V.; Xavier, R.J.; Wijmenga, C.; Joosten, L.A.B.; et al. BCG Vaccination Protects against Experimental Viral Infection in Humans through the Induction of Cytokines Associated with Trained Immunity. *Cell Host Microbe* **2018**, *23*, 89–100 e105. [CrossRef] [PubMed]
14. Blok, B.A.; Arts, R.J.W.; Van Crevel, R.; Benn, C.S.; Netea, M.G. Trained innate immunity as underlying mechanism for the long-term, nonspecific effects of vaccines. *J. Leukoc. Biol.* **2015**, *98*, 347–356. [CrossRef] [PubMed]
15. Ijaz, M.U.; Vaziri, F.; Wan, Y.-J.Y. Effects of Bacillus Calmette-Gu erin on immunometabolism, microbiome and liver diseases. *Liver Res.* **2023**, *7*, 116–123. [CrossRef] [PubMed]
16. Angelidou, A.; Pittet, L.F.; Faustman, D.; Curtis, N.; Levy, O. BCG vaccine's off-target effects on allergic, inflammatory, and autoimmune diseases: Worth another shot? *J. Allergy Clin. Immunol.* **2022**, *149*, 51–54. [CrossRef] [PubMed]
17. Kühtreiber, W.M.; Tran, L.; Kim, T.; Dybala, M.; Nguyen, B.; Plager, S.; Huang, D.; Janes, S.; Defusco, A.; Baum, D.; et al. Long-term reduction in hyperglycemia in advanced type 1 diabetes: The value of induced aerobic glycolysis with BCG vaccinations. *NPJ Vaccines* **2018**, *3*, 23. [CrossRef] [PubMed]
18. Kühtreiber, W.M.; Faustman, D.L. BCG Therapy for Type 1 Diabetes: Restoration of Balanced Immunity and Metabolism. *Trends Endocrinol. Metab.* **2019**, *30*, 80–92. [CrossRef] [PubMed]
19. Shpilsky, G.F.; Takahashi, H.; Aristarkhova, A.; Weil, M.; Ng, N.; Nelson, K.J.; Lee, A.; Zheng, H.; Kühtreiber, W.M.; Faustman, D.L. Bacillus Calmette-Guerin 's beneficial impact on glucose metabolism: Evidence for broad based applications. *iScience* **2021**, *24*, 103150. [CrossRef] [PubMed]
20. Inafuku, M.; Matsuzaki, G.; Oku, H. Intravenous Mycobacterium Bovis Bacillus Calmette-Guérin Ameliorates Nonalcoholic Fatty Liver Disease in Obese, Diabetic ob/ob Mice. *PLoS ONE* **2015**, *10*, e0128676. [CrossRef]
21. Akbarian, F.; Rahmani, M.; Tavalaee, M.; Abedpoor, N.; Taki, M.; Ghaedi, K.; Nasr-Esfahani, M.H. Effect of Different High-Fat and Advanced Glycation End-Products Diets in Obesity and Diabetes-Prone C57BL/6 Mice on Sperm Function. *Int. J. Fertil. Steril.* **2021**, *15*, 226–233. [CrossRef]
22. Folch, J.; Lees, M.; Sloane Stanley, G.H. A simple method for the isolation and purification of total lipides from animal tissues. *J. Biol. Chem.* **1957**, *226*, 497–509. [CrossRef]

23. Harada, M.; Kishimoto, Y.; Makino, S. Prevention of overt diabetes and insulitis in NOD mice by a single BCG vaccination. *Diabetes Res. Clin. Pract.* **1990**, *8*, 85–89. [CrossRef] [PubMed]
24. Shehadeh, N.; Etzioni, A.; Cahana, A.; Teninboum, G.; Gorodetsky, B.; Barzilai, D.; Karnieli, E. Repeated BCG vaccination is more effective than a single dose in preventing diabetes in non-obese diabetic (NOD) mice. *Isr. J. Med. Sci.* **1997**, *33*, 711–715.
25. Doupis, J.; Kolokathis, K.; Markopoulou, E.; Efthymiou, V.; Festas, G.; Papandreopoulou, V.; Kallinikou, C.; Antikidou, D.; Gemistou, G.; Angelopoulos, T. The Role of Pediatric BCG Vaccine in Type 1 Diabetes Onset. *Diabetes Ther.* **2021**, *12*, 2971–2976. [CrossRef] [PubMed]
26. Faustman, D.; Faustman, D. ScienceDirect. In *The Value of BCG and TNF in Autoimmunity*, 2nd ed.; Academic Press: London, UK; San Diego, CA, USA,, 2018.
27. Dias, H.F.; Mochizuki, Y.; Kühtreiber, W.M.; Takahashi, H.; Zheng, H.; Faustman, D.L. Bacille Calmette Guerin (BCG) and prevention of types 1 and 2 diabetes: Results of two observational studies. *PLoS ONE* **2023**, *18*, e0276423. [CrossRef] [PubMed]
28. Chang, Y.-C.; Lin, C.-J.; Hsiao, Y.-H.; Chang, Y.-H.; Liu, S.-J.; Hsu, H.-Y. Therapeutic Effects of BCG Vaccination on Type 1 Diabetes Mellitus: A Systematic Review and Meta-Analysis of Randomized Controlled Trials. *J. Diabetes Res.* **2020**, 8954125. [CrossRef]
29. Moghtaderi, M.; Zarei, P.; Shakerian, B.; Babaei, M.; Mostafavi, A.; Modaressi, M. The Non-Significant Benefit of BCG Vaccination for the Treatment of Iranian Patients with Type 1 Diabetes up to 48 Weeks: A Controversial Result. *Med. J. Islam. Repub. Iran* **2021**, *35*, 161. [CrossRef]
30. Allen, H.F.; Klingensmith, G.J.; Jensen, P.; Simoes, E.; Hayward, A.; Chase, H.P. Effect of Bacillus Calmette-Guerin vaccination on new-onset type 1 diabetes. A randomized clinical study. *Diabetes Care* **1999**, *22*, 1703–1707. [CrossRef] [PubMed]
31. Faustman, D.L. Benefits of BCG-induced metabolic switch from oxidative phosphorylation to aerobic glycolysis in autoimmune and nervous system diseases. *J. Intern. Med.* **2020**, *288*, 641–650. [CrossRef]
32. Faustman, D.L.; Wang, L.; Okubo, Y.; Burger, D.; Ban, L.; Man, G.; Zheng, H.; Schoenfeld, D.; Pompei, R.; Avruch, J.; et al. Proof-of-Concept, Randomized, Controlled Clinical Trial of Bacillus-Calmette-Guerin for Treatment of Long-Term Type 1 Diabetes. *PLoS ONE* **2012**, *7*, e41756. [CrossRef]
33. Keefe, R.C.; Takahashi, H.; Tran, L.; Nelson, K.; Ng, N.; Kühtreiber, W.M.; Faustman, D.L. BCG therapy is associated with long-term, durable induction of Treg signature genes by epigenetic modulation. *Sci. Rep.* **2021**, *11*, 14933. [CrossRef] [PubMed]
34. Singh, B.; Saxena, A. Surrogate markers of insulin resistance: A review. *World J. Diabetes* **2010**, *1*, 36–47. [CrossRef] [PubMed]
35. Esser, N.; Utzschneider, K.M.; Kahn, S.E. Early beta cell dysfunction vs insulin hypersecretion as the primary event in the pathogenesis of dysglycaemia. *Diabetologia* **2020**, *63*, 2007–2021. [CrossRef] [PubMed]
36. Tabak, A.G.; Jokela, M.; Akbaraly, T.N.; Brunner, E.J.; Kivimaki, M.; Witte, D.R. Trajectories of glycaemia, insulin sensitivity, and insulin secretion before diagnosis of type 2 diabetes: An analysis from the Whitehall II study. *Lancet* **2009**, *373*, 2215–2221. [CrossRef] [PubMed]
37. Matveyenko, A.V.; Gurlo, T.; Daval, M.; Butler, A.E.; Butler, P.C. Successful Versus Failed Adaptation to High-Fat Diet–Induced Insulin Resistance. *Diabetes* **2009**, *58*, 906–916. [CrossRef] [PubMed]
38. Butler, A.E.; Janson, J.; Bonner-Weir, S.; Ritzel, R.; Rizza, R.A.; Butler, P.C. Beta-cell deficit and increased beta-cell apoptosis in humans with type 2 diabetes. *Diabetes* **2003**, *52*, 102–110. [CrossRef] [PubMed]
39. Yang, H.-W.; Son, M.; Choi, J.; Oh, S.; Jeon, Y.-J.; Byun, K.; Ryu, B.M. Ishige okamurae reduces blood glucose levels in high-fat diet mice and improves glucose metabolism in the skeletal muscle and pancreas. *Fish. Aquat. Sci.* **2020**, *23*, 24. [CrossRef]
40. Hull, R.L.; Kodama, K.; Utzschneider, K.M.; Carr, D.B.; Prigeon, R.L.; Kahn, S.E. Dietary-fat-induced obesity in mice results in beta cell hyperplasia but not increased insulin release: Evidence for specificity of impaired beta cell adaptation. *Diabetologia* **2005**, *48*, 1350–1358. [CrossRef]
41. Moh Moh, M.A.; Jung, C.H.; Lee, B.; Choi, D.; Kim, B.Y.; Kim, C.H.; Kang, S.K.; Mok, J.O. Association of glucagon-to-insulin ratio and nonalcoholic fatty liver disease in patients with type 2 diabetes mellitus. *Diab. Vasc. Dis. Res.* **2019**, *16*, 186–195. [CrossRef]
42. Bang, J.; Lee, S.A.; Koh, G.; Yoo, S. Association of Glucagon to Insulin Ratio and Metabolic Syndrome in Patients with Type 2 Diabetes. *J. Clin. Med.* **2023**, *12*, 5806. [CrossRef]
43. Lee, E.E.; Sears, D.D.; Liu, J.; Jin, H.; Tu, X.M.; Eyler, L.T.; Jeste, D.V. A novel biomarker of cardiometabolic pathology in schizophrenia? *J. Psychiatr. Res.* **2019**, *117*, 31–37. [CrossRef] [PubMed]
44. Kurtz, J.; Franz, K. Innate defence: Evidence for memory in invertebrate immunity. *Nature* **2003**, *425*, 37–38. [CrossRef] [PubMed]
45. Netea, M.G.; Joosten, L.A.; Latz, E.; Mills, K.H.; Natoli, G.; Stunnenberg, H.G.; O'Neill, L.A.; Xavier, R.J. Trained immunity: A program of innate immune memory in health and disease. *Science* **2016**, *352*, aaf1098. [CrossRef] [PubMed]
46. Kleinnijenhuis, J.; Quintin, J.; Preijers, F.; Joosten, L.A.B.; Ifrim, D.C.; Saeed, S.; Jacobs, C.; Van Loenhout, J.; De Jong, D.; Stunnenberg, H.G.; et al. Bacille Calmette-Guérin induces NOD2-dependent nonspecific protection from reinfection via epigenetic reprogramming of monocytes. *Proc. Natl. Acad. Sci. USA* **2012**, *109*, 17537–17542. [CrossRef] [PubMed]
47. Cirovic, B.; de Bree, L.C.J.; Groh, L.; Blok, B.A.; Chan, J.; van der Velden, W.; Bremmers, M.E.J.; van Crevel, R.; Handler, K.; Picelli, S.; et al. BCG Vaccination in Humans Elicits Trained Immunity via the Hematopoietic Progenitor Compartment. *Cell Host Microbe* **2020**, *28*, 322–334.e325. [CrossRef] [PubMed]
48. Wu, Y.; Zhang, X.; Zhou, L.; Lu, J.; Zhu, F.; Li, J. Research progress in the off-target effects of Bacille Calmette-Guerin vaccine. *Chin. Med. J. (Engl.)* **2023**. [CrossRef] [PubMed]

49. van Puffelen, J.H.; Keating, S.T.; Oosterwijk, E.; van der Heijden, A.G.; Netea, M.G.; Joosten, L.A.B.; Vermeulen, S.H. Trained immunity as a molecular mechanism for BCG immunotherapy in bladder cancer. *Nat. Rev. Urol.* **2020**, *17*, 513–525. [CrossRef] [PubMed]
50. Atallah, A.; Grossman, A.; Nauman, R.W.; Paré, J.F.; Khan, A.; Siemens, D.R.; Cotechini, T.; Graham, C.H. Systemic versus localized Bacillus Calmette Guérin immunotherapy of bladder cancer promotes an anti-tumoral microenvironment: Novel role of trained immunity. *Int. J. Cancer* **2024**, *155*, 352–364. [CrossRef]
51. Kühtreiber, W.M.; Takahashi, H.; Keefe, R.C.; Song, Y.; Tran, L.; Luck, T.G.; Shpilsky, G.; Moore, L.; Sinton, S.M.; Graham, J.C.; et al. BCG Vaccinations Upregulate Myc, a Central Switch for Improved Glucose Metabolism in Diabetes. *iScience* **2020**, *23*, 101085. [CrossRef]

Disclaimer/Publisher's Note: The statements, opinions and data contained in all publications are solely those of the individual author(s) and contributor(s) and not of MDPI and/or the editor(s). MDPI and/or the editor(s) disclaim responsibility for any injury to people or property resulting from any ideas, methods, instructions or products referred to in the content.

Article

Diverse Strategies for Modulating Insulin Resistance: Causal or Consequential Inference on Metabolic Parameters in Treatment-Naïve Subjects with Type 2 Diabetes

Eiji Kutoh [1,2,3,4,*], Alexandra N. Kuto [1], Rumiko Okada [4], Midori Akiyama [2] and Rumi Kurihara [2]

[1] Biomedical Center, Tokyo 132-0034, Japan
[2] Division of Diabetes and Endocrinology, Department of Internal Medicine, Gyoda General Hospital, Saitama 361-0056, Japan
[3] Division of Diabetes and METABOLISM, Department of Internal Medicine, Higashitotsuka Memorial Hospital, Yokohama 244-0801, Japan
[4] Division of Diabetes, Department of Internal Medicine, Kumagaya Surgical Hospital, Kumagaya 360-0023, Japan
* Correspondence: eijikuto@gmail.com or ekuto-biomed@umin.ac.jp

Abstract: *Bacground and Objectives*: The objective of this study is to investigate how different therapies modulating insulin resistance, either causally or consequently, affect metabolic parameters in treatment-naïve subjects with T2DM. *Subjects and Methods*: A total of 212 subjects were assigned to receive either a tight Japanese diet ($n = 65$), pioglitazone at doses ranging from 15–30 mg/day ($n = 70$), or canagliflozin at doses ranging from 50–100 mg/day ($n = 77$) for a duration of three months. Correlations and changes (Δ) in metabolic parameters relative to insulin resistance were investigated. *Results*: Across these distinct therapeutic interventions, ΔHOMA-R exhibited significant correlations with ΔFBG and ΔHOMA-B, while demonstrating a negative correlation with baseline HOMA-R. However, other parameters such as ΔHbA1c, ΔBMI, ΔTC, ΔTG, Δnon-HDL-C, or ΔUA displayed varying patterns depending on the treatment regimens. Participants were stratified into two groups based on the median value of ΔHOMA-R: the lower half (X) and upper half (Y). group X consistently demonstrated more pronounced reductions in FBG compared to group Y across all treatments, while other parameters including HbA1c, HOMA-B, TC, TG, HDL-C, non-HDL-C, TG/HDL-C ratio, or UA exhibited distinct regulatory responses depending on the treatment administered. *Conclusions*: These findings suggest that (1) regression to the mean is observed in the changes in insulin resistance across these therapies and (2) the modulation of insulin resistance with these therapies, either causally or consequentially, results in differential effects on glycemic parameters, beta-cell function, specific lipids, body weight, or UA.

Keywords: insulin resistance; very low-calorie Japanese diet; pioglitazone; SGLT-2 inhibitor

1. Introduction

Insulin resistance and beta-cell function constitute pivotal components in the pathophysiology of Type 2 diabetes (T2DM). Insulin resistance denotes a state wherein cellular responsiveness to insulin diminishes, culminating in elevated blood glucose levels [1,2]. While beta-cell function involves a singular organ (the pancreas) and hormone (insulin), insulin resistance presents a complex scenario, entailing the participation of diverse molecules and signal transduction pathways in various organs such as adipose tissue, liver, or kidney [1–3]. Moreover, insulin resistance may inflict damage upon beta-cells, thereby impairing beta-cell function [3].

The mitigation of insulin resistance in obese individuals with T2DM primarily involves body weight control through dietary interventions and/or exercise. Additionally, certain drugs, such as thiazolidinedione (TZD) and SGLT-2 inhibitors, have demonstrated favorable impacts on insulin resistance [4,5].

Pioglitazone, classified as a TZD oral hypoglycemic agent, operates by activating peroxisome proliferator-activated receptor gamma (PPAR-γ), thereby regulating the expression of factors that contribute to insulin sensitivity in adipose tissue, liver, and muscle [4]. Notably, pioglitazone has exhibited the capacity to enhance beta-cell function and elicit favorable effects on lipid profiles [4,6]. However, its use has diminished due to associated adverse events, including weight gain and a suspected increase in the incidence of bladder cancer in men [4]. Despite these concerns, pioglitazone is currently under re-evaluation owing to its beneficial cardiovascular effects [7].

Canagliflozin, an SGLT-2 inhibitor, functions by impeding glucose reabsorption in the kidneys, thereby augmenting urinary glucose excretion [5]. As anticipated from their mechanism of action, SGLT-2 inhibitors induce weight loss [5]. Furthermore, SGLT-2 inhibitors are recognized for their favorable effects on insulin resistance, beta-cell function, and specific lipid profiles [5,8]. Intriguingly, it has been demonstrated that the weight loss induced by one SGLT-2 inhibitor, canagliflozin, is not inherently associated with insulin-sensitizing properties or glycemic efficacy [8].

Currently, the association between changes in insulin resistance using these methods and alterations in other diabetic parameters remains unclear. In this context, the implementation of a very low-calorie (tight) Japanese diet, pioglitazone, and canagliflozin emerges as an intriguing investigative strategy. All three approaches are acknowledged to reduce insulin resistance and glycemic parameters, yet they manifest distinct effects on other parameters such as beta-cell function, weight, and lipid profiles. While the hyperinsulinemic-euglycemic clamp and intravenous glucose tolerance test represent the most reliable methods for estimating insulin resistance, their feasibility within routine clinical settings is constrained [9]. Consequently, the HOMA-R index, a mathematical model strongly correlating with the hyperinsulinemic-euglycemic clamp procedure, has been employed to assess systemic insulin resistance across numerous studies [10]. In this study, we have selectively examined various diabetic parameters closely associated with T2DM, investigating their correlations and regulatory patterns relative to insulin resistance through the employment of three distinct therapeutic strategies.

2. Subjects and Methods

2.1. Subjects

The subjects were recruited from the outpatient divisions of the affiliated hospitals of the first author (EK). Primarily sourced from the annual health check screening system, inclusion criteria mandated that participants were either newly diagnosed with T2DM or previously diagnosed but untreated. The subjects had not received any regularly prescribed medications in the six months preceding the study. Exclusion criteria encompassed clinically significant renal impairment (creatinine > 1.5 mg/dL), hepatic dysfunction (glutamic oxalacetic transaminases/glutamic pyruvic transaminases [AST/ALT] > 70/70 IU/L), a history of heart disorders, severe hypertension (systolic blood pressure > 160 mm Hg and/or diastolic blood pressure > 100 mm Hg), Type 1 Diabetes Mellitus (T1DM), and pregnancy. The specifics of the very low-calorie/carbohydrate Japanese diet were previously elucidated by Japanese researchers [11–13]. Briefly, (1) calories do not exceeding 25 kcal/kg/day, (2) prioritize fish consumption over meat, and (3) prioritize vegetables or protein at the beginning of the meal, followed by carbohydrates such as rice, noodles, or bread. Male participants were administered a tight Japanese diet (n = 40), 30 mg/day pioglitazone (n = 53), or 100 mg/day canagliflozin (n = 59) as monotherapy. Female participants received a tight Japanese diet (n = 15), 15 mg/day pioglitazone (n = 17), or 50 mg/day canagliflozin (n = 18), owing to adverse events being more prevalent in women (e.g., edema with pioglitazone, urogenital infections with SGLT-2 inhibitors). Adherence to the study protocol was monitored during clinic visits. Participants who dropped out were excluded from the data analysis. The assignment was not strictly randomized; hence, this project entails the comparison of three observational studies. Informed consent was obtained from the participants, and the study protocol received approval from the Ethical Committee/Institutional

Review Board of Gyoda General Hospital and Kumagaya Surgery Hospital. This study adhered to the principles of the Helsinki Declaration and Good Clinical Practice.

2.2. Laboratory Measurements

The primary endpoint pertained to the changes in HOMA-R from baseline to 3 months. Secondary endpoints encompassed changes in FBG, HbA1c, insulin, HOMA-B, T-C, TG, HDL-C, TG/HDL-C, non-HDL-C, UA, and BMI over the same period. Fasting blood samples were collected in the morning. Monthly measurements of HbA1c and FBG were performed, while insulin, T-C, TG, HDL-C, and UA were measured at both the study's commencement (baseline) and conclusion (3 months). In some patients, antiglutamic acid decarboxylase (GAD) antibodies were assayed to exclude those with T1DM (Mitsubishi LSI or BML, Tokyo, Japan). HOMA-R and HOMA-B were calculated as previously described [10]: HOMA-R = insulin × FBG/405, HOMA-B = insulin × 360/(FBG-63).

2.3. Data Analyses

Statistical analysis was conducted using the PAST program developed by the University of Oslo (https://folk.uio.no/ohammer/past/ accessed through 3 January 2024 to 28 February 2024). Unpaired Student's t-tests were employed to assess baseline value differences, while paired Student's t-tests were utilized to analyze intra-group differences. Simple regression analysis was performed to investigate correlations between baseline or changes in HOMA-R and diabetic parameters. Analysis of covariance (ANCOVA) was employed to determine inter-group differences in diabetic parameter changes. Throughout the statistical analysis, significance was assigned to values of $p < 0.05$, and values within the range of $0.05 < p < 0.1$ were considered statistically insignificant but suggestive of potential differences or correlations, as per established methodology [14].

3. Results

3.1. Baseline Characteristics and Associations between Insulin Resistance and Diabetic Parameters in Newly Diagnosed, Treatment-Naïve Subjects with Type 2 Diabetes at Baseline (All Subjects)

The baseline characteristics of all the enrolled subjects are shown in Table 1.

Table 1. The baseline characteristics of all subjects encompassed in this study (n = 212).

	Baseline
F/M	49/163
age	52.3 ± 12.6
FBG (mg/dL)	202.6 ± 56.3
HbA1c (%)	9.75 ± 1.96
insulin (μL/mL)	7.44 ± 5.05
HOMA-R	3.64 ± 2.52
HOMA-B	23.23 ± 20.44
BMI	25.89 ± 4.93
T-C (mg/dL)	215.3 ± 42.2
TG (mg/dL)	181.7 ± 160.4
HDL-C (mg/dL)	52.6 ± 12.8
non-HDL-C (mg/dL)	148.0 ± 61.8
TG/HDL-C	3.86 ± 4.16
UA (mg/dL)	4.96 ± 1.33

Significant correlations were discerned between HOMA-R and various parameters at baseline, including FBG (R = 0.295), HOMA-B (R = 0.535), BMI (R = 0.466), insulin (R = 0.886), and UA (R = 0.279), whereas negative correlations manifested between HOMA-R and age (R = −0.145). TG (R = 0.127, p = 0.064). TG/HDL-C (R = 0.120, p = 0.081) exhibited a tendency towards positive correlations, while HDL-C (R = −0.131, p = 0.056) displayed a tendency towards a negative correlation with HOMA-R (Table 2).

Table 2. Correlations between insulin resistance (HOMA-R) and diabetic parameters at baseline (all the subjects).

Baseline HOMA-R vs. Baseline	R	p-Values
age	−0.145	<0.04
FBG (mg/dL)	0.295	<0.00001
HbA1c (%)	0.082	n.s.
insulin (μL/mL)	0.886	<0.00001
HOMA-B	0.535	<0.00001
BMI	0.466	<0.00001
T-C (mg/dL)	−0.019	n.s.
TG (mg/dL)	0.127	0.064
HDL-C (mg/dL)	−0.131	0.056
nonHDL-C (mg/dL)	0.019	n.s.
TG/HDL-C	0.12	0.0817
UA (mg/dL)	0.279	0.00001

Simple regression analysis was performed between HOMA-R and indicated diabetic parameters at baseline.

Subsequently, subjects were stratified into two groups based on the median baseline values of HOMA-R, yielding lower half (group A) and upper half (group B) designations. As depicted in Table 3, group B exhibited significantly elevated levels of HOMA-R, FBG, insulin, HOMA-B, BMI, and UA and concurrently lower levels of age and HDL-C in comparison to group A. TG, TG/HDL-C, and non-HDL-C displayed a propensity to be higher in group B relative to group A ($p = 0.099$ and $p = 0.051$, respectively). Conversely, HbA1c demonstrated no discernible differences between these two groups, if any.

Table 3. Comparison of baseline diabetic parameters depending on insulin resistance (all the subjects).

	A	B	p-Values
N	107	105	n.s.
age	54.1 ± 11.4	50.6 ± 13.5	<0.05
FBG (mg/dL)	191.8 ± 54.4	213.6 ± 56.4	<0.005
HbA1c (%)	9.72 ± 2.10	9.78 ± 1.82	n.s.
insulin (μL/mL)	4.13 ± 2.35	10.81 ± 4.83	<0.00001
HOMA-R	1.87 ± 1.04	5.46 ± 2.29	<0.00001
HOMA-B	14.68 ± 11.89	31.94 ± 23.48	<0.00001
BMI	24.03 ± 4.03	27.79 ± 5.07	<0.00001
T-C (mg/dL)	219.4 ± 45.5	211.2 ± 38.3	n.s.
TG (mg/dL)	163.7 ± 176.0	200.0 ± 141.3	0.099
HDL-C (mg/dL)	54.5 ± 14.4	50.7 ± 10.7	<0.04
nonHDL-C (mg/dL)	137.8 ± 74.7	160.5 ± 37.8	0.051
TG/HDL-C	3.32 ± 4.16	4.40 ± 4.09	0.058
UA (mg/dL)	4.69 ± 1.25	5.25 ± 1.36	<0.003

Unpaired Student's t-test was used to compare the baseline characteristics of the indicated diabetic parameters depending on the degree of baseline insulin resistance. The subjects were divided into two groups according to the median values of the baseline HOMA-R (lower half: group A and upper half: group B).

3.2. Alterations in Diabetic Parameters following Very Low-Calorie (Tight) Japanese Diet, Pioglitazone, or Canagliflozin Monotherapy in Treatment-Naïve Subjects with T2DM

At baseline, no significant differences in these diabetic parameters were observed among the three treatment groups (data not presented as a table).

After 3 months, significant reductions in FBG, HbA1c, and HOMA-R, along with increases in HOMA-B, were evident across all three treatment groups. Conversely, diverse regulatory patterns were observed in other parameters. Under the tight Japanese diet regimen, T-C, non-HDL-C, and BMI exhibited significant decreases, while UA increased (Table 4A; for each value and statistical significance, refer to the corresponding tables). Pioglitazone resulted in significant reductions in TG and TG/HDL, coupled with increases in HDL-C and BMI (Table 4B). Canagliflozin yielded a significant increase in HDL-C, accompanied by a decrease in BMI. TG exhibited a tendency to decrease (Table 4C).

Table 4. Changes in diabetic parameters with tight Japanese diet, pioglitazone, or canagliflozin. (Panel **A**) tight Japanese diet; (Panel **B**) pioglitazone; (Panel **C**) canagliflozin.

(A)				
	Baseline	3 Months	p-Values	% Changes
F/M	15/50			
age	50.8 ± 12.9			
FBG (mg/dL)	189.4 ± 49.3	167.6 ± 51.9	<0.0004	−11.5
HbA1c (%)	9.08 ± 1.32	7.96 ± 1.52	<0.00001	−12.3
insulin (μL/mL)	8.04 ± 5.25	7.20 ± 5.02	n.s.	−10.4
HOMA-R	3.80 ± 2.28	2.95 ± 2.18	<0.01	−22.3
HOMA-B	26.23 ± 19.94	30.60 ± 24.92	<0.05	16.6
BMI	26.20 ± 4.97	25.29 ± 4.77	<0.00001	−3.4
T-C (mg/dL)	208.0 ± 31.7	201.6 ± 33.7	$p < 0.05$	−3
TG (mg/dL)	157.2 ± 100.9	142.2 ± 85.6	n.s.	−9.5
HDL-C (mg/dL)	53.2 ± 11.7	53.2 ± 11.5	n.s.	0
nonHDL-C (mg/dL)	116.9 ± 71.7	110.8 ± 70.2	<0.03	−5.2
TG/HDL-C	3.18 ± 2.28	2.87 ± 2.05	n.s.	−9.7
UA (mg/dL)	4.84 ± 1.39	5.13 ± 1.47	<0.002	5.9
(B)				
	Baseline	3 Months	p-Values	% Changes
F/M	17/53			
age	53.0 ± 11.7			
HOMA-R	3.62 ± 2.27	2.64 ± 1.78	<0.00001	−27
FBG (mg/dL)	214.4 ± 53.1	170.2 ± 63.3	<0.00001	−20.6
HbA1c (%)	9.85 ± 1.60	8.37 ± 1.69	<0.00001	−15
insulin (μL/mL)	6.93 ± 4.36	6.61 ± 4.70	n.s.	−4.6
HOMA-B	19.00 ± 15.42	32.31 ± 38.79	<0.003	70
BMI	25.20 ± 5.23	25.64 ± 5.30	<0.00001	1.7
T-C (mg/dL)	210.3 ± 37.8	213.0 ± 36.2	n.s.	1.2
TG (mg/dL)	177.9 ± 122.6	145.4 ± 88.7	<0.0007	−18.2
HDL-C (mg/dL)	49.9 ± 11.3	56.5 ± 16.0	<0.00001	13.2
nonHDL-C (mg/dL)	160.4 ± 38.7	156.44 ± 38.8	n.s.	−2.4
TG/HDL-C	3.56 ± 10.7	2.57 ± 5.52	<0.0002	−27.8
UA (mg/dL)	4.67 ± 1.31	4.64 ± 1.21	n.s.	−0.6
(C)				
	Baseline	3 Months	p-Values	% Changes
F/M	18/59			
age	53.5 ± 12.5			
HOMA-R	3.53 ± 2.72	2.48 ± 2.05	<0.00001	−29.7
FBG (mg/dL)	203.1 ± 64.8	150.6 ± 47.9	<0.00001	−25.8
HbA1c (%)	10.24 ± 2.61	8.34 ± 1.97	<0.00001	−18.5
insulin (μL/mL)	7.38 ± 5.92	6.85 ± 6.28	n.s.	−7.1
HOMA-B	24.53 ± 26.76	33.46 ± 43.00	<0.00001	36.4
BMI	26.26 ± 5.64	25.82 ± 5.66	<0.00001	−1.6
T-C (mg/dL)	226.1 ± 49.7	224.2 ± 46.1	n.s.	−0.8
TG (mg/dL)	205.3 ± 213.0	193.1 ± 233.7	0.087	−5.9
HDL-C (mg/dL)	54.0 ± 14.4	56.2 ± 14.8	<0.03	4
nonHDL-C (mg/dL)	167.7 ± 58.7	163.7 ± 54.7	n.s.	−2.3
TG/HDL-C	4.44 ± 5.77	4.19 ± 7.24	n.s.	−5.6
UA (mg/dL)	5.33 ± 1.25	5.31 ± 1.26	n.s.	−0.3

Paired Student's t-test was used to compare the changes in the indicated parameters after 3 months of treatment with very low-calorie (tight) Japanese diet, pioglitazone, or canagliflozin. The results are expressed as the mean + SD.

3.3. Correlation between Changes in Insulin Resistance and Diabetic Parameters with Very Low Calorie (Tight) Japanese Diet, Pioglitazone or Canagliflozin

Simple regression analysis was conducted to examine the relationships between alterations in insulin resistance (ΔHOMA-R) and corresponding changes in other diabetic parameters under the three treatment strategies.

With tight Japanese diet, as delineated in Table 5A, significant correlations were observed between ΔHOMA-R and changes in ΔFBG (R = 0.599), ΔHbA1c (R = 0.256), Δinsulin (R = 0.932), or ΔHOMA-B (R = 0.452). Marked negative correlations were noted between ΔHOMA-R and baseline HOMA-R (R = −0.688, Figure 1A). Insignificant negative correlations were observed between ΔHOMA-R and ΔUA (R = −0.217, p = 0.082).

Table 5. Correlation of the changes in insulin resistance and those of other diabetic parameters. (**A**) Tight Japanese diet; (**B**) pioglitazone; (**C**) canagliflozin.

(A)		
ΔHOMA-R vs.	**R**	***p*-Values**
baseline HOMA-R	−0.688	<0.00001
ΔFBG	0.599	<0.00001
ΔHbA1c	0.256	<0.04
Δinsulin	0.932	<0.00001
ΔHOMA-B	0.452	<0.0002
ΔBMI	0.102	n.s.
ΔT-C	−0.078	n.s.
ΔTG	0.137	n.s.
ΔHDL-C	0.022	n.s.
ΔnonHDL-C	−0.091	n.s.
ΔTG/HDL-C	0.154	n.s.
ΔUA	−0.217	0.082
(B)		
ΔHOMA-R vs.	**R**	***p*-Values**
baseline HOMA-R	−0.654	<0.00001
ΔFBG	0.51	<0.00001
ΔHbA1c	0.266	<0.03
Δinsulin	0.771	<0.00001
ΔHOMA-B	0.298	<0.02
ΔBMI	−0.342	<0.04
ΔT-C	0.283	<0.02
ΔTG	0.299	<0.02
ΔHDL-C	0.087	n.s.
ΔnonHDL-C	0.26	<0.03
ΔTG/HDL-C	0.23	0.055
ΔUA	0.077	n.s.
(C)		
ΔHOMA-R vs.	**R**	***p*-Values**
baseline HOMA-R	−0.685	<0.00001
ΔFBG	0.322	<0.007
ΔHbA1c	0.118	n.s.
Δinsulin	0.849	<0.00001
ΔHOMA-B	0.365	<0.004
ΔBMI	−0.178	n.s.
ΔT-C	0.041	n.s.
ΔTG	0.215	0.072
ΔHDL-C	−0.041	n.s.
ΔnonHDL-C	0.06	n.s.
ΔTG/HDL-C	0.293	<0.02
ΔUA	−0.196	0.093

Simple regression analysis was performed between the changes in (Δ) HOMA-R and those of diabetic parameters.

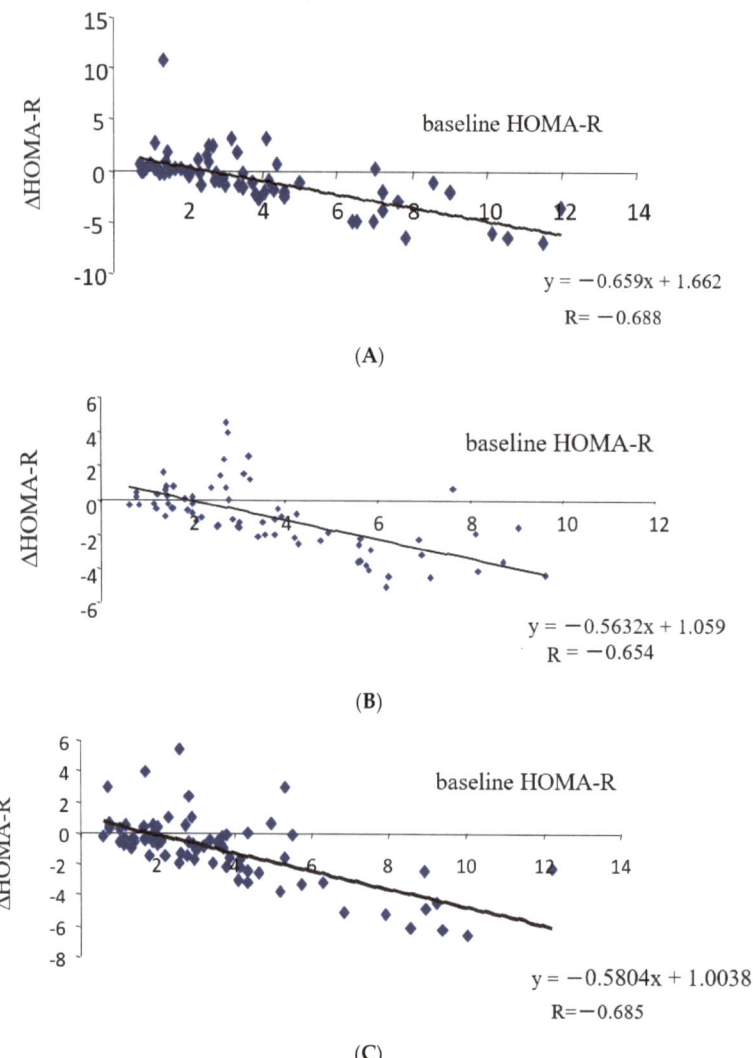

Figure 1. Baseline-dependent regulation of insulin resistance Simple regression analysis was performed between the changes in (Δ) HOMA-R and baseline HOMA-R. (**A**) Tight Japanese diet; (**B**) pioglitazone; (**C**) canagliflozin.

With pioglitazone, as illustrated in Table 5B, significant correlations were identified between ΔHOMA-R and changes (Δ) in FBG (R = 0.510), ΔHbA1c (R = 0.266), Δinsulin (R = 0.771), ΔHOMA-B (R = 0.298), ΔT-C (R = 0.283), ΔTG (R = 0.299), and Δnon-HDL-C (R = 0.260). Significant negative correlations were observed between ΔHOMA-R and baseline HOMA-R (R = −0.654, Figure 1B) and ΔBMI (R = −0.342). A tendency of correlations was observed between ΔHOMA-R and ΔTG/HDL-C (R = 0.230, p = 0.055).

With canagliflozin, as depicted in Table 5C, significant correlations were observed between ΔHOMA-R and ΔFBG (R = 0.322), Δinsulin (R = 0.849), ΔHOMA-B (R = 0.365), and ΔTG/HDL-C (R = 0.293). No correlations, if any, were observed between ΔHOMA-R and ΔHbA1c. Significant negative correlations were seen between ΔHOMA-R and baseline HOMA-R (R = −0.685, Figure 1C). Insignificant positive or negative correlations were observed between ΔHOMA-R and ΔTG (R = 0.215, p = 0.072) and ΔUA (R = −0.196, p = 0.093), respectively.

3.4. Differential Regulations of Diabetic Parameters in Two Groups with Distinct Changes in Insulin Resistance

Within each treatment group, subjects were stratified into two subgroups based on the median value of the changes (Δ) in HOMA-R: lower ΔHOMA-R (group X) and higher ΔHOMA-R (group Y), as detailed in the Subjects and Methods section. Notably, in each treatment group, baseline HOMA-R was significantly higher in group X compared to group Y (Table 6(AX,AY,BX,BY,CX,CY)).

Table 6. Effect of tight Japanese diet, pioglitazone, or canagliflozin on diabetic parameters in two groups with distinct changes in insulin resistance. (**A**) Tight Japanese diet (group X and Y); (**B**) pioglitazone (group X and Y); (**C**) canagliflozin (group X and Y).

(AX)				
	Baseline	3 Months	p-Values	% Changes
N	33			
age	51.0 ± 13.4			
HOMA-R	5.47 ± 2.89	2.83 ± 1.86	<0.00001	−48.2
FBG (mg/dL)	205.6 ± 47.8	155.9 ± 39.4	<0.00001	−24.1
HbA1c (%)	9.31 ± 1.37	7.79 ± 1.43	<0.00001	−16.3
insulin (μL/mL)	10.92 ± 5.27	7.23 ± 4.00	<0.00001	−33.7
HOMA-B	31.70 ± 21.67	20.59 ± 16.48	<0.00001	−35
BMI	27.65 ± 4.79	26.59 ± 4.88	<0.00001	−3.8
T-C (mg/dL)	208.4 ± 31.4	200.7 ± 37.5	0.094	−3.6
TG (mg/dL)	171.0 ± 81.5	153.4 ± 81.3	n.s.	−10.2
HDL-C (mg/dL)	52.0 ± 9.4	52.7 ± 11.4	n.s.	1.3
nonHDL-C (mg/dL)	156.3 ± 29.1	148.0 ± 33.8	<0.05	−5.3
TG/HDL-C	3.49 ± 1.95	3.08 ± 1.90	n.s.	−11.7
UA (mg/dL)	4.93 ± 1.45	5.40 ± 1.59	<0.0002	9.5
(AY)				
	Baseline	3 Months	p-Values	% Changes
N	32			
age	50.6 ± 12.7			
HOMA-R	2.08 ± 1.35	3.08 ± 2.50	<0.01	48
FBG (mg/dL)	172.6 ± 45.6	179.7 ± 60.4	n.s.	4.1
HbA1c (%)	8.85 ± 1.25	8.15 ± 1.60	<0.00001	−7.9
insulin (μL/mL)	5.07 ± 3.22	7.16 ± 5.96	<0.02	41.2
HOMA-B	20.59 ± 16.48	28.39 ± 26.80	<0.04	37.8
BMI	24.71 ± 4.77	23.96 ± 4.33	<0.007	−3
T-C (mg/dL)	207.6 ± 32.6	202.6 ± 29.9	n.s.	−2.4
TG (mg/dL)	143.0 ± 117.3	130.6 ± 89.6	n.s.	−8.6
HDL-C (mg/dL)	54.4 ± 13.7	53.8 ± 11.7	n.s.	−1.1
nonHDL-C (mg/dL)	153.1 ± 31.1	148.8 ± 28.8	n.s.	−2.8
TG/HDL-C	2.85 ± 2.58	2.66 ± 2.21	n.s.	−6.6
UA (mg/dL)	4.75 ± 1.35	4.86 ± 1.31	n.s.	2.3
(BX)				
	Baseline	3 Months	p-Values	% Changes
N	35			
age	52.2 ± 11.9			
HOMA-R	5.03 ± 2.15	2.52 ± 1.58	<0.00001	−49.9
FBG (mg/dL)	223.1 ± 58.1	150.7 ± 50.0	<0.00001	−32.4
HbA1c (%)	9.78 ± 1.59	7.93 ± 1.68	<0.00001	−18.9
insulin (μL/mL)	9.55 ± 4.39	6.88 ± 3.70	<0.00001	−27.9
HOMA-B	26.07 ± 18.21	35.02 ± 23.19	<0.007	34.3
BMI	26.21 ± 5.44	26.77 ± 5.58	<0.00001	2.1

Table 6. Cont.

T-C (mg/dL)	216.0 ± 42.7	211.0 ± 40.7	n.s.	−2.3
TG (mg/dL)	207.8 ± 147.4	151.4 ± 101.4	<0.0004	−27.1
HDL-C (mg/dL)	48.7 ± 10.5	54.2 ± 13.4	<0.0005	11.2
nonHDL-C (mg/dL)	167.2 ± 40.6	156.8 ± 40.9	0.052	−6.2
TG/HDL-C	4.73 ± 4.01	3.10 ± 2.53	<0.001	−34.4
UA (mg/dL)	5.12 ± 1.42	4.98 ± 1.20	n.s.	−2.7
(BY)				
	Baseline	3 Months	p-Values	% Changes
N	35			
age	53.8 ± 11.6			
HOMA-R	2.22 ± 1.33	2.76 ± 1.97	<0.02	24.3
FBG (mg/dL)	205.7 ± 46.8	189.6 ± 69.6	0.082	−7.8
HbA1c (%)	9.91 ± 1.64	8.81 ± 1.61	<0.00001	−11
insulin (μL/mL)	4.32 ± 2.30	6.35 ± 5.56	<0.01	46.9
HOMA-B	11.93 ± 6.92	29.59 ± 50.01	<0.04	148
BMI	23.89 ± 4.09	24.25 ± 4.19	<0.00001	1.5
T-C (mg/dL)	204.7 ± 31.7	215.0 ± 31.7	0.078	5
TG (mg/dL)	144.3 ± 81.1	133.5 ± 70.9	n.s.	−7.4
HDL-C (mg/dL)	51.7 ± 12.2	59.2 ± 18.2	<0.0005	14.5
nonHDL-C (mg/dL)	153.6 ± 35.9	156.0 ± 37.2	n.s.	1.5
TG/HDL-C	2.93 ± 1.83	2.60 ± 1.76	n.s.	−11.2
UA (mg/dL)	4.22 ± 1.03	4.30 ± 1.13	n.s.	1.8
(CX)				
	Baseline	3 Months	p-Values	% Changes
N	39			
age	52.46 ± 13.05			
HOMA-R	4.79 ± 2.74	2.32 ± 1.75	<0.00001	−51.5
FBG (mg/dL)	220.0 ± 61.6	143.6 ± 31.5	<0.00001	−34.7
HbA1c (%)	10.75 ± 2.74	8.59 ± 2.18	<0.00001	−20
insulin (μL/mL)	9.40 ± 6.23	6.64 ± 5.26	<0.00001	−29.3
HOMA-B	27.95 ± 28.09	33.30 ± 28.94	n.s.	19.1
BMI	26.88 ± 4.67	26.50 ± 5.01	<0.05	−1.4
T-C (mg/dL)	223.7 ± 48.4	220.0 ± 44.7	n.s.	−1.6
TG (mg/dL)	201.5 ± 133.5	175.0 ± 123.0	<0.04	−13.1
HDL-C (mg/dL)	53.0 ± 15.7	55.1 ± 15.8	n.s.	3.9
nonHDL-C (mg/dL)	170.7 ± 50.7	165.0 ± 46.0	n.s.	−3.3
TG/HDL-C	4.47 ± 4.73	3.70 ± 3.84	<0.02	−17.2
UA (mg/dL)	5.13 ± 1.14	5.33 ± 1.17	n.s.	3.8
(CY)				
	Baseline	3 Months	p-Values	% Changes
N	38			
age	53.71 ± 13.32			
HOMA-R	2.25 ± 1.30	2.65 ± 1.97	<0.05	17.7
FBG (mg/dL)	185.7 ± 59.7	157.9 ± 57.5	<0.0002	−14.9
HbA1c (%)	9.71 ± 2.14	8.09 ± 1.76	<0.0001	−16.6
insulin (μL/mL)	5.31 ± 3.52	7.07 ± 5.37	<0.01	33.1
HOMA-B	21.02 ± 19.03	33.63 ± 32.56	<0.007	59.9
BMI	25.61 ± 4.53	25.13 ± 4.27	<0.0004	−1.8
T-C (mg/dL)	228.4 ± 54.4	228.6 ± 49.6	n.s.	0
TG (mg/dL)	213.5 ± 283.5	211.6 ± 319.7	n.s.	−0.8
HDL-C (mg/dL)	55.5 ± 13.4	57.3 ± 13.8	n.s.	3.2
nonHDL-C (mg/dL)	172.8 ± 55.4	171.2 ± 51.2	n.s.	−0.9
TG/HDL-C	4.45 ± 6.72	4.69 ± 9.60	n.s.	5.3
UA (mg/dL)	5.54 ± 1.30	5.29 ± 1.37	0.079	−4.5

Paired Student's t-test was used to compare the changes in the indicated parameters in two groups with distinct changes in insulin resistance. In each treatment group, the subjects were divided into two groups based on the median value of the changes (Δ) in HOMA-R (lower half: group X and upper half: group Y). The results are expressed as the mean + SD.

Under the tight Japanese diet (X/Y = 33/32), comparable reductions in BMI were observed in both groups (Table 6(AX,AY)). In group X (Table 6(AX)), significant decreases were seen in HOMA-R, FBG, HbA1c, non-HDL-C, and insulin, while significant increases were observed in UA. T-C displayed a tendency to decrease. In group Y (Table 6(AY)), significant decreases were seen in HbA1c (not FBG), while significant increases were observed in HOMA-R, HOMA-B, and insulin. Significant inter-group differences were seen in the changes in HbA1c (greater reductions in group X versus Y, Figure 2A).

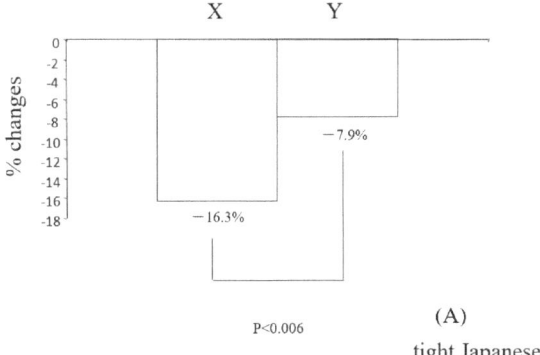

(A) tight Japanese diet

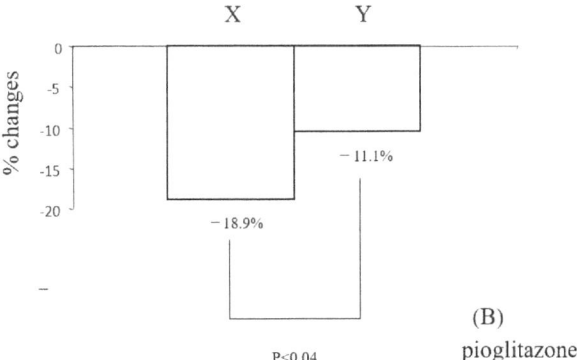

(B) pioglitazone

Figure 2. *Cont.*

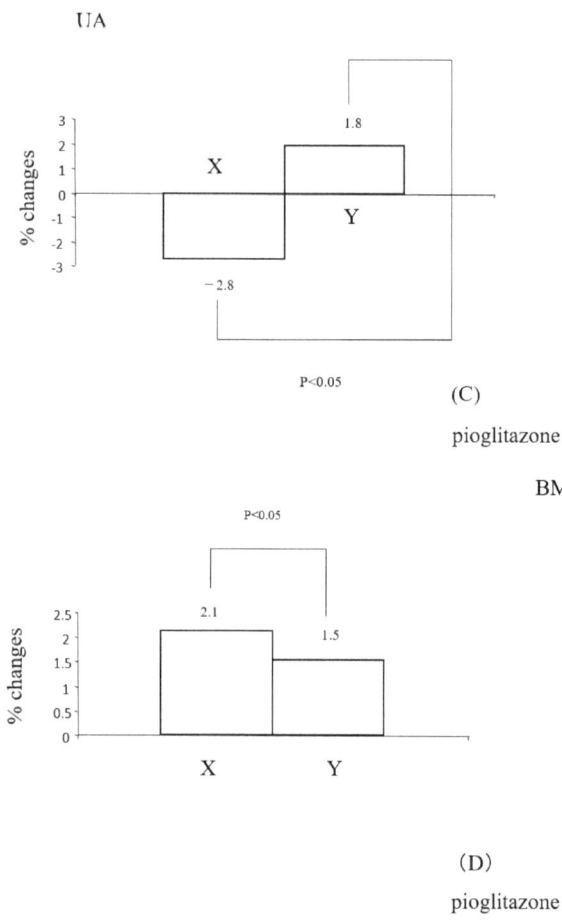

(C) pioglitazone

(D) pioglitazone

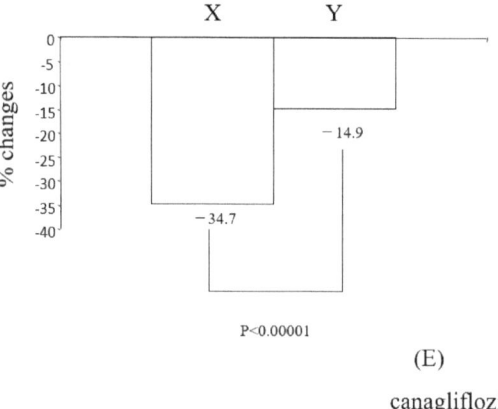

(E) canagliflozin

Figure 2. *Cont.*

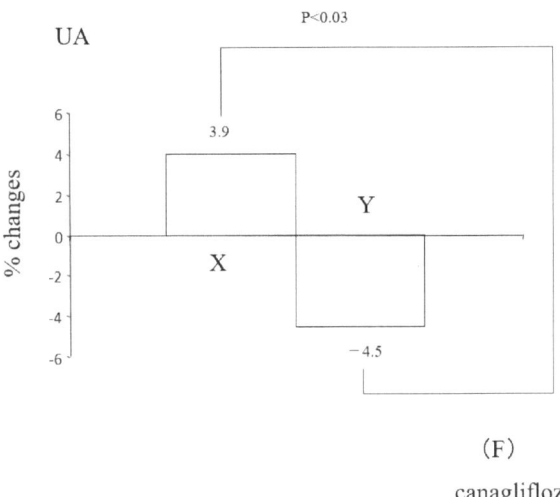

Figure 2. Differential effects on diabetic parameters by the changes in insulin resistance. ANCOVA was performed to analyze the inter-group differences in the changes in the indicated parameters in each treatment group (% changes). (**A**) HbA1c with tight Japanese diet; (**B**) HbA1c with pioglitazone; (**C**) UA with pioglitazone; (**D**) BMI with pioglitazone; (**E**) FBG with canagliflozin; (**F**) UA with canagliflozin.

With pioglitazone (X/Y = 35/35), in group X (Table 6(BX)), significant decreases were observed in HOMA-R, FBG, HbA1c, insulin, TG, and TG/HDL-C, while significant increases were seen in HOMA-B, HDL-C, and BMI. Non-HDL-C exhibited a tendency to decrease. In group Y (Table 6(BY)), significant decreases were observed in HbA1c (not FBG), while significant increases were seen in HOMA-R, HOMA-B, HDL-C, BMI, and insulin. T-C displayed a tendency to increase in this group. Significant inter-group differences were observed in the changes in HbA1c and UA (greater reductions in group X versus group Y, Figure 2B,C) and BMI (greater increases in group X versus group Y, Figure 2D).

With canagliflozin (X/Y = 39/38), in group X (Table 6(CX)), significant decreases were observed in HOMA-R, FBG, HbA1c, insulin, TG, TG/HDL-C, and BMI. In group Y (Table 6(CY)), significant decreases were observed in FBG, HbA1c, and BMI, while significant increases were seen in HOMA-R, insulin, and HOMA-B. UA exhibited a tendency to decrease. Significant inter-group differences were observed in the changes in FBG (greater reductions in group X versus Y, Figure 2E) and UA (greater reductions in group Y versus X, Figure 2F). No inter-group differences were noted in the changes in HbA1c or BMI between these two groups (Table 6(CX,CY))

4. Discussion

4.1. Characteristics of Diabetic Parameters in Newly Diagnosed, Drug-Naïve Japanese Patients with T2DM

Notably, the glycemic control of newly diagnosed, untreated Japanese patients with T2DM is considerably poor, evidenced by an elevated HbA1c close to 10% (Table 1). This poor glycemic control can be attributed, in part, to the asymptomatic nature of this disorder, often described as a "silent killer," wherein patients may not actively seek medical attention, resulting in delayed diagnosis and intervention.

HOMA-R exceeding 2.5 generally signifies the presence of insulin resistance [10,15], while HOMA-B below 30% indicates low beta-cell function [10,15]. Obesity is defined as BMI above 25 [8]. Considering these contextual factors, it is postulated that newly diagnosed, untreated Japanese patients with T2DM exhibit characteristics of high insulin resistance, relatively preserved beta-cell function, modest overweight status, and poor glycemic control. Thus, ameliorating insulin resistance is important in treating such populations. There are several strategies to reduce insulin resistance, either causally or consequentially, including a very low-calorie diet [11], pioglitazone [4,7], or SGLT-2 inhibitors [5].

Table 2 illustrates correlations between baseline levels of insulin resistance and various diabetic parameters in the overall subject cohort. These findings suggest a robust association between insulin resistance and FBG, insulin sensitivity, beta-cell function, body weight, and UA, with varying degrees of correlation with certain lipid profiles (TG, HDL-C, TG/HDL-C). The analysis in Table 3, stratifying subjects based on baseline insulin resistance levels, further supports these results. The dissociation of FBG and HbA1c concerning baseline insulin resistance may stem from the characteristic that insulin resistance may exert a lesser impact on postprandial glucose levels, while maintaining minimal glucose levels during fasting or other occasions is crucial in the early stages of diabetic history. It would be intriguing to compare data from newly diagnosed T2DM patients in other populations (e.g., Caucasians, Africans) to determine whether similar patterns emerge.

4.2. Link between Changes in Insulin Resistance and Diabetic Parameters

All three therapeutic strategies demonstrated a baseline-dependent regulation of insulin resistance, as illustrated in Figure 1A–C. After dividing subjects based on the median changes in HOMA-R values (ΔHOMA-R), group X (lower ΔHOMA-R) displayed a decrease, while group Y (higher ΔHOMA-R) exhibited an increase in HOMA-R (Table 6(AX,AY,BX,BY,CX,CY)). It is noteworthy that baseline HOMA-R was significantly higher in group X compared to group Y in each treatment group (results not presented in the table). These findings suggest that high baseline insulin resistance decreases while low baseline insulin resistance increases with each therapeutic approach. An intriguing observation is the substantial proportion of the population displaying an increase in insulin resistance. In this study, we investigated the changes in various diabetic parameters based on alterations in insulin resistance.

4.2.1. FBG

Significant correlations between changes in insulin resistance (evaluated by HOMA-R) and changes in FBG were observed in all three treatment groups. Substantiating this, when subjects were divided into two groups based on changes in HOMA-R, reductions in FBG were observed only in those with decreased HOMA-R with a tight Japanese diet and pioglitazone (group X, Table 6(AX,AY,BX,BY)). With canagliflozin, significant reductions in FBG were observed in both groups (Table 6(CX,CY)), but greater reductions were seen in those with decreased insulin resistance (group X versus group Y, Figure 2E). Collectively, these findings strongly suggest a tight connection between insulin resistance and FBG. However, the causative relationship remains undetermined. To establish causation, it is imperative to consider the temporal sequence of events.

(1) With a tight Japanese diet, initial reductions in post-meal glucose (reduced input) occur, followed by the amelioration of glucotoxicity (reduction in insulin resistance and the enhancement of beta-cell function). In the long term, reduced caloric intake leads to weight loss, subsequently decreasing insulin resistance and FBG.
(2) With pioglitazone, an initial reduction in insulin resistance occurs, followed by decreases in blood glucose levels. In the long term, enhanced insulin sensitivity may lead to weight gain and ameliorate beta-cell dysfunction.
(3) With SGLT-2 inhibitors, as anticipated from their mode of action, initial reductions in both fasting and post-meal glucose (increased output) occur. Subsequently, glucotoxicity is alleviated (reduction in insulin resistance and the enhancement of beta-cell function). In the long term, weight reduction follows, leading to decreases in insulin resistance and blood glucose levels.

While it is widely accepted that reductions in insulin resistance cause decreases in blood glucose, it is still plausible that reductions in blood glucose cause decreases in insulin resistance, as described above. Therefore, observed correlations or effects do not necessarily imply causation. Conversely, the absence of correlations or effects does not definitively rule out a causal link, as confounding factors may mask these relationships. Further well-validated basic and clinical research is required to investigate this matter.

4.2.2. HbA1c

Distinct outcomes were observed regarding HbA1c compared to FBG. While significant correlations were noted between changes in HOMA-R and HbA1c with a tight Japanese diet (R = 0.256, Table 5A) and pioglitazone (R = 0.266, Table 5B), no correlations were evident with canagliflozin (R = 0.118, Table 5C).

In a separate analysis, the SGLT-2 inhibitor canagliflozin exhibited distinct regulatory patterns compared to tight Japanese diet or pioglitazone. When subjects were stratified based on changes in insulin resistance, both tight Japanese diet and pioglitazone resulted in HbA1c reductions in both groups, with notable inter-group differences (higher reductions in HbA1c observed in those with greater reductions in insulin resistance, Table 6(AX,AY,BX,BY), Figure 2A,B). Conversely, with canagliflozin, similar, significant reductions in HbA1c were consistent regardless of changes in HOMA-R (Table 6(CX,CY)). The mechanism of action of SGLT2 inhibitors, independent of insulin secretion or action, implies that their efficacy remains unchanged irrespective of the status of insulin resistance and/or impaired beta-cell dysfunction. This could contribute to the lack of correlations between changes in HOMA-R and HbA1c with canagliflozin.

4.2.3. Beta-Cell Function

Across all three therapeutic strategies, there were notable reductions in insulin resistance and increases in beta-cell function (assessed with HOMA-B). Supporting the notion that beta-cell function is stimulated in response to insulin resistance, significant correlations were observed between changes in insulin resistance (ΔHOMA-R) and beta-cell function (ΔHOMA-B, Table 5A–C). However, beta-cell function displayed distinct regulatory patterns based on changes in insulin resistance, as elucidated below: HOMA-B was significantly up-regulated in those with elevated insulin resistance in all three strategies (group Y, Table 6(AY,BY,CY)). Conversely, it exhibited different patterns in those with reduced insulin resistance (group X). With a tight Japanese diet, HOMA-B was significantly down-regulated (Table 6(AX)). By contrast, with pioglitazone, it was up-regulated (Table 6(BX)). With canagliflozin, there was a tendency to increase (Table 6(CX)). The mechanisms and implications of this divergent regulation in this subgroup are presently under investigation.

4.2.4. Weight

Body weight management is crucial for obese patients with diabetes. It is well-established that excess weight exacerbates glucose control through deteriorated insulin resistance, and conversely, weight control positively impacts insulin sensitivity [16]. However, controversies surround this issue in pharmacotherapies. For instance, certain diabetes drugs like insulin or sulphonylurea have no impact on insulin resistance but induce weight gain [17]. DPP-4 inhibitors are considered neutral in weight or insulin sensitivity, but individuals responding efficiently to these drugs may experience weight gain [18,19]. SGLT-2 inhibitors reduce both weight and insulin resistance. However, previous findings have suggested that specific populations treated with SGLT-2 inhibitors may not experience weight loss, and correlations between changes in insulin resistance and weight are not consistently observed [8]. A TZD drug, such as pioglitazone, reduces insulin resistance but contributes to weight gain [4,6]. These complexities imply that weight loss (or gain) does not consistently correlate with decreased (or increased) insulin resistance during pharmacotherapies.

In this study, we explored the relationship between changes in insulin resistance and weight across three distinct therapeutic strategies, all of which aim to reduce insulin resistance.

(a) With a tight Japanese diet, similar weight reductions were observed irrespective of changes in insulin resistance (Table 6(AX,AY)). No correlations were identified between changes in insulin resistance and weight (Table 5A).

(b) With pioglitazone, on the contrary, more significant weight increases were noted in individuals with reduced insulin resistance (Figure 2D). Reductions in insulin resistance correlated with increased weight (Table 5B). The precise mechanism behind weight gain with pioglitazone remains unclear, but it has been hypothesized that the activation of PPARγ leads to an increase in the number and size of fat cells, resulting in increased fat storage [20]. In addition, the improvement in insulin sensitivity with this drug (referred to as group X in this paper) is accompanied by an increase in weight due to heightened lipogenesis [8]. These could contribute to an overall gain in body fat, leading to increased body weight.

(c) With canagliflozin, changes in insulin resistance were not associated with changes in weight (Table 5C). Irrespective of changes in insulin resistance with this drug, similar and significant reductions in weight were observed (Table 6(CX,CY)).

These findings challenge the conventional notion that increased weight worsens insulin resistance, while weight reduction improves it. There are several assumptions to explain these discrepancies. In human physiology, feedback mechanisms operate in many instances. In our results, body weight reduction with a tight Japanese diet and/or SGLT-2 inhibitor may activate feedback mechanisms that attempt to increase insulin resistance and conserve glucose.

4.2.5. Lipids

Diabetic dyslipidemia is typically characterized by increased TG and reduced HDL-C [21]. Non-HDL-C is frequently increased and considered a better parameter for atherogenic lipid than LDL-C [21]. In this study, changes in insulin resistance with these three strategies resulted in differential correlations or regulations among the lipid parameters in relation to insulin resistance, as indicated below.

(a) With a tight Japanese diet, favorable effects on T-C or non-HDL-C were observed, as expected from the components of the Japanese diet (Table 4A). However, no correlations or changes in lipid parameters were noted, irrespective of changes in insulin resistance (Tables 5A and 6(AX,AY)).

(b) With pioglitazone, the significant down-regulation of TG and up-regulation of HDL-C were observed (Table 4B), consistent with other reports [22]. Changes in insulin resistance correlated with changes in T-C, TG, and non-HDL-C (Table 5B) but not with HDL-C (Table 5B). Significant reductions in TG and TG/HDL-C were observed in individuals with reduced insulin resistance (group 6BX). Collectively, pioglitazone appears to have favorable effects on certain lipid parameters, and the reductions in these lipids seem to be linked to reductions in insulin resistance.

(c) Effects on lipids with SGLT-2 inhibitors are controversial [23]. In this study, with canagliflozin, the insignificant down-regulation of TG and significant up-regulation of HDL-C were observed (Table 4C). Changes in insulin resistance had no correlation with changes in HDL-C but showed a tendency to correlate with changes in TG (Table 5C). TG significantly decreased only in individuals with reduced insulin resistance (group 6CX). Thus, it appears that the modulation of insulin resistance with this SGLT-2 inhibitor is somewhat associated with TG but is not clearly associated with other lipid parameters.

4.2.6. UA

In comparison to other diabetic parameters such as weight or lipids, UA is less well studied regarding its involvement in T2DM or insulin resistance. UA can impair insulin signaling pathways and interfere with insulin's ability to regulate glucose metabolism [24,25]. Besides this, UA can promote inflammation, oxidative stress, and endothelial dysfunction, which contribute to the development of insulin resistance and beta-cell dysfunction [24,25]. However, the exact nature of this relationship is complex and not fully understood. In this study, the baseline UA had significant correlations with that of insulin resistance (Table 2). Another analysis showed that UA is more elevated in those with higher vs. lower baseline insulin resistance (Table 3). These results strongly argue that insulin resistance and UA are linked. However, it remains unclear whether insulin resistance causes the elevation of UA or the other way around. The therapeutic strategies in this present study all reduced insulin resistance; however, distinct UA regulatory patterns were seen as described below. (1) Weight reductions resulted in reduced insulin resistance and UA [26]. However, with a tight Japanese diet, unexpectedly and surprisingly, UA was significantly increased though reductions in body weight and insulin resistance (Table 4A). Further, reductions in insulin resistance appear to be negatively correlated to UA (Table 5A). Those with reduced insulin resistance had increased UA, though these subjects had reduced weight (Table 6(AX)). This may be due to the fact that the Japanese diet contains high UA [11,12]. It is of interest to evaluate this using other diets (e.g., Mediterranean). (2) With pioglitazone or canagliflozin, changes in insulin resistance may not have significant correlations or effects on UA (Table 5B,C and Table 6(BX,BY,CX,CY)). However, UA regulatory patterns depending on the changes in insulin resistance are distinct between these two drugs; relative reductions or increases in UA were observed in those with reduced insulin resistance in pioglitazone or canagliflozin, respectively (Figure 2C,F). Some diabetes drugs including DPP-4 inhibitors are known to elevate UA [13]. It is possible that reduced blood glucose levels per se somehow increase UA through reduced excretion or increased re-absorption in the kidneys. Taken together, these results indicate that UA regulation is rather complex, and in addition to insulin resistance, other mechanisms may be involved in the regulation of UA during therapies in T2DM. Basic research is required to investigate this issue.

4.3. Limitations of This Study

Certain drawbacks or limitations exist in this study. It is an observational study with a relatively small number of subjects and short study duration. Additionally, there is a gender disparity in the number of subjects and dosing of the drugs. The insulin resistance-lowering mechanisms of the therapies in this study (low-calorie diet, pioglitazone, or canagliflozin) are distinct. Therefore, it may not be appropriate to directly compare them. It remains to be investigated whether similar or different results would be observed in other populations. Thus, the results presented in this study might only be considered "hypothesis-generating". To prove the credibility of these results, randomized controlled trials in different diabetic populations are required. However, this could be expensive, time consuming, and, on some occasions, unethical. In observational studies, randomization may naturally occur. Further, based on the design of the protocol (monotherapy in drug-naïve subjects), the observed results were most probably caused exclusively by the treatments undertaken (tight Japanese diet, pioglitazone, or canagliflozin).

5. Conclusions

In conclusion, the investigation into the impacts of a very low-calorie (tight) Japanese diet, the thiazolidinedione (TZD) pioglitazone, and the sodium-glucose cotransporter-2 (SGLT-2) inhibitor canagliflozin unveiled a collective reduction in insulin resistance. The significant negative correlations observed between the changes in insulin resistance, as assessed by HOMA-R, and the baseline insulin resistance indicate a tendency for individuals with high insulin resistance to experience a decrease, while those with low insulin resistance may exhibit an increase (regression to the means). Unexpectedly, a noteworthy proportion of subjects demonstrated an increase in insulin resistance with these therapeutic strategies. The analyses conducted in this study revealed that, while significant correlations were identified between the changes in insulin resistance and FBG, insulin levels, or beta-cell function, other parameters such as HbA1c, body weight, some lipids, or uric acid (UA) displayed distinct regulatory patterns contingent upon the type of therapy employed. Stratifying subjects into two groups based on the median value of the changes in HOMA-R in each group-lower half (group X) and upper half (group Y)-revealed divergent regulations in FBG, beta-cell function, certain lipids, body weight, and UA.

Author Contributions: E.K. and A.N.K. participated in the design of the study and acquisition of the data, performed the statistical analysis, and drafted the manuscript. R.O., M.A. and R.K. made substantial contributions to the conception and design of the study and helped draft the manuscript. Writing—review and editing: E.K. and A.N.K. All authors have read and agreed to the published version of the manuscript.

Funding: This research received no external funding.

Institutional Review Board Statement: This study was reviewed and approved by the Institutional Review Board of Gyoda General Hospital (approval Code: DM Nr 3 2016, approval Date: 8 May 2016 and Kumagaya Surgery Hospital (approval Code: EK2/2018, approval Date: 8 December 2018).

Informed Consent Statement: Informed consent was obtained from all patients and stored in the electronic medical record system.

Data Availability Statement: The data that support the findings of this study are available from the corresponding author (E.K.) upon reasonable request.

Acknowledgments: The authors express gratitude to Kenji Kawashima, Kazuki Dohi, Hiromichi Suzuki, Tetsuro Sugawara, and Takashi Niida for their insightful discussions and valuable feedback. The authors also appreciate Keiko Saido-Ozawa for her administrative support.

Conflicts of Interest: The authors declare that no conflicts of interest exist regarding this manuscript.

Abbreviations

T2DM	type 2 diabetes
SGLT-2	sodium-glucose co-transporter
BMI	body mass index
FBG	fasting blood glucose
HOMA-R	homeostasis model assessment-R
HOMA-B	homeostasis model assessment-B
T-C	total cholesterol
TG	triglyceride
HDL-C	high density lipoprotein cholesterol
UA	uric acid

References

1. DeFronzo, R.A.; Ferrannini, E. Insulin resistance. A multifaceted syndrome responsible for NIDDM, obesity, hypertension, dyslipidemia, and atherosclerotic cardiovascular disease. *Diabetes Care* **1991**, *14*, 173–194. [CrossRef] [PubMed]
2. Lebovitz, H.E. Insulin resistance: Definition and consequences. *Exp. Clin. Endocrinol. Diabetes* **2001**, *109* (Suppl. S2), S135–S148. [CrossRef] [PubMed]
3. Cerf, M.E. Beta cell dysfunction and insulin resistance. *Front. Endocrinol.* **2013**, *4*, 37. [CrossRef]
4. Lebovitz, H.E. Thiazolidinediones: The Forgotten Diabetes Medications. *Curr. Diab Rep.* **2019**, *19*, 151. [CrossRef] [PubMed]
5. Scheen, A.J.; Paquot, N. Metabolic effects of SGLT-2 inhibitors beyond increased glucosuria: A review of the clinical evidence. *Diabetes Metab.* **2014**, *40*, S4–S11. [CrossRef]
6. Kutoh, E. Differential regulations of lipid profiles between Japanese responders and nonresponders treated with pioglitazone. *Postgrad. Med.* **2011**, *123*, 45–52. [CrossRef]
7. DeFronzo, R.A.; Inzucchi, S.; Abdul-Ghani, M.; Nissen, S.E. Pioglitazone: The forgotten, cost-effective cardioprotective drug for type 2 diabetes. *Diab Vasc. Dis. Res.* **2019**, *16*, 133–143. [CrossRef] [PubMed]
8. Kutoh, E.; Kuto, A.N.; Ozawa, E.; Kurihara, R.; Akiyama, M. Regulation of Adipose Tissue Insulin Resistance and Diabetic Parameters in Drug Naïve Subjects with Type 2 Diabetes Treated with Canagliflozin Monotherapy. *Drug Res.* **2023**, *73*, 279–288. [CrossRef] [PubMed]
9. Gastaldelli, A. Measuring and estimating insulin resistance in clinical and research settings. *Obesity* **2022**, *30*, 1549–1563. [CrossRef]

10. Matthews, D.R.; Hosker, J.P.; Rudenski, A.S.; Naylor, B.A.; Treacher, D.F.; Turner, R.C. Homeostasis model assessment: Insulin resistance and beta-cell function from fasting plasma glucose and insulin concentrations in man. *Diabetologia* **1985**, *28*, 412–419. [CrossRef]
11. Sakamoto, N.; Koh, N. Very low calorie diet therapy of patients with diabetes mellitus. *Nihon Rinsho* **1990**, *48*, 894–898. (In Japanese) [PubMed]
12. Sakata, T. A very-low-calorie conventional Japanese diet: Its implications for prevention of obesity. *Obes. Res.* **1995**, *3* (Suppl. S2), 233s–239s. [CrossRef] [PubMed]
13. Kutoh, E.; Ukai, Y. Alogliptin as an initial therapy in patients with newly diagnosed, drug naïve type 2 diabetes: A randomized, control trial. *Endocrine* **2012**, *41*, 435–441. [CrossRef] [PubMed]
14. Amrhein, V.; Korner-Nievergelt, F.; Roth, T. The earth is flat ($p > 0.05$): Significance thresholds and the crisis of unreplicable research. *PeerJ* **2017**, *5*, e3544. [CrossRef] [PubMed]
15. Ghasemi, A.; Tohidi, M.; Derakhshan, A.; Hasheminia, M.; Azizi, F.; Hadaegh, F. Cut-off points of homeostasis model assessment of insulin resistance, beta-cell function, and fasting serum insulin to identify future type 2 diabetes: Tehran Lipid and Glucose Study. *Acta Diabetol.* **2015**, *52*, 905–915. [CrossRef] [PubMed]
16. Reaven, G.M. Importance of identifying the overweight patient who will benefit the most by losing weight. *Ann. Intern. Med.* **2003**, *138*, 420–423. [CrossRef] [PubMed]
17. Apovian, C.M.; Okemah, J.; O'Neil, P.M. Body Weight Considerations in the Management of Type 2 Diabetes. *Adv. Ther.* **2019**, *36*, 44–58. [CrossRef] [PubMed]
18. Yokoh, H.; Kobayashi, K.; Sato, Y.; Takemoto, M.; Uchida, D.; Kanatsuka, A.; Kuribayashi, N.; Terano, T.; Hashimoto, N.; Sakurai, K.; et al. Efficacy and safety of the dipeptidyl peptidase-4 inhibitor sitagliptin compared with alpha-glucosidase inhibitor in Japanese patients with type 2 diabetes inadequately controlled on metformin or pioglitazone alone (Study for an Ultimate Combination Therapy to Control Diabetes with Sitagliptin-1): A multicenter, randomized, open-label, non-inferiority trial. *J. Diabetes Investig.* **2015**, *6*, 182–191.
19. Kutoh, E.; Kuto, A.N.; Wada, A.; Hayashi, J.; Kurihara, R. Sitagliptin as an Initial Therapy and Differential Regulations of Metabolic Parameters Depending on its Glycemic Response in Subjects with Type 2 Diabetes. *Drug Res.* **2021**, *71*, 157–165. [CrossRef]
20. Kubota, N.; Terauchi, Y.; Miki, H.; Tamemoto, H.; Yamauchi, T.; Komeda, K.; Satoh, S.; Nakano, R.; Ishii, C.; Sugiyama, T.; et al. PPAR gamma mediates high-fat diet-induced adipocyte hypertrophy and insulin resistance. *Mol. Cell.* **1999**, *4*, 597–609. [CrossRef] [PubMed]
21. Hirano, T. Pathophysiology of Diabetic Dyslipidemia. *J. Atheroscler. Thromb.* **2018**, *25*, 771–782. [CrossRef] [PubMed]
22. Rosenblatt, S.; Miskin, B.; Glazer, N.B.; Prince, M.J.; Robertson, K.E.; Pioglitazone 026 Study Group. The impact of pioglitazone on glycemic control and atherogenic dyslipidemia in patients with type 2 diabetes mellitus. *Coron. Artery Dis.* **2001**, *12*, 413–423. [CrossRef] [PubMed]
23. Premji, R.; Nylen, E.S.; Naser, N.; Gandhi, S.; Burman, K.D.; Sen, S. Lipid Profile Changes Associated with SGLT-2 Inhibitors and GLP-1 Agonists in Diabetes and Metabolic Syndrome. *Metab. Syndr. Relat. Disord.* **2022**, *20*, 321–328. [CrossRef] [PubMed]
24. Tassone, E.J.; Cimellaro, A.; Perticone, M.; Hribal, M.L.; Sciacqua, A.; Andreozzi, F.; Sesti, G.; Perticone, F. Uric Acid Impairs Insulin Signaling by Promoting Enpp1 Binding to Insulin Receptor in Human Umbilical Vein Endothelial Cells. *Front. Endocrinol.* **2018**, *9*, 98. [CrossRef] [PubMed]
25. Zhu, Y.; Hu, Y.; Huang, T.; Zhang, Y.; Li, Z.; Luo, C.; Luo, Y.; Yuan, H.; Hisatome, I.; Yamamoto, T.; et al. High uric acid directly inhibits insulin signalling and induces insulin resistance. *Biochem. Biophys. Res. Commun.* **2014**, *447*, 707–714. [CrossRef]
26. Wasada, T.; Katsumori, K.; Saeki, A.; Iwatani, M. Hyperuricemia and insulin resistance. *Nihon Rinsho* **1996**, *54*, 3293–3296.

Disclaimer/Publisher's Note: The statements, opinions and data contained in all publications are solely those of the individual author(s) and contributor(s) and not of MDPI and/or the editor(s). MDPI and/or the editor(s) disclaim responsibility for any injury to people or property resulting from any ideas, methods, instructions or products referred to in the content.

Article

Regular Physical Activity in the Prevention of Post-Transplant Diabetes Mellitus in Patients after Kidney Transplantation

Karol Graňák [1,2], Matej Vnučák [1,2,*], Monika Beliančinová [1], Patrícia Kleinová [1,2], Tímea Blichová [1,2], Margaréta Pytliaková [3] and Ivana Dedinská [1,2]

1. Transplant-Nephrology Department, University Hospital Martin, Kollárova 2, 036 01 Martin, Slovakia; granak.k@gmail.com (K.G.)
2. Department of I. Internal Medicine, University Hospital Martin, Jessenius Faculty of Medicine of Comenius University, 03601 Martin, Slovakia
3. Department of Gastroenterological Internal Medicine, University Hospital Martin, Jessenius Faculty of Medicine of Comenius University, 03601 Martin, Slovakia
* Correspondence: vnucak.matej@gmail.com; Tel.: +421-43-4203-184

Abstract: *Background and Objectives*: Post-transplant diabetes mellitus (PTDM) is a significant risk factor for the survival of graft recipients and occurs in 10–30% of patients after kidney transplant (KT). PTDM is associated with premature cardiovascular morbidity and mortality. Weight gain, obesity, and dyslipidemia are strong predictors of PTDM, and by modifying them with an active lifestyle it is possible to reduce the incidence of PTDM and affect the long-term survival of patients and grafts. The aim of our study was to determine the effect of regular physical activity on the development of PTDM and its risk factors in patients after KT. *Materials and Methods*: Participants in the study had to achieve at least 150 min of moderate-intensity physical exertion per week. The study group ($n = 22$) performed aerobic or combined (aerobic + strength) types of sports activities. Monitoring was provided by the sports tracker (Xiaomi Mi Band 4 compatible with the Mi Fit mobile application). The control group consisted of 22 stable patients after KT. Each patient underwent an oral glucose tolerance test (oGTT) at the end of the follow-up. The patients in both groups have the same immunosuppressive protocol. The total duration of the study was 6 months. *Results*: The patients in the study group had significantly more normal oGTT results at 6 months compared to the control group ($p < 0.0001$). In the control group, there were significantly more patients diagnosed with PTDM ($p = 0.0212$) and with pre-diabetic conditions (impaired plasma glucose and impaired glucose tolerance) at 6 months ($p = 0.0078$). *Conclusions*: Regular physical activity after KT provides significant prevention against the development of pre-diabetic conditions and PTDM.

Keywords: post-transplant diabetes mellitus; physical activity; kidney transplantation

Citation: Graňák, K.; Vnučák, M.; Beliančinová, M.; Kleinová, P.; Blichová, T.; Pytliaková, M.; Dedinská, I. Regular Physical Activity in the Prevention of Post-Transplant Diabetes Mellitus in Patients after Kidney Transplantation. *Medicina* **2024**, *60*, 1210. https://doi.org/10.3390/medicina60081210

Academic Editors: Yuzuru Ohshiro, Kunimasa Yagi and Yasuhiro Maeno

Received: 15 June 2024
Revised: 15 July 2024
Accepted: 22 July 2024
Published: 26 July 2024

Copyright: © 2024 by the authors. Licensee MDPI, Basel, Switzerland. This article is an open access article distributed under the terms and conditions of the Creative Commons Attribution (CC BY) license (https://creativecommons.org/licenses/by/4.0/).

1. Introduction

Diabetes mellitus (DM), which occurs after organ transplantation, is a common and serious metabolic complication [1]. Its incidence has increased over the past few decades and remains high. PTDM develops in 10–30% of the cases in the first year after transplantation [2]. The risk depends on the type of transplanted organ, the recipient's genetic predisposition, and their age. Its occurrence is most often monitored after KT, where its incidence is around 15–30%. Dedinska et al., in screening patients after KT in the Slovak Republic in 2014, identified PTDM in 38.3% of patients according to the valid American Diabetes Association (ADA) criteria [3]. The current definition of PTDM is based on the recommendations of the International Consensus of 2014, and the diagnosis follows the criteria of the ADA and the World Health Organization (WHO) for the diagnosis of type 2 DM and pre-diabetic conditions: fasting blood glucose > 126 mg/dL (7 mmol/L) in more than one case, random glucose > 200 mg/dL (11.1 mmol/L) with symptoms, and glycemia two hours after the administration of 75 g of glucose in the oral glucose tolerance test (oGTT) > 200 mg/dL (11.1 mmol/L) [4,5].

PTDM is a serious risk factor for the survival of recipients and grafts after KT. It is strongly linked to the occurrence of infections and diseases of the cardiovascular system. It is cardiovascular morbidity and mortality that significantly limit the long-term survival of patients after KT [1,2,6,7].

The major risk factors for the development of PTDM are the metabolic side effects of immunosuppressive drugs, post-transplant viral infections, and hypomagnesemia, following traditional risk factors, typical of patients with type 2 DM [8]. In a study of more than 600 patients after primary KT, the authors confirmed that metabolic syndrome (MS) before KT is an independent risk factor for PTDM [9]. In the multicenter study by Dedinska et al., the strongest predictor of PTDM was insulin resistance before KT [3]. Central obesity, another aspect of MS, is associated with high triglycerides, adipocyte-controlled cytokine release, and subclinical inflammation. These all lead to insulin resistance, which increases the risk of developing PTDM [10]. Low adiponectin levels are closely associated with insulin resistance, and significantly increase the risk of developing PTDM regardless of gender, age, and type of immunosuppression [11,12]. In contrast, in obese patients, leptin production increases significantly in the post-transplant period and is significantly associated with the development of PTDM [11]. It is possible to reduce the high incidence of PTDM by influencing only modifiable risk factors, including obesity, associated insulin resistance, and the other components of MS. Low levels of physical activity have even been identified as a major modifiable risk factor for mortality in patients with end-stage chronic kidney disease (CKD) [13]. Many studies have confirmed that regular physical activity has a significant effect on patients with type 2 DM, not only in prevention but also in treatment regimens [14,15]. Therefore, we can expect a similar effect of active lifestyle modification in patients after KT, despite several differences in the pathogenesis of type 2 DM. However, the current literature lacks sufficient data on the importance of physical activity in the prevention of PTDM and, in particular, an objective assessment of its effect.

Recently, in a sample of patients after KT, we confirmed the positive impact of physical activity on the development of insulin resistance and the parameters of glucose and lipid metabolism. Therefore, our aim was to follow up on these favorable results. In this study, we investigated the effect of regular physical activity on the development of PTDM in patients after KT.

2. Materials and Methods

We created a pilot prospective intervention-controlled study that included patients after primary deceased-donor or living-donor KT. All the enrolled patients underwent three outpatient check-ups at the Transplant–nephrology department in Martin during the follow-up: at the beginning, at the third, and at the sixth month. The observation period lasted six months.

2.1. Inclusion and Exclusion Criteria

The study included collaborating patients who reached the limit at least 3 months after KT. Other inclusion criteria included a good and stable graft function, defined as an estimated glomerular filtration rate (eGFR) of less than 60 mL/min/1.73 m^2, calculated using the chronic kidney disease—epidemiology collaboration index (CKD-EPI) formula, and stabilized comorbidities. Patients with confirmed DM, PTDM, or known pre-diabetic status (impaired glucose tolerance and elevated fasting glucose); hemoglobin levels < 100 g/L; and age over 65 years were excluded from the study. Each member of the cohort was set up for the same prophylactic immunosuppression (tacrolimus, mycophenolic acid, and prednisone).

2.2. Physical Activity and Its Monitoring

After the primary screening according to inclusion criteria (age, graft function, time after KT, hemoglobin level, and maintenance immunosuppression), we randomly chose patients for the intervention group ($n = 22$). We randomly selected participants from a relatively small pool to ensure the equal representation of women and men, considering the size of the transplanted population at our center. Patient selection was not performed by software but by a third party who was not active in the study. Based on the latest ADA recommendations from 2016 for the prevention and treatment of DM and pre-diabetic conditions, the patients in the intervention group were prescribed a limit of at least 150 min of medium-intensity physical activity per week. The second condition was not to have a break for more than 2 days between activities. The patients had to perform an aerobic type of sport (running, brisk walking, cycling, or swimming) or they could combine it with strength training. Each patient received a Xiaomi Mi Band 4 sports bracelet for the detailed monitoring of sports performance parameters, with a compatible Mi Fit mobile application collecting data on activity type, heart rate, energy expenditure, activity duration, and frequency throughout the monitoring period. Power intensity was determined by the percentage of the maximum heart rate (HRmax). Medium-intensity physical activity includes a range of 64 to 76% HRmax, as well as a high-intensity range of 77 to 93% [16]. At each clinic check-up, the investigator revised and sent the data from the Mi Fit mobile application to his or her email. The investigator was available to answer the patients' study questions by phone or email between checkups. The patients in the control group also met all of the study's inclusion and exclusion criteria. They were selected to match the patients in the intervention group according to gender and age. They were also checked out at the same intervals during the follow-up, a total of three times. The patients were instructed not to perform any sports activities during the observed period. The activities allowed were to perform basic work at home, around the house, and walk to work or shop. The goal of not reaching the level of physical activity that was prescribed for the intervention group was controlled using The International Physical Activity Questionnaire (IPAQ) at each clinic follow-up. However, neither a continuous monitoring of physical activity nor a prescribed minimum weekly level of training was in place for this patient group.

2.3. Recorded Characteristics

For each patient in both groups, we recorded at the baseline of the follow-up: age (years), body weight (kg), waist circumference (cm), body mass index—BMI (kg/m^2), the underlying cause of renal failure, family history of DM, smoking, history of acute rejection and its type, time since KT (months), type of induction immunosuppression (antithymocyte immunoglobulin or basiliximab), the presence of delayed graft function, significant cytomegalovirus (CMV) replication (cut-off 10,000), and the need for transient discontinuation of mycophenolic acid and treatment with valganciclovir for 21 days. We recorded the following lab values: serum creatinine (umol/L), eGFR according to CKD-EPI (ml/min/1.73 m^2), vitamin D (ug/L), hemoglobin (g/L), glycemia (mmol/L), glycated hemoglobin—HbA1c (%), immunoreactive insulin (uIU/mL), c-peptide (pmol/L), cholesterol (mmol/L), triacylglycerols (mmol/L), and proteinuria (g/day). Proteinuria was examined from a 24 h urine sample. We used the homeostatic model assessment of insulin resistance (HOMA-IR) index to determine IR. All the patients maintained a stable serum tacrolimus level between 3.0 and 6.0 ng/L during the follow-up period. We recorded the daily dose of prednisone for each control. At 6 months, at the end of the follow-up, each patient underwent oGTT by drinking a solution of 75 g of glucose in 200 mL of water. We measured plasma glucose levels both during fasting and 30 and 120 min after administering the glucose solution. During the test, the examinee sat in calm conditions.

2.4. Statistical Analysis

We used MedCalc version 13.1.2, a certified statistical program (MedCalc Software VAT registration number BE 0809 344,640, Member of the International Association of

Statistical Computing, Ostend, Belgium). We used parametric (*t*-test) or non-parametric (Mann–Whitney) tests to compare the continuous variables between the groups, and the t2 test and Fisher's exact test, as appropriate, to analyze associations between the categorical variables. We used logistic regression for multivariate analysis to identify the independent predictors of PTDM. We identified independent risk factors by means of the Cox proportional hazard model. A *p*-value < 0.05 was considered to be statistically significant.

We did not perform sample size calculations because this was a pilot study, and there was no historical data available.

3. Results

The study enrolled 44 patients, with 22 patients each in the intervention and control groups. The basic characteristics of the group and laboratory parameters are shown in Tables 1 and 2.

Table 1. Basic group characteristics and anthropometric data.

Group Characteristics	Monitored Group $n = 22$	Control Group $n = 22$	*p*-Value
	Basic group characteristics		
Gender—men (%)	50	54.5	0.7677
Age (years)	42.6 ± 8.8	42.8 ± 13.2	0.9531
Time after KT (M)	60.6 ± 50	15.8 ± 9	**0.0002**
Basiliximab in induction (%)	36.4	31.8	0.7504
Delayed graft function (%)	4.5	4.5	1.0000
DM positive family history (%)	41	45.5	0.7658
Smokers (%)	9	13.6	0.6338
Anamnesis of CMV (%)	13.6	9.1	0.6418
Anamnesis of acute rejection	18	4.5	0.1613
Average prednisone dose (mg/day)	5.9 ± 2.4	5.5 ± 1	0.4745
	Anthropometric data		
Body weight (kg) baseline	75 ± 14.3	77 ± 16.4	0.6686
Body weight (kg) 3 M	74.9 ± 13.8	78.3 ± 16.4	0.4610
Body weight (kg) 6 M	75.1 ± 13.4	79.7 ± 17	0.3246
BMI (kg/m^2) baseline	25.5 ± 3.2	25.5 ± 3.8	1.0000
BMI (kg/m^2) 3 M	25.4 ± 3	26 ± 3.7	0.5578
BMI (kg/m^2) 6 M	25.5 ± 2.9	26.4 ± 4	0.3977
Waist circumference (cm) baseline	90.6 ± 12.4	94.1 ± 12.2	0.3507
Waist circumference (cm) 3 M	89.3 ± 11.5	96.7 ± 12.1	**0.0437**
Waist circumference (cm) 6 M	89.1 ± 11.1	96.7 ± 12.3	**0.0372**
Body height (cm)	171 ± 8.2	173 ± 11.7	0.5150

KT—kidney transplant; DM—diabetes mellitus; CMV—cytomegalovirus; BMI—body mass index; M—month.

The age ($p = 0.9531$) and gender ($p = 0.7677$) structures of both groups were not significantly different. The patients forming the control group had an overall shorter time after KT compared to the intervention group ($p = 0.0002$). The mean interval from KT was 15.8 +- 9 months, thus meeting the inclusion criterion (3 months from KT). We did not observe a difference in the daily dose of prednisone or in the use of basiliximab in the induction protocol between the two groups. Vitamin D levels were significantly lower in the control group at the beginning of the follow-up, but this disparity leveled out during the follow-up, and we did not notice it at 6 months. The hemoglobin level was also significantly lower at the beginning, but on average it was in the zone of very mild anemia and thus met the inclusion criterion. In addition, it was saturated during follow-up, and these differences leveled off. Differences in glucose metabolism were also observed. In the intervention group, the fasting plasma glucose levels were significantly lower at baseline ($p = 0.0045$), 3 months ($p = 0.0016$), and 6 months ($p = 0.0003$). However, the higher baseline glycemia in the control group was within the normo-glycemic range and did not represent a pathological condition. We primarily recorded the magnesium levels, but since all the study participants were taking magnesium replacement, their levels were within the physiological range, and there was no difference between the groups. Figure 1 shows the file distribution based on the underlying cause of renal failure. Chronic glomerulonephritis and chronic tubulointerstitial nephritis accounted for the majority.

Table 2. Basic group characteristics—laboratory findings.

	Laboratory Parameters—Graft Function		
Creatinine (μmol/L) baseline	95.1 ± 17.4	101.4 ± 23.3	0.3154
Creatinine (μmol/L) 3 M	98.2 ± 19.2	114.5 ± 30.1	**0.0381**
Creatinine (μmol/L) 6 M	94.5 ± 22.1	110.1 ± 27.1	**0.0425**
eGFR CKD-EPI (ml/min) baseline	76.4 ± 15.5	72.7 ± 18.9	0.4816
eGFR CKD-EPI (ml/min) 3 M	74.1 ± 16	63.4 ± 16.3	**0.0036**
eGFR CKD-EPI (ml/min) 6 M	78.2 ± 17.5	65.7 ± 14.6	**0.0137**
Quantitative proteinuria (g/L) baseline	0.236 ± 0.15	0.220 ± 0.18	0.7503
Quantitative proteinuria (g/L) 3 M	0.220 ± 0.19	0.267 ± 0.29	0.5383
Quantitative proteinuria (g/L) 6 M	0.187 ± 0.13	0.347 ± 0.42	0.0952
Vitamin D (μg/L) baseline	31.9 ± 10.2	23.2 ± 7.2	**0.0022**
Vitamin D (μg/L) 3 M	29.5 ± 8.9	24 ± 8	**0.0369**
Vitamin D (μg/L) 6 M	27.9 ± 8.8	26.5 ± 10	0.6246
Hemoglobin (g/L) baseline	143 ± 12.1	133 ± 15	**0.0193**
Hemoglobin (g/L) 3 M	145 ± 10.3	137 ± 16.5	0.0605
Hemoglobin (g/L) 6 M	146 ± 11.5	142 ± 16.7	0.3601
	Laboratory parameters—glucose metabolism		
Fasting glucose (mmol/L) baseline	4.7 ± 0.6	5.2 ± 0.3	**0.0045**
Fasting glucose (mmol/L) 3 M	4.8 ± 0.6	5.7 ± 1.1	**0.0016**
Fasting glucose (mmol/L) 6 M	4.8 ± 0.6	5.7 ± 0.9	**0.0003**
C—peptide (μg/L) baseline	2.5 ± 1	3.1 ± 1.2	0.0788
C—peptide (μg/L) 3 M	2.2 ± 0.8	2.8 ± 1.1	**0.0447**
C—peptide (μg/L) 6 M	2.4 ± 0.9	4 ± 2	**0.0014**
Immunoreactive insulin—IRI (mU/l) baseline	8.4 ± 6.2	7.8 ± 3.2	0.6887
Immunoreactive insulin—IRI (mU/l) 3 M	8.3 ± 6.8	7.9 ± 3.4	0.8063
Immunoreactive insulin—IRI (mU/l) 6 M	8.7 ± 4.7	9.7 ± 3.7	0.4374
HOMA-IR baseline	1.7 ± 1.3	1.8 ± 0.8	0.7601
HOMA-IR 3 M	1.8 ± 1.5	2.4 ± 1.2	0.1504
HOMA-IR 6 M	1.9 ± 1	2.5 ± 1.1	0.0653
Glycated hemoglobin—HbA1c (%) baseline	3.6 ± 0.4	3.8 ± 0.7	0.2512
Glycated hemoglobin—HbA1c (%) 3 M	3.6 ± 0.5	3.6 ± 0.5	1.0000
Glycated hemoglobin—HbA1c (%) 6 M	3.6 ± 0.5	3.9 ± 0.7	0.1094
	Laboratory parameters—lipid profile		
Cholesterol (mmol/L) baseline	5 ± 0.8	5.3 ± 1	0.2781
Cholesterol (mmol/L) 3 M	4.9 ± 0.8	5.1 ± 1	0.4679
Cholesterol (mmol/L) 6 M	4.8 ± 0.8	5.2 ± 1	0.1504
LDL—low-density lipoprotein (mmol/L) baseline	2.8 ± 0.7	3.2 ± 0.8	0.0848
LDL—low-density lipoprotein (mmol/L) 3 M	2.8 ± 0.7	3 ± 0.9	0.4153
LDL—low-density lipoprotein (mmol/L) 6 M	2.7 ± 0.8	3.2 ± 0.8	**0.0444**
HDL—high-density lipoprotein (mmol/L) baseline	1.5 ± 0.4	1.3 ± 0.4	0.1047
HDL—high-density lipoprotein (mmol/L) 3 M	1.5 ± 0.5	1.4 ± 0.5	0.5107
HDL—high-density lipoprotein (mmol/L) 6 M	1.5 ± 0.5	1.4 ± 0.4	0.4679
TAG—triglycerides (mmol/L) baseline	2 ± 1.4	2.2 ± 1	0.5885
TAG—triglycerides (mmol/L) 3 M	1.5 ± 0.9	1.8 ± 0.7	0.2240
TAG—triglycerides (mmol/L) 6 M	1.7 ± 0.8	2 ± 0.9	0.2492

M—month; eGFR—estimated glomerular filtration rate; CKD-EPI—chronic kidney disease—epidemiology collaboration; HOMA-IR—homeostatic model assessment for insulin resistance, LDL—low-density lipoprotein; HDL—high-density lipoprotein; TAG—triglycerides.

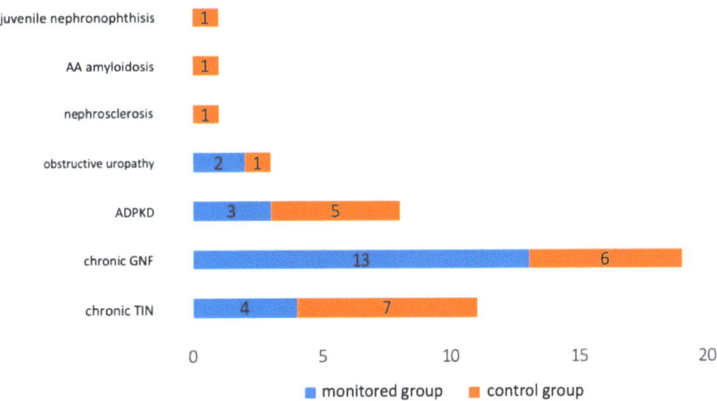

Figure 1. The distribution of the study file according to the cause of renal failure. ADPKD—autosomal dominant polycystic kidney disease; GNF—glomerulonephritis; TIN—tubulointerstitial nephritis.

In an intervention group of 22 patients, 15 practiced isolated aerobic training and 7 combined types of training. Cycling, running, brisk walking, swimming, hiking, and weight training have been practiced.

In addition to comparing the observed parameters between the two groups at the specified time points, we also evaluated the development of these parameters in the individual groups during the study period. We found no significant difference in the development of body weight, BMI, waist circumference, creatinine level, eGFR, proteinuria, vitamin D, hemoglobin, c-peptide, immunoreactive insulin, HbA1c, cholesterol, LDL, HDL, and triacylglycerols for a 6-month follow-up in both groups. On the other hand, we found a significant difference in the fasting glucose (p = 0.0227) and HOMA-IR index (p = 0.0202) in the control group.

By analyzing the oGTT at the end of the follow-up, we observed differences in the incidence of glucose metabolism disorders (Figure 2).

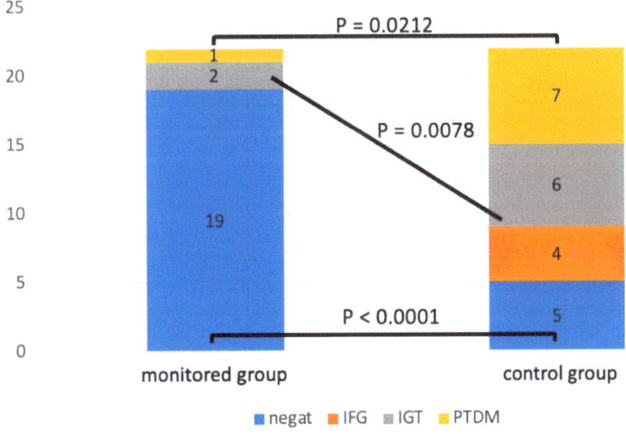

Figure 2. Oral glucose tolerance test results. IFG—impaired fasting glucose, IGT—impaired glucose tolerance, and PTDM—post-transplant diabetes mellitus.

The patients in the control group had significantly fewer physiological oGTT results compared to the intervention group ($p < 0.0001$). In the control group, we also identified a significantly higher incidence of pre-diabetic conditions: impaired glucose tolerance, fasting hyperglycemia ($p = 0.0078$), and diagnosed PTDM ($p = 0.0212$). Figure 3 shows a more detailed analysis of oGTT in both groups. In the control group, the blood glucose level was significantly higher at 30 min of the test ($p = 0.0034$) as well as after 120 min ($p = 0.0011$) compared to the intervention group.

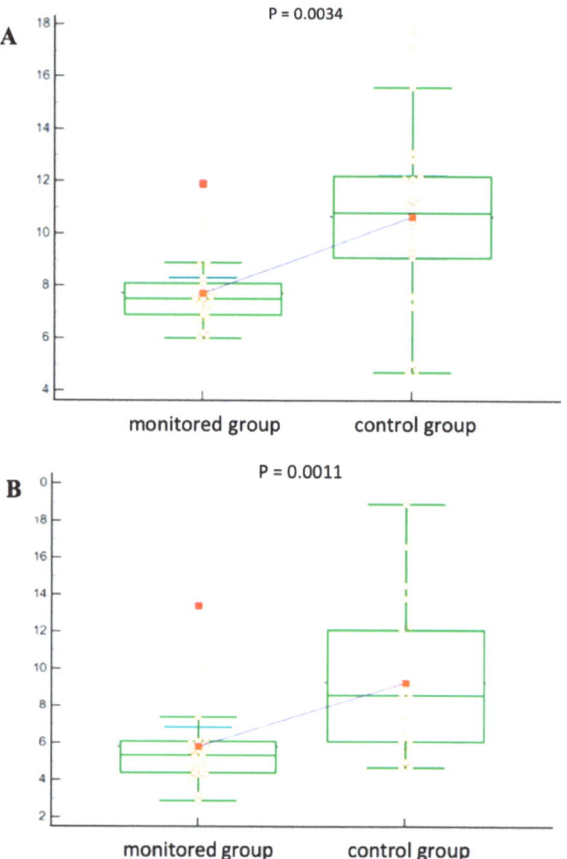

Figure 3. Glycemia during oGTT in both groups, (**A**): 30 min and (**B**): 120 min.

4. Discussion

In our study, we discovered that regular physical activity of at least 150 min of moderate intensity per week helped the patients in the intervention group prevent PTDM development. After 6 months of follow-up, the patients in the control group who did not practice such a level of regular physical activity developed significantly more often PTDM, or a pre-diabetic condition (impaired glucose tolerance and fasting hyperglycemia).

Even though there is a lot of clear evidence that exercise can help prevent type 2 diabetes and keep blood sugar levels under control in people who already have it, there is not a lot of information on how important regular physical activity is for patients after KT [16]. In the first intervention study in 2008, Sharif et al. pointed out that active lifestyle modification, including physical training, resulted in improved 2 h postprandial blood glucose levels in patients with already confirmed glucose intolerance [17]. In this case, however, it is more of a therapeutic aspect of using physical activity than a preventive one. In 2015, the authors

Karelis et al. conducted a randomized pilot study in patients 6–8 weeks after KT who underwent strength training for 16 weeks. A total of 20 patients, divided 1:1 between the intervention and passive groups, had their physical and cardiometabolic attributes assessed, including the risk of developing PTDM, and completed follow-up. After 16 weeks of follow-up, the group of subjects showed no significant differences in metabolic or anthropometric parameters, and oGTT found no lower incidence of PTDM. This work's limitation lies in its use of isolated strength training, despite the recommendation for aerobic or combined activity to prevent metabolic disorders. These patients underwent three trainings per week, with only one under the supervision of a gym trainer and the other two undergoing treatment at home using an elastic band, thus lacking an objective of sufficient intensity and duration of training. Additionally, the study included patients who were very shortly after KT, did not have stable serum levels, and were on daily doses of immunosuppressive drugs with frequent adjustments, potentially impacting glucose metabolism parameters [18]. Under the proactive supervision of a team of experts, the randomized prospective study CAVIAR (Comparing Glycemic Benefits of Active Versus Passive Lifestyle Intervention in Kidney Allograft Recipients) demonstrated that post-KT recipients can modify their lifestyle to improve their cardiometabolic risk profile. The study randomly assigned 130 patients to the active and passive intervention groups. A renal dietitian led the active intervention using a behavioral change technique, specifically focusing on adjusting eating habits. Additionally, the dietitian promoted an exercise program to enhance physical activity and encouraged participants to maintain an exercise diary to track their progress. We identified no effect of active intervention on the primary endpoints—insulin secretion, sensitivity, and disposition index—during the 6-month follow-up period. On the other hand, this group showed improvement in secondary endpoints, such as a significant reduction in body weight and fat mass. In this study, the development of PTDM showed a clinically declining trend in the intervened recipients after KT but did not reach statistical significance [19]. In contrast to our study, we did not actively intervene in the dietary regime, instead placing more emphasis on it. Simultaneously, these patients lacked the minimum required level (duration and intensity) of physical training and did not undergo objective monitoring using physical performance-sensing devices. In a prospective monocentric study of a sample of 650 recipients, Byambasukh et al. assessed the effect of physical activity on the development of PTDM, cardiovascular mortality, and overall mortality at least one year after KT. The total follow-up was 5.3 years. Mild (4.0–6.5 MET) to intense ($\geq$6.5 MET) sports activities were applied to the subjects. Mild to intense extracurricular physical activity has been associated with a reduced risk of developing PTDM and cardiovascular and overall mortality, regardless of age, gender, baseline renal parameters, transplant characteristics, and other lifestyle habits. However, after adjusting for immunosuppressive therapy, metabolic and anthropometric parameters, and baseline blood glucose levels, this physical load was no longer associated with PTDM. The authors used a validated SQUASH questionnaire to determine regular physical performance during follow-up. Only leisure sports and activities performed while commuting to work in minutes per week were evaluated. We did not include physical activity that occurred during work. The results also found a relatively high group of inactive patients, up to 38%. The limitation in this case, as in other previous studies, is the patients subjectively assessed physical activity. On the other hand, the strong point is a sufficiently long follow-up [20].

In our study, we followed the 2012 Kidney Disease: Improving Global Outcome (KDIGO) guidelines for CKD investigation and management, which recommend that these patients perform at least 30 min of moderate aerobic physical activity five times per week, adjusting to their cardiovascular health and tolerance [21]. Our findings on how to stop PTDM are in line with what the American Diabetes Association (ADA) says should be done for all high-risk patients and people with prediabetes to stop or delay the development of DM [22]. These PTDM prevention results complement and support the first phase of our study's evaluation, where we demonstrated a significant effect of regular physical activity on reducing the development of pre-diabetic conditions, insulin resistance, and fasting hyperglycemia. A multivariate analysis using the Cox proportional hazard model demonstrated a significantly higher incidence of insulin resistance, as measured by the HOMA-IR index, in the control group after 6 months of follow-up ($p = 0.0202$), and fasting hyperglycemia after 3 months of follow-up ($p = 0.0279$). Only a univariate comparison revealed the impact on anthropometric parameters (waist circumference) and graft function, as expressed by eGFR, between the two groups [23]. One potential contributing factor to the significant difference in the incidence of PTDM and pre-diabetic conditions could have been the significantly different time since KT. As part of the inclusion criteria, we determined a time of at least 3 months from KT, since by that time all the patients have stable levels, a dose of maintenance immunosuppression, and normal graft function. A recent international consensus of experts in the field of PTDM recommends performing a diagnosis of PTDM 10–13 weeks after KT, which our criteria agree with [24]. On the other hand, a proportion of patients diagnosed early after KT could have a potentially reversible disorder and, in the future, will require confirmation at least 1 year after KT. Upon analyzing the patients, we found that five patients in the control group received a diagnosis within a year, while three others received a diagnosis within 6 months of KT. The second factor could be some degree of predisposition, which we did not identify before the study because of the higher basal glycemia in the control group compared with the intervention group. However, we could not identify potential risks before inclusion because all the patients had physiological fasting glycemia at baseline, and there was no difference in anthropometric parameters or other glucose metabolism parameters (HOMA-IR, immunoreactive insulin, c-peptide, and HbA1c). Furthermore, prior to the study entry, all the enrolled patients had a negative oGTT test, although the study protocol did not include this test. However, the patients did not have this test performed at the same time point, which ranged from 3 to 12 months. The patients who were shortly post-transplant had it performed just before the KT; conversely, some patients who were longer post-transplant may have had an oGTT 12 months ago. Another potential factor may be the unequal level of the physical activity of the patients between the groups before the start of the study, as we did not objectively ascertain their exact average activity before the study. Only at the baseline visit did we ascertain the level of physical activity through an interview, concluding that three-quarters of the participants in both groups did not meet the threshold of 150 min of physical activity per week. The patients in the study group were determined to persevere for the entire period; the investigators kept in regular contact with them and motivated them. Those who did not take up a particular sport (e.g., running, swimming, or cycling) practiced daily brisk walking.

Our study has limitations in terms of the total length of follow-up and the size of the examined sample. On the other hand, given that this is a pilot study evaluating physical activity in a transplanted population by objective measurement using sports bracelets, we consider the number of subjects involved to be appropriate. We need larger, ideal multicenter and randomized studies in the future to confirm these findings. With our findings, we aim to reduce the high incidence of PTDM and other metabolic complications after KT by strictly implementing these measures into the regimen early in the post-transplant period, ideally within 3 months of the operation. Furthermore, we require specific data to formulate lifestyle recommendations for kidney recipients, encompassing not only physical activity but also dietary measures.

5. Conclusions

Our study's findings emphasize the importance of incorporating regular physical activity into preventable measures for postoperative glucose metabolism disorders. Carrying out regular aerobic or combined training lasting at least 150 min a week provided our intervening recipients after KT with significant prevention of PTDM or pre-diabetic conditions after only 6 months.

Post-transplant diabetes mellitus is a serious complication after KT with a negative impact on long-term graft survival as well as cardiovascular and overall mortality. When potential recipients arrive at a KT, they already show signs of metabolic syndrome, which is a significant risk factor for the later onset of PTDM, especially in the field of combination immunosuppressive therapy. It is, therefore, necessary to include this type of non-pharmacological intervention in combination with dietary measures as soon as the patient's general condition, surgical wound healing, and the development of graft function allow, ideally 3 months after KT.

Author Contributions: Conceptualization, K.G. and M.V.; methodology, K.G.; software, I.D.; validation, K.G. and I.D.; formal analysis, M.B. and P.K.; investigation, K.G., T.B. and M.P.; resources, M.V. and M.P.; data curation, K.G.; writing—original draft preparation, K.G.; writing—review and editing, I.D. and M.V.; visualization, P.K. and M.P.; supervision, I.D.; project administration, K.G. and M.V.; funding acquisition, K.G. and I.D. All authors have read and agreed to the published version of the manuscript.

Funding: This study was supported by grant VEGA-1/0238/21: Continual glucose monitoring and glycemic variability in early post-transplantation period as a predictor of complications after kidney transplantation.

Institutional Review Board Statement: The study was conducted in accordance with the Declaration of Helsinki, and approved by the Institutional Review Board of University Hospital Martin and Jessenius Faculty of Medicine (EK 33/2018, 25 April 2018).

Informed Consent Statement: Informed consent was obtained from all subjects involved in the study.

Data Availability Statement: The data that support the findings of this study are available from the first author upon reasonable request.

Conflicts of Interest: The authors declare no conflicts of interest.

Abbreviations

ADA	American Diabetes Association
BMI	body mass index
CKD	chronic kidney disease
CKD-EPI	chronic kidney disease—epidemiology collaboration index
CMV	cytomegalovirus
DM	diabetes mellitus
eGFR	estimated glomerular filtration rate
FPG	fasting plasma glucose
FPI	fasting plasma insulin
HbA1c	glycated hemoglobin
HOMA-IR	homeostatic model assessment of insulin resistance
HRmax	maximum heart rate
KDIGO	Kidney Disease: Improving Global Outcomes
PTDM	post-transplant diabetes mellitus
KT	kidney transplantation
WHO	World Health Organization
oGTT	oral glucose tolerance test
MS	metabolic syndrome
MET	metabolic equivalent task

References

1. Shivaswamy, V.; Boerner, B.; Larsen, J. Post Transplant Diabetes Mellitus: Causes, Treatment, and Impact on Outcomes. *Endocr. Rev.* **2016**, *37*, 37–61.
2. Senior, P.A.; Almehthel, M.; Miller, A.; Paty, B.W. Diabetes and Transplantation. *Can J. Diabetes* **2018**, *42* (Suppl. S1), S145–S149. [CrossRef]
3. Dedinská, L.; Baltesová, T.; Beňa, Ĺ.; Čellár, M.; Galajda, P.; Chrastina, M.; Mokáň, M. Incidence of Diabetes Mellitus After Kidney Transplantation in Slovakia: Multicentric, Prospective Analysis. *Transplant. Proc.* **2016**, *48*, 3292–3298. [CrossRef]
4. American Diabetes Association. 2 Classification and diagnosis of diabetes. *Diabetes Care* **2015**, *38* (Suppl. S1), S8–S16. [CrossRef] [PubMed]

5. World Health Organization. *Definition and Diagnosis of Diabetes Mellitus and Intermediate Hyperglycaemia*; Report of a WHO/IDF Consultation; WHO: Geneva, Switzerland, 2006.
6. Rangaswami, J.; Mathew, R.O.; Parasuraman, R.; Tantisattamo, E.; Lubetzky, M.; Rao, S.; Dadhania, D.M. Cardiovascular disease in the kidney transplant recipient: Epidemiology, diagnosis and management strategies. *Nephrol. Dial. Transplant.* **2019**, *34*, 760–773. [CrossRef] [PubMed]
7. Devine, P.A.; Courtney, A.E.; Maxwell, A.P. Cardiovascular risk in renal transplant recipients. *J. Nephrol.* **2019**, *32*, 389–399. [CrossRef] [PubMed]
8. Jenssen, T.; Hartmann, A. Post-transplant diabetes mellitus in patients with solid organ transplants. *Nat. Rev. Endocrinol.* **2019**, *15*, 172–188. [CrossRef] [PubMed]
9. Cai, R.; Wu, M.; Xing, Y. Pretransplant metabolic syndrome and its components predict post-transplantation diabetes mellitus in Chinese patients receiving a first renal transplant. *Ther. Clin. Risk Manag.* **2019**, *15*, 497–503. [CrossRef] [PubMed]
10. Despres, J.P.; Lemieux, I. Abdominal obesity and metabolic syndrome. *Nature* **2006**, *444*, 881–887. [CrossRef] [PubMed]
11. Dedinská, I.; Mačková, N.; Kantárová, D.; Kováčiková, L.; Graňák, K.; Laca, Ľ.; Mokáň, M. Leptin—A new marker for development of post-transplant diabetes mellitus? *J. Diabetes Its Complicat.* **2018**, *32*, 863–869. [CrossRef]
12. Bayés, B.; Granada, M.L.; Pastor, M.C.; Lauzurica, R.; Salinas, I.; Sanmartí, A.; Romero, R. Obesity, adiponectin and inflammation as predictors of new-onset diabetes mellitus after kidney transplantation. *Am. J. Transplant.* **2007**, *7*, 416–422. [CrossRef]
13. O'Hare, A.M.; Tawney, K.; Bacchetti, P.; Johansen, K.L. Decreased survival among sedentary patients undergoing dialysis: Results from the dialysis morbidity and mortality study wave 2. *Am. J. Kidney Dis.* **2003**, *41*, 447–454. [CrossRef]
14. Kosaka, K.; Noda, M.; Kuzuya, T. Prevention of type 2 diabetes by lifestyle intervention: A Japanese trial in IGT males. *Diab. Res. Clin. Pract.* **2005**, *67*, 152–162. [CrossRef]
15. Balk, E.M.; Earley, A.; Raman, G.; Avendano, E.A.; Pittas, A.G.; Remington, P.L. Combined diet and physical activity promotion programs to prevent type 2 diabetes among persons at increased risk: A systematic review for the community preventive services task force. *Ann. Intern. Med.* **2015**, *163*, 437–451. [CrossRef]
16. De Smet, S.; Van Craenenbroeck, A.H. Exercise training in patients after kidney transplantation. *Clin. Kidney J.* **2021**, *14* (Suppl. S2), ii15–ii24. [CrossRef]
17. Sharif, A.; Moore, R.; Baboolal, K. Influence of lifestyle modification in renal transplant recipients with postprandial hyperglycemia. *Transplantation* **2008**, *85*, 353–358. [CrossRef] [PubMed]
18. Karelis, A.D.; Hébert, M.J.; Rabasa-Lhoret, R.; Räkel, A. Impact of Resistance Training on Factors Involved in the Development of New-Onset Diabetes After Transplantation in Renal Transplant Recipients: An Open Randomized Pilot Study. *Can. J. Diabetes* **2016**, *40*, 382–388. [CrossRef]
19. Wilcox, J.; Waite, C.; Tomlinson, L.; Driscoll, J.; Karim, A.; Day, E.; Sharif, A. Comparing glycaemic benefits of Active Versus passive lifestyle Intervention in kidney Allograft Recipients (CAVIAR): Study protocol for a randomised controlled trial. *Trials* **2016**, *17*, 417. [CrossRef]
20. Byambasukh, O.; Osté, M.C.; Gomes-Neto, A.W.; van den Berg, E.; Navis, G.; Bakker, S.J.; Corpeleijn, E. Physical activity and the development of post-transplant diabetes mellitus, and cardiovascular-and all-cause mortality in renal transplant recipients. *J. Clin. Med.* **2020**, *9*, 415. [CrossRef] [PubMed]
21. Stevens, P.E.; Levin, A. KDIGO 2012 Clinical Practice Guideline for the Evaluation and Management of Chronic Kidney Disease. Chapter 3: Management of progression and complications of CKD. *Kidney Int. Suppl.* **2013**, *3*, 73–90.
22. Colberg, S.R.; Sigal, R.J.; Yardley, J.E.; Riddell, M.C.; Dunstan, D.W.; Dempsey, P.C.; Tate, D.F. Physical Activity/Exercise and Diabetes: A Position Statement of the American Diabetes Association. *Diabetes Care* **2016**, *39*, 2065–2079. [CrossRef]
23. Granak, K.; Vnucak, M.; Beliancinova, M.; Dedinska, I. Effect of regular physical activity and lifestyle changes on insulin resistance in patients after kidney transplantation. *Bratisl Lek Listy* **2024**, *125*, 250–257. [CrossRef] [PubMed]
24. Sharif, A.; Chakkera, H.; de Vries, A.P.; Eller, K.; Guthoff, M.; Haller, M.C.; Hecking, M. International consensus on post-transplantation diabetes mellitus. *Nephrol Dial Transpl.* **2024**, *39*, 531–549. [CrossRef] [PubMed]

Disclaimer/Publisher's Note: The statements, opinions and data contained in all publications are solely those of the individual author(s) and contributor(s) and not of MDPI and/or the editor(s). MDPI and/or the editor(s) disclaim responsibility for any injury to people or property resulting from any ideas, methods, instructions or products referred to in the content.

Article

Ultrasonographic Achilles Tendon Measurements and Static and Dynamic Balance in Prediabetes

Fulya Bakılan [1,*], Sultan Şan Kuşcu [1], Burcu Ortanca [1], Fezan Şahin Mutlu [2], Pınar Yıldız [3] and Onur Armağan [1]

[1] Department of Physical Medicine and Rehabilitation, Eskişehir Osmangazi University, 26040 Eskişehir, Turkey; sultann1905@gmail.com (S.Ş.K.); burcu-ayik@hotmail.com (B.O.); dronurarmagan@hotmail.com (O.A.)
[2] Department of Biostatistics, Eskişehir Osmangazi University, 26040 Eskişehir, Turkey; fsahin@ogu.edu.tr
[3] Department of Internal Medicine, Eskişehir Osmangazi University, 26040 Eskişehir, Turkey; pinaresogu@gmail.com
* Correspondence: fulyabakilan@gmail.com

Abstract: *Background and Objectives:* There is a lack of studies examining balance problems and Achilles tendon thickness in prediabetes despite their common occurrence in diabetes mellitus. The aim of this study was to evaluate Achilles tendon size and static and dynamic balance, as well as the role of the Achilles tendon in balance, in prediabetic patients. *Materials and Methods:* A total of 96 participants were divided into three groups: (1) the control group, consisting of participants without diabetes mellitus; (2) the prediabetes group; and (3) the diabetes mellitus group. Ultrasonographic measurements of Achilles tendon sizes (thickness, width and area) were performed. Dynamic balance was assessed using the Berg Balance Scale, and static balance (the Fall and Stability Indices) was assessed using a Tetrax device. The Self-Leeds Assessment of Neuropathic Symptoms and Signs was utilized to identify neuropathic pain. *Results:* In the prediabetes group, the median dynamic balance scores [54.0 (51.0–56.0)] were lower than those of the control group [55.0 (54.0–56.0)] but higher than those of the patients with diabetes mellitus [52.50 (49.0–54.25)]; however, this difference did not reach statistical significance. The ultrasonographic measurements of the Achilles tendon size were similar among the three groups. On the other hand, in the prediabetes group, a positive correlation was observed between the bilateral Achilles tendon anterior–posterior thickness and Fall Index score ($p = 0.045$), while a negative correlation was found between the left Achilles tendon anterior–posterior thickness and the Berg Balance Score ($p = 0.045$). *Conclusions:* In prediabetes, neither Achilles tendon size nor static or dynamic balance appears to be significantly affected. However, in prediabetic patients, increased Achilles tendon thickness appears to be associated with increased risk of falls and decreased balance.

Keywords: Achilles tendon; balance; dynamic balance; prediabetes; static balance

Citation: Bakılan, F.; Kuşcu, S.Ş.; Ortanca, B.; Mutlu, F.Ş.; Yıldız, P.; Armağan, O. Ultrasonographic Achilles Tendon Measurements and Static and Dynamic Balance in Prediabetes. *Medicina* **2024**, *60*, 1349. https://doi.org/10.3390/medicina60081349

Academic Editors: Yuzuru Ohshiro, Kunimasa Yagi and Yasuhiro Maeno

Received: 24 July 2024
Revised: 9 August 2024
Accepted: 17 August 2024
Published: 19 August 2024

Copyright: © 2024 by the authors. Licensee MDPI, Basel, Switzerland. This article is an open access article distributed under the terms and conditions of the Creative Commons Attribution (CC BY) license (https://creativecommons.org/licenses/by/4.0/).

1. Introduction

Multiple mechanisms contribute to controlling balance, including reactive, anticipatory, sensory and dynamic factors; the limits of the balance system; and physiological factors, such as the vestibular, visual and proprioceptive systems; muscle strength; reaction time; and the ankle and foot complex [1]. All these play crucial roles in maintaining balance [2]. Type 2 diabetes mellitus (DM) frequently induces changes that impact the somatosensory, vestibular and visual systems and lead to a high incidence of falls because of inadequate balance [3–5].

Previous studies have shown that patients with DM have had several changes in the ankle and foot complex [1,6], and the biomechanical characteristics of the ankle and foot complex have been reported to correlate with postural control in individuals with DM and be associated with vestibular inputs [1]. The Achilles tendon has been shown to be thicker in DM [1,7,8], while increases in the thickness of the Achilles tendon have been associated with the capacity of patients with DM to rely primarily on vestibular inputs for maintaining balance in the event of disrupted somatosensory input [1].

Prediabetes affects a wide section of the population and typically occurs for a significant period before the onset of DM. At the point of diagnosis, many patients already exhibit microvascular and macrovascular complications [9,10]. Individuals with prediabetes experience not only elevated peak glucose levels but also prolonged hyperglycemia due to their concurrent insulin resistance [11]. The latter triggers numerous metabolic processes, leading to endoneurial hypoxia and altering nerve perfusion, especially in glucose-dependent tissues such as the peripheral nerves and vestibular system. Furthermore, persistent hyperglycemia can result in muscle weakness, joint stiffness and premature degenerative alterations in the brain. On the other hand, it has been suggested that the thickening of the collagenous component, attributed to non-enzymatic glycation due to hyperglycemia, may be a factor contributing to the thickness of the Achilles tendon, which has been reported to be associated with vestibular inputs [1]. If persistent hyperglycemia disrupts balance through all the mechanisms described above, it is conceivable that impairment of balance and thickness of the Achilles tendon may also occur during the prediabetic period. Early detection of complications is important to reduce negative effects on patients' quality of life not only in the DM stage but also at the prediabetic stage. In the literature, other musculoskeletal problems, such as carpal tunnel syndrome, have been reported to be more common in prediabetes [12]. Curiosity has been expressed about the Achilles tendon, particularly whether it is affected during the prediabetic period and its role in balance. However, to our knowledge, there has, as of now, been no study that has investigated balance, changes to the Achilles tendon or the role of the Achilles tendon in balance in the prediabetic period.

An emerging hypothesis states that Achilles tendon thickening and balance impairment may occur during the prediabetic period, which is considered to be the early stages of DM, and that the former may disrupt balance.

The primary aim of this study was to evaluate Achilles tendon size and static and dynamic balance in the prediabetic stage by comparing these factors in an appropriate group with those in the DM and control groups, while the secondary aim was to explore the correlation between the Achilles tendon size and static and dynamic balance.

2. Materials and Methods

2.1. Ethical Considerations

This trial was conducted in accordance with the ethical principles outlined in the 1964 Declaration of Helsinki, ensuring the protection of participants' rights and well-being. Approval was obtained from the local ethics committee, with decision number 45, dated 17.01.23. Prior to participation, written informed consent was obtained from all participants after they were fully informed about this study's purpose, procedures, potential risks and benefits. Ethical considerations, including the confidentiality of participant data and the voluntary nature of participation, were strictly upheld throughout this study.

2.2. Participants and Measurements

This cross-sectional study was conducted with 96 participants who were admitted to a Physical Medicine and Rehabilitation outpatient clinic between February and November 2023. They were divided into three groups: (1) the control group (those who were neither prediabetic nor diabetic; (2) the prediabetes group (prediabetic patients); and (3) the DM group (patients with Type 2 DM).

The demographic data of the participants were collected, including age, sex, body mass index (BMI) and duration of the disease. The laboratory data included HbA1c (%) and fasting plasma glucose (FPG) levels during the most recent three months.

Diagnosis of Type 2 DM was confirmed through patients' medical records. Patients with an FPG of ≥ 126 mg/dL or 75 g oral glucose tolerance test that resulted in second-hour plasma glucose of ≥ 200 mg/dL and HbA1c $\geq 6.5\%$ were included in the DM group.

Diagnosis of prediabetes was also confirmed through patients' medical records. Patients with an FPG of 100–125 mg/dL or 75 g oral glucose tolerance test that resulted in second-hour plasma glucose of 140–199 mg/dL and HbA1c values of 5.7–6.4% were included in the prediabetes group.

The patients in the control group had an FPG of <100 mg/dL or 75 g oral glucose tolerance test that resulted in second-hour plasma glucose of <140 mg/dL and HbA1c values of <5.7% [13].

The final inclusion criterion for participation in this study was being between 40 and 65 years old. This age range was selected to focus on the adult population, as balance values can differ significantly between younger and geriatric populations.

Patients meeting any of the following criteria were excluded from this study: extremity amputation; vitamin B12 deficiency (all participants had documented B12 levels in the patient file system); Type 1 DM, prior exposure to neurotoxic agents; peripheral neuropathy for reasons such as chronic kidney failure, liver failure or hypothyroidism; hereditary or inflammatory peripheral neuropathies; neuromuscular diseases; malignancies; anti-neuropathic drug usage; radiculopathy; nerve trauma or surgery; vasculitis and autoimmune disorders; peripheral vascular disease; pregnancy; vestibular and cerebellar problems; history of lower extremity surgery; presence of medication affecting balance; history of alcoholism; and presence of visual impairment.

The Self-Leeds Assessment of Neuropathic Symptoms and Signs (S-LANSS) is a questionnaire consisting of seven items with Turkish validity and reliability. It is utilized to identify neuropathic pain. The S-LANSS was applied to all participants. The original S-LANSS, with a cut-off of ≥ 12, was considered indicative of the presence of neuropathic pain [14,15].

Ultrasonographic measurements of the Achilles tendon were taken using a Samsung Sonoace X7 ultrasound system (Samsung Medison Co., Ltd., Seoul, Republic of Korea) equipped with an 8–13 MHz linear transducer. After assuming a prone position on the examination table, each participant placed their feet against the wall and flexed their ankles to ensure optimal contact between the probe and the tendon. Measurements were taken separately on the right and left sides for each participant. Initially, the probe was positioned perpendicularly to the long axis of the tendon for an axial plane assessment, followed by measurements of the thickness (anterior posterior), width (medial–lateral) and area at the level of the medial malleolus. The thickness and width measurements utilized the tendon's major axes, while the area measurements were automatically calculated by the device through continuous tracing of the tendon circumference in the same section [16]. All sonographic measurements were performed by a sonographer with 5 years of experience in musculoskeletal ultrasound (Figure 1).

Dynamic balance was evaluated with the Berg Balance Test, which consists of 14 different questions evaluating the maintenance of static positions during changes in the center of body mass. Each participant was observed by a physician while performing the relevant activities, and the participants were given scores from 0 to 4. A score of 4 represented the activity being performed without any support, while 0 indicated the need for full support or the inability to perform the activity at all. The highest total score obtainable was 56, which represented excellent balance [17,18]. The test was performed only once per participant. To ensure accurate and reliable results, the following considerations were made: Participants were given sufficient rest before the test to avoid fatigue, which could have impacted balance performance. The test was conducted in a comfortable and distraction-free environment to facilitate optimal performance. Additionally, the purpose and instructions thereof were clearly explained to the participants to ensure they fully understood the procedure.

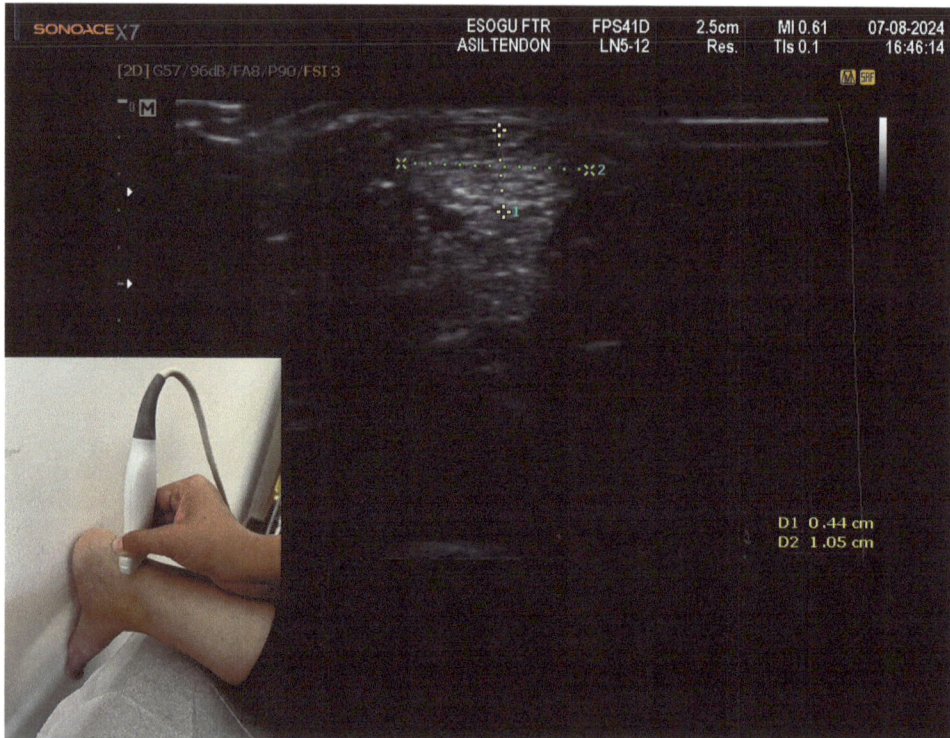

Figure 1. Ultrasound Imaging of the Achilles Tendon. The small image demonstrates the positioning of the ultrasound probe, placed perpendicular to the long axis of the Achilles tendon to obtain an axial plane at the level of the medial malleolus. The larger image depicts the measurements of the Achilles tendon, including thickness (anterior-posterior dimension, labeled as 1) and width (medial-lateral dimension, labeled as 2).

A Tetrax device (Tetrax-Sunlight Medical Ltd., Ramat Gan, Israel) was used to assess static balance. After the device was calibrated, each participant was positioned on the platform and subjected to tests in eight different positions. For each position, a test measurement was made for a duration of 32 s, totaling approximately 5 min. The normal eyes-open position was taken as a reference. The effects of vision on balance were observed in the eyes-closed position. When a participant was in the eyes-open position on a pillow, foam rubber pads limited the somatosensory system. In the eyes-closed position on a pillow, only the vestibular system functioned and was tested. With the head turned to the right and left in the eyes-closed position, both the vestibular and somatosensory systems were examined. In the eyes-closed position with the head tilted 30 degrees backward, the effects of central and peripheral vestibular disorders were observed. Balance was dependent on the back of the heels and the lower vertebrae in this position. Conversely, in the eyes-closed position with the head tilted 30 degrees forward, there was a load on the upper vertebrae and neck. Following the measurements, the same device was used to calculate the Fall and Stability Indices [19].

2.3. Statistical Analysis

A power analysis was conducted to determine the minimum required sample size for this study using the G*Power software, version 3.1.9.4 (Franz Faul, Universität Kiel, Düsseldorf, Germany). Based on the statistical findings from the reference publication, when the effect size (d) for the percentage change in the left Achilles tendon thickness parameter was taken as 5.0 and the standard deviation (SD) was taken as 0.1, the number of samples determined for power = 1.00 and α = 0.05 had minimums of $n1$= 22, $n2$ = 23 and $n3$ = 30 for each subgroup, respectively [20]. Considering the exclusion criteria, 30 participants were planned to be enrolled in each group.

In this study, the Shapiro–Wilk normality test was applied to the continuous variables. Descriptive statistics were presented as mean ± standard deviation (SD) and median (25th–75th percentiles), and categorical variables were presented as frequency and percentage. The independent samples t-test and one-way analysis of variance (ANOVA) were conducted for normally distributed variables. The Tukey and Tamhane Multiple Comparison tests were utilized to test differences among groups. The Mann–Whitney U and Kruskal–Wallis tests were applied for non-normally distributed variables, with Dunn's Multiple Comparison Test used for intergroup differences. The Chi-square test was applied for categorical variables. The Spearman correlation coefficient was employed to examine correlation (a coefficient of <0.1: negligible correlation; 0.1–0.39: weak; 0.4–0.69: moderate; 0.7–0.89: strong; and ≥0.9: very strong) [21]. The significance level was set at $p < 0.05$. Statistical analyses were performed using the IBM SPSS Statistics 21.0 program (SPSS Inc., Chicago, IL, USA).

3. Results

Two patients were excluded due to chronic kidney failure, one patient was excluded due to a history of malignancy and three patients were excluded due to anti-neuropathic drug usage. Considering the inclusion and exclusion criteria, 90 patients, 72 females and 18 males, with a mean age of 53.06 ± 7.33, were accepted to our study.

The demographic characteristics of the three groups are shown in Table 1. Seven (23.3%) of the 30 patients with DM were using insulin, while twenty-eight (93.3%) of them were using oral anti-diabetic drugs. In the prediabetes group, only eight (26.7%) patients were using oral anti-diabetic drugs. None of the patients had a foot ulcer.

Table 1. Demographic and laboratory data of all groups.

	Control Group n = 30 Median (25–75%)	Prediabetes Group n = 30 Median (25–75%)	DM Group n = 30 Median (25–75%)	p-Value
Demographic data				
Age (years)	52.0 (44.75–58.25)	55.0 (45.75–59.0)	56.0 (48.75–60.0)	0.495
Sex (Female/Male) n %	24 (80%)/6 (20%)	25 (83.3%)/5 (16.7%)	23 (76.7%)/7 (23.3%)	0.812
BMI [1] (kg/m^2)	27.16 ± 4.11	31.67 ± 6.01 *	29.66 ± 4.18	**0.002**
Disease duration (months)	0 (0–0)	27.0 (3.0–48.0) [a]	84.0 (36.0–171.0)	**<0.001**
Laboratory data				
HbA1c (%)	5.55 (5.37–5.60)	5.95 (5.77–6.10) °	7.0 (6.57–8.72)	**<0.001**
Fasting glucose levels (mg/dL)	89.5 (83.0–95.0)	96.0 (91.5–101.75) ±	133.0 (118.0–145.0)	**<0.001**

[1] Mean ± SD (DM: Diabetes Mellitus). * The Prediabetes Group was significantly different from the Control Group ($p = 0.004$). [a] The Prediabetes Group was significantly different from the DM Group ($p < 0.001$). ° The Prediabetes Group was significantly different from the DM Group ($p < 0.001$) and the Control Group ($p < 0.001$). The DM Group was significantly different from the Control Group ($p < 0.001$). ± The Prediabetes Group was significantly different from the DM Group ($p < 0.001$) and the Control Group ($p = 0.019$). The DM Group was significantly different from the Control Group ($p < 0.001$).

None of the control group participants, only one patient in the prediabetes group and only one patient in the DM group had an S-LANSS score greater than 12. This means that almost all the patients had an S-LANSS score lower than 12. A comparison of the static and dynamic balance parameters among the three groups showed that only the Berg Balance Score was significantly lower in the DM group than in the control group ($p = 0.001$) (Table 2). In the patients with prediabetes, the median dynamic balance scores [54.0 (51.0–56.0)] were lower than those of the control group [55.0 (54.0–56.0)] but higher than those of the patients with DM [52.50 (49.0–54.25)]; however, this difference did not reach statistical significance.

Moreover, the ultrasonographic measurements of Achilles tendon size was similar among the three groups (Table 3).

Table 2. Comparison of S-LANSS, the static and dynamic balance parameters among prediabetes, diabetes and control groups.

	Control Group n = 30 Median (25–75%)	Prediabetes Group n = 30 Median (25–75%)	DM Group n = 30 Median (25–75%)	p-Value
S-LANSS [1]	0 (0–0)	0 (0–6) *	5.5 (1–8)	<0.001
Static Balance				
Stability Index (eyes open)	10.15 (8.77–12.37)	11.70 (9.15–14.42)	12.01 (10.45–15.77)	0.075
Stability Index (eyes closed)	17.10 (11.82–21.59)	17.85 (14.10–23.40)	19.25 (14.42–23.90)	0.251
Fall Index	21 (7.5–32.5)	24.0 (14.0–37.5)	21.0 (9.50–42.50)	0.624
Dynamic Balance				
Berg Balance Score	55.0 (54.0–56.0)	54.0 (51.0–56.0)	52.50 (49.0–54.25) °	**0.001**

[1] S-LANSS: The Self-Leeds Assessment of Neuropathic Symptoms and Signs (DM: Diabetes Mellitus). * The Prediabetes Group was significantly different from the Control Group ($p = 0.006$) and the DM Group ($p = 0.008$). The DM Group was significantly different from the Control Group ($p < 0.001$). ° The DM Group was significantly different from the Control Group ($p = 0.001$).

Table 3. Comparison of the ultrasonographic measurement of the Achilles tendon among prediabetes, diabetes, and control groups.

	Control Group n = 30 Median (25–75%)	Prediabetes Group n = 30 Median (25–75%)	DM Group n = 30 Median (25–75%)	p-Value
Ultrasonographic Measurements of Achilles Tendon (mm)				
Thickness [1] (A-P) (R)	0.48 ± 0.08	0.50 ± 0.08	0.48 ± 0.07	0.547
Thickness (A-P) (L)	0.45 (0.40–0.50)	0.45 (0.41–0.53)	0.47 (0.43–0.50)	0.400
Width (M-L) (R)	1.24 (1.17–1.35)	1.31 (1.21–1.48)	1.30 (1.19–1.38)	0.327
Width [1] (M-L) (L)	1.28 ± 0.12	1.32 ± 0.17	1.28 ± 0.16	0.586
Area (R)	0.51 (0.44–0.60)	0.56 (0.46–0.70)	0.51 (0.47–0.58)	0.179
Area [1] (L)	0.51 ± 0.11	0.55 ± 0.13	0.53 ± 0.09	0.433

[1] Mean ± SD, (DM: Diabetes Mellitus, R: Right, L: Left, A-P: Anterior–posterior, M-L: Medio–lateral, mm: millimeters).

In the DM group, a weak positive correlation was found between BMI and the bilateral Achilles tendon medio–lateral width (right side: $p = 0.034$, $r = 0.388$; left side: $p = 0.042$, $r = 0.374$). On the other hand, in the control group, a moderate negative correlation was found between BMI and Berg Balance Score values ($p < 0.001$, $r = -0.658$). In the prediabetes group, a moderate negative correlation was found among age ($p = 0.002$, $r = -0.549$), BMI ($p = 0.003$, $r = -0.527$) and the Berg Balance Score. In the same group, a weak positive correlation was observed among age, the bilateral Achilles tendon anterior–posterior thickness and the Fall Index score (right: $p = 0.045$, $r = 0.368$; left: $p = 0.045$, $r = 0.369$). Additionally, a weak negative correlation was found between the Berg Balance Score and the left Achilles tendon anterior–posterior thickness ($p = 0.045$, $r = -0.369$). In the DM group, a moderate negative correlation was observed between age and the Berg Balance Score ($p = 0.025$, $r = -0.409$), while a moderate positive correlation was observed between age and the Fall Index score ($p = 0.007$, $r = 0.478$) (Table 4).

Table 4. Correlation analysis among age, body mass index, ultrasonographic measurements of the Achilles tendon and static/dynamic balance scores.

	Control Group n = 30		Prediabetes Group n = 30		DM Group n = 30	
	Dynamic Balance (Berg Balance Score)	Static Balance (Fall Index)	Dynamic Balance (Berg Balance Score)	Static Balance (Fall Index)	Dynamic Balance (Berg Balance Score)	Static Balance (Fall Index)
Age (years)	$p = 0.122$ $r = -0.288$	$p = 0.372$ $r = 0.169$	**$p = 0.002$** **$r = -0.549$**	**$p = 0.038$** **$r = 0.381$**	**$p = 0.025$** **$r = -0.409$**	**$p = 0.007$** **$r = 0.478$**
Body Mass Index (kg/m^2)	**$p < 0.001$** **$r = -0.658$**	$p = 0.930$ $r = 0.017$	**$p = 0.003$** **$r = -0.527$**	$p = 0.181$ $r = 0.251$	$p = 0.840$ $r = -0.039$	$p = 0.456$ $r = 0.141$
Ultrasonographic Measurements of Achilles Tendon (mm)						
Thickness (A-P) (R)	$p = 0.515$ $r = -0.124$	$p = 0.404$ $r = 0.158$	$p = 0.330$ $r = -0.184$	**$p = 0.045$** **$r = 0.368$**	$p = 0.889$ $r = -0.027$	$p = 0.774$ $r = 0.055$
Thickness (A-P) (L)	$p = 0.423$ $r = -0.152$	$p = 0.676$ $r = 0.080$	**$p = 0.045$** **$r = -0.369$**	**$p = 0.045$** **$r = 0.369$**	$p = 0.902$ $r = -0.023$	$p = 0.422$ $r = 0.152$
Width (M-L) (R)	$p = 0.388$ $r = -0.164$	$p = 0.987$ $r = 0.003$	**$p = 0.394$** $r = 0.161$	$p = 0.594$ $r = -0.101$	$p = 0.925$ $r = -0.018$	$p = 0.478$ $r = 0.135$
Width (M-L) (L)	$p = 0.297$ $r = -0.197$	$p = 0.972$ $r = -0.007$	$p = 0.593$ $r = 0.102$	$p = 0.813$ $r = -0.045$	$p = 0.595$ $r = -0.101$	$p = 0.489$ $r = 0.131$
Area (R)	$p = 0.518$ $r = -0.123$	$p = 0.707$ $r = 0.072$	$p = 0.965$ $r = 0.008$	$p = 0.527$ $r = 0.120$	$p = 0.889$ $r = 0.027$	$p = 0.889$ $r = 0.027$
Area (L)	$p = 0.503$ $r = -0.127$	$p = 0.970$ $r = -0.007$	$p = 0.874$ $r = -0.030$	$p = 0.585$ $r = 0.104$	$p = 0.736$ $r = -0.064$	$p = 0.889$ $r = 0.027$

DM: Diabetes Mellitus, R: Right, L: Left, A-P: Anterior–posterior, M-L: Medio–lateral, mm: millimeters.

4. Discussion

To the best of our knowledge, there is currently no documentation on whether the significant impairment of balance and Achilles tendon thickness observed in the diabetic stage are also present in the prediabetic stage.

The present study showed that in patients with prediabetes, the median dynamic balance scores were lower than those of the control group but higher than those of patients with DM; however, this difference did not reach statistical significance. This finding was consistent with the characteristic features of prediabetes, which typically include blood glucose levels that are above normal but below the thresholds for diabetes. Considering that the dynamic balance scores of the DM and prediabetes groups were found to be similar in our study, it cannot be concluded that dynamic balance is unimpaired in the prediabetes period. Further research with more sensitive assessment parameters is needed in this regard.

The static balance measurements, including the Fall and Stability Indices, with eyes both open and closed, were found to be similar across the three groups. Static balance did not appear to be affected in the patients with prediabetes or diabetes, who primarily presented without neuropathic pain. Nearly all the participants included in this study had an S-LANSS score below 12, indicating a minimal presence of neuropathic pain. This finding aligns with the existing literature. Previous studies have shown that neuropathy, which plays a role in the deterioration of balance, occurs not only in the diabetic stage but also in the prediabetic stages [22]. Palma et al. [23] conducted a comparison of static balance in Type 2 DM patients, both with and without neuropathy, and reported that static balance was worse in the former. In addition, Lim et al. [24] reported worse static balance scores in patients with diabetic polyneuropathy than in patients with diabetes but no polyneuropathy. In diabetic polyneuropathy, the presence of factors such as decreased proprioception, weakened muscles and reduced somatosensory function explains the deterioration of balance. If the S-LANSS scores of most of the patients in our DM and prediabetes groups had not been below 12, or if neuropathic patients had been predominant in our study, the static balance values of the diabetes and prediabetes groups could have been worse.

In the present study, neither the DM group nor the prediabetes group exhibited thickening of the Achilles tendon, which is the part of the ankle and foot complex that correlates to postural control and vestibular inputs [1]. Most of the Achilles tendon size parameters were higher in the prediabetes group than in the DM and control groups; however, this result did not reach statistical significance. If the number of participants in this study had been higher, we believe that a statistically significant difference could have been found. This result may have contributed to higher BMIs in prediabetes patients, as the increase in mechanical load on the foot, often observed in individuals with DM due to their higher body masses [25], is thought to play a role in the thickness of the Achilles tendon. In addition, BMI was positively correlated with Achilles tendon width in the DM group. In contrast to the findings of the present study, Achilles tendon thickness has been consistently associated with Type 2 DM in many studies [1,7,8]. Most have clarified this association by attributing it to neuropathy [8,26]. Evranos et al. [8] reported that Achilles tendon thickness values were significantly higher in diabetic patients with foot ulcers than in diabetic patients without foot ulcers and the control group. They also reported similar results to our findings between diabetic patients without foot ulcers and the control group. In our study, none of the patients had a foot ulcer, and the number of patients showing neuropathy symptoms was minimal.

In the literature, Achilles tendon size has been reported to be correlated with postural control in DM [1]. To our knowledge, this issue has not yet been investigated in prediabetes. In our study results, for the prediabetes patients, a positive correlation was observed between the bilateral Achilles tendon anterior–posterior thickness and the Fall Index score (which shows impaired static balance). Furthermore, in prediabetes patients, a negative correlation was found between the left Achilles tendon anterior–posterior thickness and the Berg Balance Score (which shows impaired dynamic balance). These results showed that not only age and BMI but also Achilles tendon thickness affect balance in the prediabetic stage. On the other hand, no correlation was found between static and dynamic balance and the Achilles tendon parameters in the control and DM groups. In DM, polypharmacy, muscle weakness of ankle dorsiflexion, and micro- and macrovascular complications, such as neuropathy and retinopathy [27], are all factors that affect balance and increase the risk of falls [28]. Without these other risk factors, the thickening of the Achilles tendon may not be significant or substantial enough to disrupt balance in overt diabetes by itself. However, in prediabetes, Achilles tendon thickness was found to affect both dynamic and static balance. Preventing the transition of patients from prediabetes to diabetes and ensuring that they adhere to the correct diets and exercise regimens are invaluable in preventing balance from being impaired. Monitoring Achilles tendon thickness using ultrasonography during the prediabetic period may provide guidance in monitoring balance.

5. Conclusions

The main limitation of this study is the absence of neuropathic diabetic and prediabetic groups due to the low S-LANSS scores of the patients participating in this study and the small sample size. The lack of assessment of dynamic balance using the Biodex Balance System or similar objective measurement tools is another. The biggest strength of the present study is that it is, to the best of our knowledge, the first study that has evaluated balance, Achilles tendon size and their relationship in the prediabetic stage. Balance was assessed both statically and dynamically, and moreover, static balance was measured using an objective tool. While the Berg Balance Scale offers advantages such as cost-effectiveness, ease of administration and minimal training requirements, its limitations, including ceiling effects and inadequate prediction of falls during active movement, may affect its ability to objectively measure balance in prediabetic patients. Therefore, more precise methods are needed to accurately assess balance in this population. Future research should focus on evaluating balance in prediabetic patients using advanced methodologies, such as electromyographic studies, to explore the relationship among balance, Achilles tendon size and neuropathy. Additionally, investigating alternative balance assessment tools and interventions could provide further insights into improving fall risk prediction and management in these patients.

In conclusion, in prediabetes, neither Achilles tendon size nor static or dynamic balance appears to be significantly affected. In prediabetic patients, increased Achilles tendon thickness appears to be associated with an increased risk of falls and decreased balance.

Author Contributions: Conceptualization, F.B.; methodology, O.A., B.O., S.Ş.K. and P.Y.; software, F.Ş.M.; validation, F.B., F.Ş.M. and S.Ş.K.; formal analysis, F.B. and B.O.; investigation, F.B. and S.Ş.K.; resources, F.B.; data curation, F.B., S.Ş.K., B.O., P.Y. and O.A.; writing—original draft preparation, F.B.; writing—review and editing, F.B., S.Ş.K., B.O., P.Y., F.Ş.M. and O.A.; visualization, F.B.; supervision, O.A., P.Y. and F.Ş.M.; project administration, none; funding acquisition, none. All authors have read and agreed to the published version of the manuscript.

Funding: This research received no external funding.

Institutional Review Board Statement: This study was conducted in accordance with the Declaration of Helsinki and approved by the Ethics Committee of Eskişehir Osmangazi University (protocol code: 45 and date of approval: 17 January 2023).

Informed Consent Statement: Written informed consent was obtained from all subjects involved in the study.

Data Availability Statement: The data that support the findings of this study are available from the corresponding author upon reasonable request.

Conflicts of Interest: The authors declare no conflicts of interest.

References

1. Cheing, G.L.; Chau, R.M.; Kwan, R.L.; Choi, C.H.; Zheng, Y.P. Do the biomechanical properties of the ankle–foot complex influence postural control for people with Type 2 diabetes? *Clin. Biomech.* **2013**, *28*, 88–92. [CrossRef] [PubMed]
2. Choi, S.; Jun, H.-P. Effects of rehabilitative exercise and neuromuscular electrical stimulation on muscle morphology and dynamic balance in individuals with chronic ankle instability. *Medicina* **2024**, *60*, 1187. [CrossRef]
3. Tander, B.; Atmaca, A.; Ulus, Y.; Tura, Ç.; Akyol, Y.; Kuru, Ö. Balance performance and fear of falling in older patients with diabetics: A comparative study with non-diabetic elderly. *Turk. J. Phys. Med. Rehabil.* **2016**, *62*, 314–322. [CrossRef]
4. Dixon, C.J.; Knight, T.; Binns, E.; Ihaka, B.; O'Brien, D. Clinical measures of balance in people with type two diabetes: A systematic literature review. *Gait Posture* **2017**, *58*, 325–332. [CrossRef] [PubMed]
5. Choi, J.H.; Kim, H.R.; Song, K.H. Musculoskeletal complications in patients with diabetes mellitus. *Korean J. Intern. Med.* **2022**, *37*, 1099–1110. [CrossRef]
6. Tuna, H.; Birtane, M.; Güldiken, S.; Soysal, N.A.; Taşpinar, Ö.; Süt, N.; Taştekin, N. The Effect of Disease Duration on Foot Plantar Pressure Values in Patients with Type 2 Diabetes Mellitus. *Turk. J. Phys. Med. Rehabil.* **2014**, *60*, 231–235. [CrossRef]
7. Papanas, N.; Courcoutsakis, N.; Papatheodorou, K.; Daskalogiannakis, G.; Maltezos, E.; Prassopoulos, P. Achilles tendon volume in type 2 diabetic patients with or without peripheral neuropathy: MRI study. *Exp. Clin. Endocrinol. Diabetes* **2009**, *117*, 645–648. [CrossRef] [PubMed]
8. Evranos, B.; Idilman, I.; Ipek, A.; Polat, S.B.; Cakir, B.; Ersoy, R. Real-time sonoelastography and ultrasound evaluation of the Achilles tendon in patients with diabetes with or without foot ulcers: A cross sectional study. *J. Diabetes Complicat.* **2015**, *29*, 1124–1129. [CrossRef] [PubMed]
9. Baranowska-Jurkun, A.; Matuszewski, W.; Bandurska-Stankiewicz, E. Chronic microvascular complications in prediabetic states—An overview. *J. Clin. Med.* **2020**, *9*, 3289. [CrossRef]
10. Parakash, S.S. Hyperinsulinemia, obesity, and diabetes mellitus. *Int. J. Diabetes Dev. Ctries.* **2023**, *43*, 289–290. [CrossRef]
11. Sharafi, M.; Eftekhari, M.H.; Mohsenpour, M.A.; Afrashteh, S.; Baeradeh, N.; Fararouei, M.; Pezeshki, B. Progression of prediabetes to diabetes and its associated factors: The Fasa Adult Cohort Study (FACS). *Int. J. Diabetes Dev. Ctries.* **2023**, *43*, 908–915. [CrossRef]
12. Erol, K.; Topaloğlu, U.S.; Göl, M.F. Frequency of carpal tunnel syndrome and hand dysfunction in prediabetes: A cross-sectional, controlled study. *Turk. J. Phys. Med. Rehabil.* **2022**, *68*, 62. [CrossRef] [PubMed]

13. American Diabetes Association Professional Practice Committee. 2. Diagnosis and Classification of Diabetes: Standards of Care in Diabetes—2024. *Diabetes Care* **2024**, *47*, 20–42. [CrossRef]
14. Koç, R.; Erdemoglu, A.K. Validity and reliability of the Turkish Self-administered Leeds Assessment of Neuropathic Symptoms and Signs (S-LANSS) questionnaire. *Pain Med.* **2010**, *11*, 1107–1114. [CrossRef]
15. Bennett, M.I.; Smith, B.H.; Torrance, N.; Potter, J. The S-LANSS score for identifying pain of predominantly neuropathic origin: Validation for use in clinical and postal research. *J. Pain* **2005**, *6*, 149–158. [CrossRef] [PubMed]
16. Canbolat, M.; Özbağ, D.; Özdemir, Z.; Demirtaş, G.; Kafkas, A.Ş. Effects of physical characteristics, exercise and smoking on morphometry of human Achilles tendon: An ultrasound study. *Anatomy* **2015**, *9*, 128–134. [CrossRef]
17. Berg, K.; Wood-Dauphinee, S.; Williams, J.I. The Balance Scale: Reliability assessment with elderly residents and patients with an acute stroke. *Scand. J. Rehabil. Med.* **1995**, *27*, 27–36.
18. Sahin, F.; Yilmaz, F.; Ozmaden, A.; Kotevoglu, N.; Sahin, T.; Kuran, B. Reliability and validity of the Turkish version of the Berg Balance Scale. *J. Geriatr. Phys. Ther.* **2008**, *31*, 32–37. [CrossRef]
19. Bozbaş, G.T.; Gürer, G. Does the lower extremity alignment affect the risk of falling? *Turk. J. Phys. Med. Rehabil.* **2018**, *64*, 140. [CrossRef]
20. İyidir, Ö.T.; Rahatlı, F.K.; Bozkuş, Y.; Ramazanova, L.; Turnaoğlu, H.; Nar, A.; Tütüncü, N.B. Acoustic radiation force impulse elastography and ultrasonographic findings of Achilles tendon in patients with and without diabetic peripheral neuropathy: A cross-sectional study. *Exp. Clin. Endocrinol. Diabetes* **2021**, *129*, 99–103. [CrossRef]
21. Schober, P.; Boer, C.; Schwarte, L.A. Correlation coefficients: Appropriate use and interpretation. *Anesth. Analg.* **2018**, *126*, 1763–1768. [CrossRef] [PubMed]
22. Koçer, A.; Domac, F.M.; Boylu, E.; Us, Ö.; Tanridağ, T. A comparison of sural nerve conduction studies in patients with impaired oral glucose tolerance test. *Acta Neurol. Scand.* **2007**, *116*, 399–405. [CrossRef] [PubMed]
23. Palma, F.H.; Antigual, D.U.; Martínez, S.; Monrroy, M.A.; Gajardo, R.E. Static balance in patients presenting diabetes mellitus type 2 with and without diabetic polyneuropathy. *Arq. Bras. Endocrinol. Metab.* **2013**, *57*, 722–726. [CrossRef] [PubMed]
24. Lim, K.B.; Kim, D.J.; Noh, J.H.; Yoo, J.; Moon, J.W. Comparison of balance ability between patients with type 2 diabetes and with and without peripheral neuropathy. *PM&R* **2014**, *6*, 209–214. [CrossRef]
25. Duffin, A.C.; Lam, A.; Kidd, R.; Chan, A.K.F.; Donaghue, K.C. Ultrasonography of plantar soft tissue thickness in young people with diabetes. *Diabet. Med.* **2002**, *19*, 1009–1013. [CrossRef] [PubMed]
26. Giacomozzi, C.; D'ambrogi, E.; Uccioli, L.; Macellari, V. Does the thickening of Achilles tendon and plantar fascia contribute to the alteration of diabetic foot loading? *Clin. Biomech.* **2005**, *20*, 532–539. [CrossRef] [PubMed]
27. Kafa, N.; Citaker, S.; Tuna, Z.; Guney, H.; Kaya, D.; Guzel, N.A.; Basar, S.; Yetkin, I. Is plantar foot sensation associated with standing balance in type 2 diabetes mellitus patients. *Int. J. Diabetes Dev. Ctries.* **2015**, *35*, 405–410. [CrossRef]
28. Reeves, N.D.; Orlando, G.; Brown, S.J. Sensory-motor mechanisms increasing falls risk in diabetic peripheral neuropathy. *Medicina* **2021**, *57*, 457. [CrossRef]

Disclaimer/Publisher's Note: The statements, opinions and data contained in all publications are solely those of the individual author(s) and contributor(s) and not of MDPI and/or the editor(s). MDPI and/or the editor(s) disclaim responsibility for any injury to people or property resulting from any ideas, methods, instructions or products referred to in the content.

Article

Hyperornithinemia–Hyperammonemia–Homocitrullinuria Syndrome in Vietnamese Patients

Khanh Ngoc Nguyen [1,2], Van Khanh Tran [3], Ngoc Lan Nguyen [3], Thi Bich Ngoc Can [1], Thi Kim Giang Dang [1], Thu Ha Nguyen [1], Thi Thanh Mai Do [1], Le Thi Phuong [3], Thinh Huy Tran [4], Thanh Van Ta [4], Nguyen Huu Tu [5] and Chi Dung Vu [1,2,*]

[1] Center of Endocrinology, Metabolism, Genetic/Genomics and Molecular Therapy, Vietnam National Children's Hospital, 18/879 La Thanh, Dong Da, Hanoi 11512, Vietnam; khanhnn@nch.gov.vn (K.N.N.); ngocctb@nch.gov.vn (T.B.N.C.); giangdk@nch.gov.vn (T.K.G.D.); thuha@nch.gov.vn (T.H.N.); maidtt@nch.gov.vn (T.T.M.D.)

[2] Department of Paediatrics, Hanoi Medical University, 1st Ton That Tung Street, Hanoi 11521, Vietnam

[3] Center for Gene and Protein Research, Hanoi Medical University, 1st Ton That Tung Street, Hanoi 11521, Vietnam; tranvankhanh@hmu.edu.vn (V.K.T.); nguyenngoclan@hmu.edu.vn (N.L.N.); phuongle@hmu.edu.vn (L.T.P.)

[4] Biochemistry Department, Hanoi Medical University, 1st Ton That Tung Street, Hanoi 11521, Vietnam; tranhuythinh@hmu.edu.vn (T.H.T.); tathanhvan@hmu.edu.vn (T.V.T.)

[5] Hanoi Medical Univerity Hospital, Hanoi Medical University, 1st Ton That Tung Street, Hanoi 11521, Vietnam; nguyenhuutu@hmu.edu.vn

* Correspondence: dungvu@nch.gov.vn; Tel.: +84-98-2706-899

Abstract: *Background and Objectives*: Hyperornithinemia–hyperammonemia–homocitrullinuria syndrome (HHH; OMIM 238970) is one of the rare urea cycle disorders. Ornithine carrier 1 deficiency causes HHH syndrome, characterized by failure of mitochondrial ornithine uptake, hyperammonemia, and accumulation of ornithine and lysine in the cytoplasm. The initial presentation and time of diagnosis in HHH highly varies. Genetic analysis is critical for diagnosis. *Materials and Methods*: This study encompassed retrospective and prospective analyses of four unrelated Vietnamese children diagnosed with HHH syndrome. *Results*: The age of diagnosis ranged from 10 days to 46 months. All four cases demonstrated hyperornithinemia and prolonged prothrombin time. Three out of four cases presented with hyperammonemia, elevated transaminases, and uraciluria. No homocitrulline was detected in the urine. Only one case depicted oroticaciduria. Genetic analyses revealed three pathogenic variants in the *SLC25A15* gene, with the c.535C>T (p.Arg179*) variant common in Vietnamese patients. The c.562_564del (p.Phe188del) and c.408del (p.Met137Cysfs*10) variants were detected in one case. The latter variant has yet to be reported in the literature on HHH patients. After intervention with a protein-restricted diet, ammonia-reducing therapy, and L-carnitine supplementation, hyperammonemia was not observed, and liver enzyme levels returned to normal. *Conclusions*: Our results highlighted the clinical and biochemical heterogeneity of HHH syndrome and posed that HHH syndrome should be considered when individuals have hyperammonemia, elevated transaminase, and decreased prothrombin time.

Keywords: HHH syndrome; *SLC25A15* variant; Vietnamese patients; p.Arg179*; p.Phe188del; p.Met137Cysfs*10

1. Introduction

Hyperornithinemia–hyperammonemia–homocitrullinuria syndrome (HHH; OMIM 238970) is a rare urea cycle disorder [1]. HHH syndrome elucidates 1–3.8% of urea cycle disorders [2]. Ornithine carrier 1 deficiency causes HHH syndrome, characterized by failure of mitochondrial ornithine uptake, hyperammonemia, and accumulation of ornithine and lysine in the cytoplasm [1,3]. The carbamylation of increased cytoplasmic lysine forms homocitrulline (ε-amino-carbamoyl-lysine). Patients display protein intolerance, episodic vomiting, growth retardation, hepatomegaly, liver failure, coagulopathy, and neurological symptoms, including loss of consciousness, seizures, pyramidal signs, and cognitive impairment with or without behavioral problems [4]. Camacho summarized data from 122 patients with HHH syndrome and reported lethargy (62%), coma (33%), increased liver enzymes (52%), and coagulation disorders (49%) [5]. A protein-restricted diet did not affect long-term neurological impairments with pyramidal tract signs (75%), intellectual disability (66%), and myoclonic epilepsy (34%). However, clinical manifestations and age of disease onset can broadly vary among individuals, even in the same family [5]. The age of onset ranges from the neonatal period to adulthood. Those with neonatal onset seem normal for the first 24–48 h, followed by the onset of symptoms associated with hyperammonemia (poor feeding, vomiting, lethargy, low temperature, and rapid breathing). Those with later onset may display chronic neurocognitive deficits and/or unexplained seizures, spasticity, acute encephalopathy secondary to hyperammonemic crisis, or chronic liver dysfunction.

Hyperornithinemia–hyperammonemia–homocitrullinuria (HHH) syndrome can be diagnosed by mutations in the *SLC25A15* gene and marked by elevated ammonia, homocitrulline, and ornithine levels [6]. HHH syndrome differs from other defects due to high urinary homocitrulline and ornithine [7]. Hyperornithinemia is present in almost all patients; however, a small proportion does not exhibit hyperammonemia and homocitrullinuria. Therefore, genetic testing for *SLC25A15* variants accompanied by at least one of three metabolic traits, hyperornithinemia, hyperammonemia, and homocitrullinuria, is pivotal for a definite diagnosis of HHH syndrome [8]. Early diagnosis improves clinical outcomes [8]. Acute treatment of HHH syndrome is similar to other urea cycle disorders, whereas long-term treatment of HHH syndrome is similar to carbamoyl phosphate synthetase I and ornithine transcarbamylase deficiency [7]. Protein restriction, citrulline, arginine, supplementation of essential amino acids, and sodium benzoate/sodium phenylbutyrate are required.

The pathogenic variants in the *SLC25A15* gene, an autosomal recessive inheritance pattern, cause hyperornithinemia–hyperammonemia–homocitrullinuria (HHH) syndrome [2]. The *SLC25A15* gene is located on chromosome 13q14.11 and comprises seven exons encoding for isoform 1 of the ornithine carrier ORC1 with a length of 301 amino acids [9]. The substrate binding of the ORC1 includes Glu77, Arg179, and Glu180 residues, and the Asn74 and Asn78 are situated in the substrate binding pocket [10]. HHH syndrome has been more frequently reported in French-Canadian, Italian, Japanese [11], and Palestinian [12]. The major mutant alleles in the *SLC25A15* gene included p.Phe188del and p.Arg179*, identified in 45% of HHH patients [13]. The pathogenic variant, p.Arg179*, was identified in a 5-year-old Vietnamese boy who migrated to the USA [3,14]. In this study, we report the phenotype, genotype, treatment, and outcome of four Vietnamese patients diagnosed with HHH syndrome. To our knowledge, our study is the first report of HHH patients in Vietnam.

2. Materials and Methods

2.1. Individuals

The study involved retrospective and prospective analyses of four unrelated Vietnamese children diagnosed with HHH syndrome at the Center of Endocrinology, Metabolism, Genetic/Genomics and Molecular Therapy, Vietnam National Children's Hospital. Clinical symptoms, including seizure, poor feeding, vomiting, lethargy, pyramidal signs, and intellectual disability, were followed.

2.2. Biochemical Parameters

Biochemical investigations included plasma ammonia, amino and acyl acid profiles, liver function, and coagulation.

Urinary organic acid analysis was conducted using gas chromatography–mass spectrometry (GC/MS). The urine specimen was soaked into two 2.5 cm × 2 cm blotting papers and subsequently dried naturally at room temperature for 4 h [15,16]. Then, the dried blotting papers were folded into a 2.5 mL syringe and placed in a 10 mL test tube. Distilled water (1.2 mL) was added to each tube and kept for 5 min. After centrifugation at 3000 rpm for 5 min, the supernatant was collected and transferred into a new 1.5 mL Eppendorf tube. We used two internal standards: margarate (MGA) and tetracosane (C24) (Sigma-Aldrich, Saint Louis, MO, USA). The 20 μg of each internal standard was added to the urine sample. Then, the distilled water was added to yield a final volume of 2 mL. One milliliter of 5% hydroxylamine hydrochloride (Sigma-Aldrich, Saint Louis, MO, USA) was added after adjusting the pH to 12–14 with 2N NaOH (Sigma-Aldrich, Saint Louis, MO, USA) and kept at room temperature for 60 min to oximize 2-ketoacids. Then, the resulting solution was acidified to pH 1.0 with 0.2 mL of 6N HCl (Sigma-Aldrich, Saint Louis, MO, USA), and 1 g of NaCl (Sigma-Aldrich, Saint Louis, MO, USA) was added. The urinary organic acids were extracted twice with 6 mL of ethyl acetate (Sigma-Aldrich, Saint Louis, MO, USA) and once with 6 mL of diethyl ether (Sigma-Aldrich, Saint Louis, MO, USA). After centrifugation, the organic layers were combined and dehydrated with 5 g of anhydrous sodium sulfate (Sigma-Aldrich, Saint Louis, MO, USA). Then, the supernatant was removed and dried with nitrogen at 60 °C. The extracted organic acids were derivatized by adding 100 μL of a mixture of N, O-bis(trimethylsilyl) trifluoroacetamide (BSTFA), and trimethylchlorosilane (TMCS) (10:1, v/v) (Sigma-Aldrich, Saint Louis, MO, USA). Afterward, they were kept at 80 °C for 30 min. Then, the derivatized solution was analyzed using a capillary GC/MS system (Shimadzu model QP 5000, Kyoto, Japan). The capillary column was a silica-coated DB-5 (30 m in length × 0.25 mm inside diameter) with a 1 μm film thickness of 5% phenyl methyl silicone (Shimadzu, Kyoto, Japan). The mass spectrum was formed by standard electron impact ionization scanning from the ion fragments with mass numbers from 50 m/z to 600 m/z at 0.4 s/rev. The initial temperature was 100 °C for 4 min, then increased to 290 °C at 4 °C/min and held constant for 10 min. The temperature of the injection port and transfer line was 280 °C. The flow rate of the helium gas was 1.5 mL/min, and the maximum velocity was 40.2 m/s. One microliter of the final derivatized solution was injected into the GC/MS for split-flow analysis. The urinary organic acids were analyzed using an automated data analysis system.

2.3. Genetic Testing and In Silico Analysis

Total genomic DNA samples were extracted from the whole blood samples using the QIAamp DNA Blood Mini Kit (Qiagen, Hilden, Germany). Genomic DNA samples were enriched for targeted regions employing a hybridization-based protocol and sequenced utilizing Illumina technology. The gene panel comprised 10 genes associated with urea cycle disorders. These genes were *ALDH18A1*, *ARG1*, *ASL*, *ASS1*, *CPS1*, *NAGS*, *OAT*, *OTC*, *SLC25A13*, and *SLC25A15*. Bioinformatics analyses and variant interpretation of the patients were performed at the Center for Gene-Protein Research, Hanoi Medical University. Reads were aligned to a reference sequence (GRCh37). Variants were identified and interpreted using relevant reference transcripts of NM_002860.3 for *ALDH18A1*, NM_000045.3 for *ARG1*, NM_000048.3 for *ASL*, NM_000050.4 for *ASS1*, NM_001875.4 for *CPS1*, NM_153006.2 for *NAGS*, NM_000274.3 for *OAT*, NM_000531.5 for *OTC*, NM_014251.2 for *SLC25A13*, and NM_014252.3 for *SLC25A15*.

The effect of variants was predicted using the Mutation Taster tool [17]. The pathogenicity of variants was determined using the ClinVar database. The frequency of variants was checked based on the information in the dbSNP154 database (https://www.ncbi.nlm.nih.gov/snp/, accessed on 2 May 2024) and gnomAD v4.1.0 database (https://gnomad.broadinstitute.org/, accessed on 2 May 2024).

2.4. Management

The patient management was based on the previous guidelines [7]. In managing acute hyperammonemia, the patients stopped all protein intake for 24–48 h and were given glucose infusion at 8–10 mg/kg/min, supplemental arginine 300–500 mg/kg/day,

L-carnitine 100 mg/kg/day, and sodium benzoate 200–300 mg/kg/day and underwent extracorporeal detoxification. The long-term treatment for the patients involved a low protein diet, arginine of 300–500 mg/kg/day, L-carnitine of 100 mg/kg/day, and sodium benzoate of 200–300 mg/kg/day.

3. Results

3.1. Clinical Findings

One male and three females were involved in this study (Table 1). One neonatal onset case and three late-onset cases existed (Table 1). The age of onset was 7 days and 18–31 months in the neonatal and late-onset forms, respectively. Cases 2 and 4 were diagnosed immediately after onset; however, Cases 1 and 3 were diagnosed 15–17 months after onset. At the diagnosis, two cases exhibited poor feeding and episodic vomiting; three cases had lethargy, with four cases not displaying any pyramidal signs (Table 1). Hyperammonemia, elevated transaminases, and uraciluria were observed in three cases (Table 1). Biochemical analyses revealed hyperornithinemia and liver failure with elevated international normalized ratio (INR) but not homocitrulline in the urine in four cases. Four cases demonstrated biochemical heterogeneity (Table 1).

Table 1. Clinical features and biochemical profile of the four cases.

	Case 1	Case 2	Case 3	Case 4
Sexuality	Female	Female	Male	Female
Age of onset	31 months	7 days	31 months	18 months
	Elevated transaminases and prolonged prothrombic time	Poor feeding, vomiting, jaundice, lethargy	Fever, lethargy, irritability, elevated transaminases, and prolonged prothrombic time	Vomiting and seizure
Age of diagnosis	46 months	10 days	48 months	19 months
Poor feeding	+	+	−	−
Episodic vomiting	+	+	−	−
Seizure	−	−	−	−
Lethargy	+	+	+	−
Hepatosplenomegaly	−	−	+	−
History of fibrile seizure	−	−	−	+
Biochemical profile at the diagnosis				
Hyperammonemia (µmol/L)	+ 98.0	+ 192.0	+ 76.8	−49.5
Hyperornithinemia (µmol/L)	+ 258.8	+ 256.5	+ 329.2	+ 193.7
Elevated transaminases	+	−	+	+
✓ Alanine transaminase (UI/L)	118	30	1055	69
✓ Aspartate transaminase (UI/L)	56	19	1033	58
Prolonged Prothrombin time (%)	32	46	48	46
International normalized ratio (INR)	1.67	1.52	1.55	1.74
Uraciluria	+	−	+	+
Oroticaciduria	−	−	−	+
Homocitrullinuria	−	−	−	−

(+), present; (−) not present.

Case 1 is the second child of family 1, born by vaginal delivery. She had normal physical development. At 31 months of age, she had elevated transaminases and prolonged prothrombin time in the context of fever and fatigue episodes. Diagnosed with unknown liver failure, she was treated with arginine and vitamin K supplements for 14 months. At 46 months of age, she displayed fatigue and lethargy and was admitted to our department. After analyzing for plasma ammonia and amino and acyl acid analyses, she indicated hyperammonemia (98.0 μmol/L) and hyperornithinemia (258.8 μmol/L) with elevated transaminases and uraciluria. The family history illustrated normal conditions.

Case 2 is the second child of family 2, also born by vaginal delivery. After the birth, she had symptoms of poor feeding, vomiting, jaundice, and lethargy and was admitted to our department. Biochemical investigations yielded hyperammonemia (192.0 μmol/L), hyperornithinemia (256.5 μmol/L), and normal levels of transaminases. The urine analysis revealed normal uracil, orotic acid, and homocitrulline levels. Her brother exhibited normal development.

Born by vaginal delivery, Case 3 is the second child of family 3. At 31 months of age, he exhibited fever, cough, fatigue, irritability, lethargy, hypertransaminase, and prolonged prothrombin time. Hence, he was diagnosed with liver failure and treated with glucose infusion, arginine, and vitamin K. At 48 months of age, he exhibited fatigue and lethargy and was admitted to our department. Biochemical investigations indicated slight hyperammonemia (76.8 μmol/L), hyperornithinemia (329.2 μmol/L), and high transaminases (Alanine transaminase 1055 UI/L and aspartate transaminase 1033 UI/L). Increased uracil was detected in the urine sample.

Case 4 is the first child of family 4, born by vaginal delivery. She displayed normal physical development. At 18 months of age, she had a seizure (under 10 s) and vomiting and was admitted to our department. Biochemical investigations revealed hypertransaminases (Alanine transaminase 69 UI/L and aspartate transaminase 58 UI/L), hyperornithinemia (193.7 μmol/L), and prolonged prothrombin time, but normal levels of ammonia (49.5 μmol/L).

3.2. Molecular Findings

Through molecular analyses of the four patients, we identified three variants in the *SLC25A15* gene classified as pathogenic variants according to ClinVar (Table 2). The variant c.535C>T (p.Arg179*) was a common variant, which was present in all four patients. Cases 1 and 2 were homozygous for the c.535C>T (p.Arg179*) variant. Case 3 harbored compound heterozygosity for c.408delC (p.Met137Cysfs*10) and c.535C>T (p.Arg179*). Case 4 had compound heterozygous variants of c.535C>T (p.Arg179*) and c.562_564delTTC (p.Phe188del).

Table 2. Molecular analyses of the four cases.

Patient	Gene	Variant	dbSNP152	ClinVar
Case 1	SLC25A15	c.535C>T (p.Arg179*)/c.535C>T (p.Arg179*)	rs104894429	5994 (Pathogenic)
Case 2	SLC25A15	c.535C>T (p.Arg179*)/c.535C>T (p.Arg179*)	rs104894429	5994 (Pathogenic)
Case 3	SLC25A15	c.535C>T (p.Arg179*)/ c.408delC (p.Met137Cysfs*10)	rs104894429 rs780201405	5994 (Pathogenic) 851,641 (Pathogenic)
Case 4	SLC25A15	c.535C>T (p.Arg179*)/ c.562_564delTTC (p.Phe188del)	rs104894429 rs202247803	5994 (Pathogenic) 5992 (Pathogenic)

The variant c.408delC (p.Met137Cysfs*10) has not been reported in the literature on HHH patients. The variant was reported by Invitae and Baylor Genetics in the ClinVar database (ID 85164) as a pathogenic or likely pathogenic variant. The variant c.408delC was observed in an East Asian at the heterozygous state (https://gnomad.broadinstitute.org/variant/13-40805209-AC-A?dataset=gnomad_r4, accessed on 2 May 2024). The variant was predicted as a deleterious variant in the Mutation Taster tool.

3.3. Treatment and Outcome

Cases 1 and 3 were misdiagnosed with unknown liver failure and treated with vitamin K1 (phytok 5 mg/day) and arginine supplement. However, their liver functions only improved once they were diagnosed accurately. After an accurate diagnosis, four cases were subjected to a protein-restricted diet (1–1.5 g/kg/day), L-carnitine (100 mg/kg/day), arginine (300–500 mg/kg/day), and sodium benzoate (100–250 mg/kg/day) (Table 3). All cases responded well to the treatment, depicting no acute hyperammonemia. Cases 1, 2, 3, and 4 were discharged after 10, 10, 7, and 5 days of treatment, respectively. After 3 days of treatment, Cases 1 and 2 rapidly returned to normal levels of transaminase and coagulation. Nevertheless, in Case 4, transaminases decreased gradually to normal levels after 2 months of treatment. Meanwhile, in Case 3, liver enzyme and blood clotting index slowly improved and returned to normal after 2 years of treatment. In the last visit, all cases depicted normal levels of ammonia, normal brain magnetic resonance imaging, and normal physical development. Cases 1, 2, and 3 had slightly increased levels of transaminases and international normalized ratio (INR), with Case 4 demonstrating normal levels. Case 3 displayed attention deficit hyperactivity disorder (Table 3). At the age of five, Case 3 was diagnosed with mild deficit hyperactivity disorder and received psychotherapy intervention at home. Case 3 is six years old now and has improved and increased focus on learning.

Table 3. Treatment and outcome of the four HHH cases.

	Case 1	Case 2	Case 3	Case 4
Treatment before accurate diagnosis				
Age of treatment	31 months	None	31 months	None
Vitamin K1				
Arginine supplement (mg/kg/day)	300–500	None	300–500	None
Treatment after accurate diagnosis				
Age of treatment	46 months	10 days	48 months	19 months
Low protein diet (g/kg/day)				
L-carnitine supplement (mg/kg/day)	100	100	100	100
Arginine supplement (mg/kg/day)	300–500	300–500	300–500	300–500
Sodium benzoate supplement at the Acute episodes (mg/kg/day)	100–250	100–250	100–250	100–250
Outcomes				
Treatment time to achieve normal transaminase and coagulation	3 days	3 days	2 years	2 months
Current age	9 years	4 years	6 years	4 years
Physical development	Normal	Normal	Normal	Normal
✓ Height	−0.6 SD	−1.8 SD	−0.4 SD	−1.5 SD
✓ Weight	−1.4 SD	−1.8 SD	−0.4 SD	−0.6 SD
Attention deficit hyperactivity disorder	None	None	Yes	None
Brain magnetic resonance imaging	Normal	Normal	Normal	Normal
Relapse	None	None	None	None
Biochemical profile at the last visit				
Plasma ammonia level (µmol/L) Transaminases	29.4	28.2	21.0	10.2
✓ Alanine transaminase (UI/L)	34	50	47	39.3
✓ Aspartate transaminase (UI/L)	47	44	55	32.4
Prothrombin time (%) International normalized ratio (INR)	73 / 1.25	75 / 1.22	61 / 1.43	81 / 1.15

4. Discussion

Hyperornithinemia–hyperammonemia–homocitrullinuria (HHH) syndrome is a sporadic disorder of the urea cycle in Vietnam. Until now, only four cases have been diagnosed in our center. The diagnosis was based on clinical, biochemical, and molecular analyses.

In our study, we observed diverse onset ages and delayed diagnoses. The age of onset ranged from neonatal to toddler. Case 2 was presented at 10 days old, which aligned with the findings of Camacho and colleagues [5], who posed that the neonatal onset rate was about 8% in people with HHH syndrome. This onset rate usually appears 24–48 h after breastfeeding with acute symptoms [5]. Martinelli and colleagues (2015) reported symptoms in the neonatal period for 2% of patients, 24% from 1 month to under 1 year old, and 44% from 1–12 years old [2]. Even though up to 1/3 of children have symptoms of neonatal onset, diagnosis is often delayed, with an average diagnostic delay of 6.3 ± 10.1 years (range 0–37 years) [2]. In our study, Cases 1 and 3 were misdiagnoses of unknown liver failure, with the accurate diagnosis delayed 15 months and 17 months, respectively. The clinical symptoms of children with HHH syndrome are diverse and nonspecific [2,6,12]. For example, due to hyperammonemia, HHH patients exhibited acute neurological symptoms, including seizures, poor appetite, vomiting, and lethargy. Therefore, these HHH patients can easily be misdiagnosed with encephalitis, epilepsy, cerebral hemorrhage, or poisoning. Additionally, hepatosplenomegaly can not immediately suggest a metabolic disorder. Other symptoms encompass acute encephalopathy, chronic liver disease, or cognitive impairment/learning disability/seizures.

Four cases also depicted biochemical heterogeneity. Three of them were presented with hyperammonemia. The median blood ammonia concentration usually ranged from 100 to 300 µmol/L, with newborns having higher average blood ammonia concentrations than older children and adults [5]. Wild et al. reported that a premature infant diagnosed with HHH syndrome had a blood ammonia concentration of 1300 µmol/L when they received intravenous nutrition and 623 µmol/L after stopping intravenous nutrition [1]. In our study, Case 3, which was early onset at 7 days, had the highest ammonia level (192 µmol/L). Case 4 displayed a normal level of ammonia but positive oroticaciduria at the diagnosis, contrasting with the other patients. The normal level of ammonia and elevated levels of transaminases caused Case 4 to be mistaken for liver failure of unknown cause or autoimmune hepatitis. Therefore, we recommend that patients with unexplained liver failure be repeatedly tested for ammonia primary when a change in consciousness and inborn errors of metabolism screening exist, such as MS/MS and plasma amino acids.

In our study, four cases had hyperornithinemia, which is observed in HHH patients [2,18]. Martinelli and colleagues reported that blood ornithine concentrations of patients with HHH syndrome increased from 216 to 1915 µmol/L (Normal: 30–110 µmol/L) [2]. Despite treatment with medication and a protein-restricted diet, blood ornithine levels remained elevated, and only a few patients were reported to have normal levels upon long-term follow-up [2]. Thus, the blood ornithine index was reliable in newborn screening to help with early diagnosis of HHH syndrome. No homocitrulline was detected in the urine of the four cases. Homocitrullineuria is a characteristic sign of the disease; however, patients who may have no or only marginal homocitrulline excretion in the urine are present [12,19]. Especially in neonates, homocitrulline may be obscured by plasma amino acid profile and abnormal aminoaciduria in liver dysfunction [5]. Such factors may have impacted the detection of urinary homocitrulline in our four cases.

Our study involved six of eight alleles of c.535C>T (p.Arg179*), suggesting that it was a "hot spot" variant in Vietnamese patients with HHH syndrome. The c.535C>T (p.Arg179*) variant was one of the most common variants observed in various HHH patients, including Japanese, Italian, Senegal, Morocco, Han Chinese, Korean, and Thailand [2]. Therefore, the c.535C>T (p.Arg179*) variant might have a broader carrier distribution in Vietnamese and could be used in population screening programs. Another common variant, c.562_564delTTC (p.Phe188del), which was reported in French-Canadian [11,18,20], Italian [21], Korean [22], and Pakistan [23], was identified in one of our patients. These two variants are located in exon 4 of the SLC25A15 gene. Thirteen pathogenic or likely pathogenic variants are reported in exon 4 in the ClinVar database. The third variant, c.408delC (p.Met137Cysfs*10), is situated in exon 3. This variant causes a frameshift; methionine at the position of amino acid 137 changes to cysteine and early termination after mutation of 10 amino acids. The identified variants in our study are nonsense or deletion variants aligning with the ClinVar database (accessed on 23 July 2024). Figure 1 depicts that the nonsense and deletion variants contribute 18% and 33% of the total pathogenic or likely pathogenic variants in the *SLC25A15* gene, respectively.

Both Cases 1 and 2 harbored the same nonsense variant c.535C>T (p.Arg179*) at the homozygous state. However, they posed clinical heterogeneity. Case 2 depicted neonatal onset at 7 days, whereas Case 1 demonstrated the late-onset form at 31 months. Table 4 displays thirteen previously reported patients harboring c.535C>T/c.535C>T with the details of clinical and biochemical characteristics selected for comparison. Overall, 17 patients with c.535C>T/c.535C>T variant exhibited diverse phenotypes with the typical characteristics of lethargy, hyperornithinemia, hyperammonemia, and elevated homocitruline (Table 4). Seven patients were diagnosed at very late onset (>10 years), whereas three presented symptoms during the neonatal period. Nevertheless, they depicted different phenotypes in coma, coagulopathy, seizure, pyramidal signs, and levels of ornithine and ammonia [23]. Table 4 illustrates that our patient (P17) had normal ammonia levels, aligning with five Turkish patients. However, the five Turkish patients did not depict any medical history of hyperammonia. The probands P8 and P11 were referred for gait disturbances [6], and genetic analysis revealed that they harbored homozygotes of c.535C>T (p.Arg179*). Then, their siblings were screened for the pathogenic variant. Results demonstrated that the siblings also carried the pathogenic variant and presented with elevated ornithine and homocitruline. However, they did not exhibit any symptoms of HHH syndrome. No correlation between the genotype and phenotype in HHH patients seemed to exist [12,20,23]. To our knowledge, no published data to date of patients harboring compound heterozygous variants *SLC25A15*: c.535C>T/c.408delC or c.535C>T/c.562_564delTTC are present.

Before obtaining accurate diagnoses, the liver functions of Cases 1 and 3, managed with vitamin K1 and arginine supplement, had not improved over 15 and 17 months. After a protein-restricted diet, L-carnitine, arginine, and sodium benzoate, liver functions in all cases became normal. For Cases 1 and 2, liver enzyme levels returned to normal after 3 days of treatment. Nonetheless, it took 2 years and 2 months for Cases 3 and 4, respectively. All cases had not occurred with the acute crisis of hyperammonemia for 2–5 years because they were prevented by enough energy support (glucose infusion) in the case of stresses (fever, vomiting, vaccination, etc.), aligning with other studies. All of our patients exhibited normal physical development. Therefore, early, accurate diagnosis of HHH syndrome is necessary more than ever, especially in newborn screening. Diet therapy, controlling plasma ammonia levels, and preventing acute crises from stresses provide better outcomes.

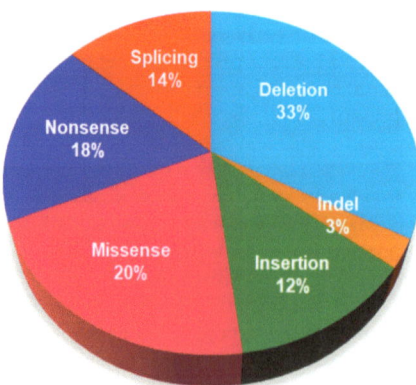

Figure 1. Distribution of pathogenic or likely pathogenic variants in the *SLC25A15* in the ClinVar database. All 73 pathogenic or likely pathogenic variants were listed in ClinVar database (accessed on 23 July 2024).

Table 4. Clinical characteristics of patients with homozygote c.535C>T (p.Arg179*).

Cases	Sex	Ethnicity	Onset Age	Diagnosis Age	Lethargy	Coma	Elevated Transaminases	Coagulopathy	Intellectual Disability	Seizures, Myoclonic	Pyramidal Signs	Ornithine (μmol/L)	Ammonia (μmol/L)	Elevated Homoc-itrulline	Ref.
P1	m	Japanese	3.0 y	10.0 y	+	–	–	–	Severe	–	+	↑419	↑204	+	[24]
P2	m	Japanese	11.0 y	41.0 y	+	+	–	–	Mild	–	+	↑586	↑98	+	[25]
P3	f	Japanese	NR	52.0 y	–	–	–	–	–	+	+	↑	↑242	+	[26]
P4	m	Italian	12.0 y	26.0 y	+	–	–	+	Mild	+	+	NR	↑	NR	[27]
P5	f	Senegal	Birth	Birth	+	+	–	+	Mild	+	+	↑509	↑700	+	[23]
P6	f	Senegal	Birth	Birth	+	–	–	–	Mild	–	+	↑290	↑100	+	[23]
P7	m	Morocco	1.7 y	1.7 y	+	–	+	+	–	–	–	↑493	↑96	+	[23]
P8	m	Turkish	16.0 y	16.0 y	–	–	–	–	–	–	+	↑367	44	+	[6]
P9	f	Turkish	10.4 y	10.4 y	–	–	–	–	–	–	-	↑468	37	+	[6]
P10	m	Turkish	2.3 y	2.3 y	–	–	–	–	–	–	–	↑305	50	+	[6]
P11	m	Turkish	13.1 y	14.0 y	–	–	–	–	–	–	+	↑446	40	+	[6]
P12	f	Turkish	17.5 y	17.5 y	–	–	–	–	–	–	-	↑248	27	+	[6]
P13	f	Vietnamese	1.0 y	5.0 y	+	+	+	+	Moderate	+	+	↑1439	↑353	–	[3]
P14	f	Vietnamese	2.6 y	3.8 y	+	–	+	+	–	–	–	↑259	↑98	–	This study
P15	f	Vietnamese	7 d	10 d	+	–	–	+	–	–	–	↑256	↑192	–	This study
Summary					9/15 (60%)	3/15 (20%)	3/15 (20%)	6/15 (40%)	6/15 (40%)	4/15 (27%)	9/15 (60%)	14/15 (93%)	10/15 (67%)	11/15 (73%)	

m, male; f, female; y, year; NR, not recorded; +, present; –, not present; ↑, elevated.

No brain abnormalities were detected in the brain MRI of our four cases. Neuroradiological abnormalities usually occur after 20 years of age [28]. In our study, brain MRI was performed at 4 to 9 years of age. Hence, neurological abnormalities have not been illustrated in the MRI findings. Additionally, after age 15, HHH patients developed spastic paraparesis, which was not related to dietary intake and sodium benzoate treatment [12]. Therefore, HHH patients should be carefully monitored by clinical examination and spinal and brain MRI for the long term. We suggest that an MRI be performed annually after 15 years of age.

Our study's limitations include a small sample size and lack of functional testing to demonstrate the effect of pathogenic variants on protein functions in vitro or in vivo. Further studies would be needed to reach conclusive conclusions.

5. Conclusions

Hyperornithinemia–hyperammonemia–homocitrullinuria (HHH) syndrome is a clinical and biochemical heterogeneity disorder. The clinical spectrum is diverse, and biochemical changes are nonspecific. HHH syndrome should be considered when evaluating individuals with unexplained hyperammonemia or persistently elevated liver enzymes and decreased prothrombin ratio, especially in newborn screening. The hot spot at residue 179 of SLC25A15 in Vietnamese cases with HHH syndrome may be used in the screening of individuals suspected of HHH syndrome.

Author Contributions: Conceptualization, K.N.N. and V.K.T.; methodology, K.N.N. and C.D.V.; validation, V.K.T. and C.D.V.; formal analysis, K.N.N.; investigation, K.N.N., T.B.N.C., T.K.G.D., T.H.N. and T.T.M.D.; data curation, N.H.T. and K.N.N.; writing—original draft preparation, K.N.N. and N.L.N.; writing—review and editing, V.K.T., N.L.N.; L.T.P., T.H.T., T.V.T., N.H.T. and C.D.V.; visualization, T.B.N.C., T.K.G.D., T.H.N., T.T.M.D. and C.D.V.; supervision, C.D.V. All authors have read and agreed to the published version of the manuscript.

Funding: This research was funded by the Vietnam Ministry of Science and Technology under grant number ĐTĐL.CN-133/21.

Institutional Review Board Statement: The study was conducted in accordance with the Declaration of Helsinki and approved by the Ethics Committee of Hanoi Medical University Institutional Ethical Review Board (Approval No: 940/GCN-HĐĐĐNCYSH-ĐHYHN on 30 June 2023).

Informed Consent Statement: Written informed consent has been obtained from the parents of patients to publish this paper.

Data Availability Statement: The raw data supporting the conclusions of this article will be made available by the authors on request.

Conflicts of Interest: The authors declare no conflicts of interest.

References

1. Wild, K.T.; Ganetzky, R.D.; Yudkoff, M.; Ierardi-Curto, L. Hyperornithinemia, hyperammonemia, and homocitrullinuria syndrome causing severe neonatal hyperammonemia. *JIMD Rep.* **2018**, *44*, 103–107. [CrossRef]
2. Martinelli, D.; Diodato, D.; Ponzi, E.; Monné, M.; Boenzi, S.; Bertini, E.; Fiermonte, G.; Dionisi-Vici, C. The hyperornithinemia-hyperammonemia-homocitrullinuria syndrome. *Orphanet J. Rare Dis.* **2015**, *10*, 29. [CrossRef]
3. Kim, S.Z.; Song, W.J.; Nyhan, W.L.; Ficicioglu, C.; Mandell, R.; Shih, V.E. Long-term follow-up of four patients affected by HHH syndrome. *Clin. Chim. Acta Int. J. Clin. Chem.* **2012**, *413*, 1151–1155. [CrossRef]
4. Lee, H.H.C.; Poon, K.H.; Lai, C.K.; Au, K.M.; Siu, T.S.; Lai, J.P.S.; Mak, C.M.; Yuen, Y.P.; Lam, C.W.; Chan, A.Y.W. Hyperornithinaemia-hyperammonaemia-homocitrullinuria syndrome: A treatable genetic liver disease warranting urgent diagnosis. *Hong Kong Med. J. Xianggang Yi Xue Za Zhi* **2014**, *20*, 63–66. [CrossRef]
5. Camacho, J.; Rioseco-Camacho, N. Hyperornithinemia-hyperammonemia-homocitrullinuria syndrome. 2012 May 31 [Updated 2020 Feb 13]. In *GeneReviews®*; Adam, M.P., Feldman, J., Mirzaa, G.M., Pagon, R.A., Wallace, S.E., Bean, L.J., Gripp, K.W., Amemiya, A., Eds.; University of Washington: Seattle, WA, USA, 1993–2024.
6. Bilgin, H.; Bilge, S.; Binici, M.; Tekes, S. Clinical, biochemical, and genotypical characteristics in urea cycle mitochondrial transporter disorders. *Eur. Rev. Med. Pharmacol. Sci.* **2024**, *28*, 1873–1880. [CrossRef]

7. Häberle, J.; Burlina, A.; Chakrapani, A.; Dixon, M.; Karall, D.; Lindner, M.; Mandel, H.; Martinelli, D.; Pintos-Morell, G.; Santer, R.; et al. Suggested guidelines for the diagnosis and management of urea cycle disorders: First revision. *J. Inherit. Metab. Dis.* **2019**, *42*, 1192–1230. [CrossRef]
8. Martinelli, D.; Fiermonte, G.; Häberle, J.; Boenzi, S.; Goffredo, B.M.; Travaglini, L.; Agolini, E.; Porcelli, V.; Dionisi-Vici, C. CUGC for hyperornithinemia-hyperammonemia-homocitrullinuria (HHH) syndrome. *Eur. J. Hum. Genet. EJHG* **2020**, *28*, 982–987. [CrossRef]
9. Indiveri, C.; Tonazzi, A.; Palmieri, F. Identification and purification of the ornithine/citrulline carrier from rat liver mitochondria. *Eur. J. Biochem.* **1992**, *207*, 449–454. [CrossRef]
10. Monné, M.; Miniero, D.V.; Daddabbo, L.; Robinson, A.J.; Kunji, E.R.S.; Palmieri, F. Substrate specificity of the two mitochondrial ornithine carriers can be swapped by single mutation in substrate binding site. *J. Biol. Chem.* **2012**, *287*, 7925–7934. [CrossRef]
11. Camacho, J.A.; Obie, C.; Biery, B.; Goodman, B.K.; Hu, C.-A.; Almashanu, S.; Steel, G.; Casey, R.; Lambert, M.; Mitchell, G.A.; et al. Hyperornithinaemia- hyperammonaemia- homocitrullinuria syndrome is caused by mutations in a gene encoding a mitochondrial ornithine transporter. *Nat. Genet.* **1999**, *22*, 151–158. [CrossRef]
12. Dweikat, I.; Khalaf-Nazzal, R. Clinical heterogeneity of hyperornithinemia-hyperammonemia-homocitrullinuria syndrome in thirteen palestinian patients and report of a novel variant in the SLC25A15 gene. *Front. Genet.* **2022**, *13*, 1004598. [CrossRef]
13. Palmieri, F.; Scarcia, P.; Monné, M. Diseases caused by mutations in mitochondrial carrier genes SLC25: A review. *Biomolecules* **2020**, *10*, 655. [CrossRef]
14. Nyhan, W.L. Hyperornithinemia hyperammonemia homocitrullinemia (HHH) syndrome. In *Atlas of Metabolic Diseases*, 2nd ed.; Oxford University Press: New York, NY, USA, 2005; pp. 228–232.
15. Kimura, M.; Yamamoto, T.; Yamaguchi, S. Automated metabolic profiling and interpretation of GC/MS data for organic acidemia screening: A personal computer-based system. *Tohoku J. Exp. Med.* **1999**, *188*, 317–334. [CrossRef]
16. Yamaguchi, S.; Kimura, M.; Iga, M.; Fu, X.W.; Ohie, T.; Yamamoto, T. Automated, simplified GC/MS data processing system for organic acidemia screening and its application. *Southeast Asian J. Trop. Med. Public Health* **1999**, *30* (Suppl. S2), 174–180.
17. Schwarz, J.M.; Cooper, D.N.; Schuelke, M.; Seelow, D. MutationTaster2: Mutation prediction for the deep-sequencing age. *Nat. Methods* **2014**, *11*, 361–362. [CrossRef]
18. Sokoro, A.A.H.; Lepage, J.; Antonishyn, N.; McDonald, R.; Rockman-Greenberg, C.; Irvine, J.; Lehotay, D.C. Diagnosis and high incidence of hyperornithinemia-hyperammonemia-homocitrullinuria (HHH) syndrome in northern Saskatchewan. *J. Inherit. Metab. Dis.* **2010**, *33* (Suppl. S3), 275–281. [CrossRef]
19. Korman, S.H.; Kanazawa, N.; Abu-Libdeh, B.; Gutman, A.; Tsujino, S. Hyperornithinemia, hyperammonemia, and homocitrullinuria syndrome with evidence of mitochondrial dysfunction due to a novel *SLC25A15 (ORNT1)* gene mutation in a Palestinian family. *J. Neurol. Sci.* **2004**, *218*, 53–58. [CrossRef]
20. Debray, F.-G.; Lambert, M.; Lemieux, B.; Soucy, J.F.; Drouin, R.; Fenyves, D.; Dubé, J.; Maranda, B.; Laframboise, R.; Mitchell, G.A. Phenotypic variability among patients with hyperornithinaemia-hyperammonaemia-homocitrullinuria syndrome homozygous for the delF188 mutation in SLC25A15. *J. Med. Genet.* **2008**, *45*, 759–764. [CrossRef]
21. Filosto, M.; Alberici, A.; Tessa, A.; Padovani, A.; Santorelli, F.M. Hyperornithinemia-hyperammonemia-homocitrullinuria (HHH) syndrome in adulthood: A rare recognizable condition. *Neurol. Sci. Off. J. Ital. Neurol. Soc. Ital. Soc. Clin. Neurophysiol.* **2013**, *34*, 1699–1701. [CrossRef]
22. Jang, K.M.; Hyun, M.C.; Hwang, S.K. A novel SLC25A15 mutation causing Hyperornithinemia-Hyperammonemia-Homocitrullinuria syndrome. *J. Korean Child Neurol. Soc.* **2017**, *25*, 204–207. [CrossRef]
23. Tessa, A.; Fiermonte, G.; Dionisi-Vici, C.; Paradies, E.; Baumgartner, M.R.; Chien, Y.-H.; Loguercio, C.; de Baulny, H.O.; Nassogne, M.-C.; Schiff, M.; et al. Identification of novel mutations in the SLC25A15 gene in hyperornithinemia-hyperammonemia-homocitrullinuria (HHH) syndrome: A clinical, molecular, and functional study. *Hum. Mutat.* **2009**, *30*, 741–748. [CrossRef] [PubMed]
24. Nakajima, M.; Ishii, S.; Mito, T.; Takeshita, K.; Takashima, S.; Takakura, H.; Inoue, I.; Saheki, T.; Akiyoshi, H.; Ichihara, K. Clinical, biochemical and ultrastructural study on the pathogenesis of hyperornithinemia-hyperammonemia-homocitrullinuria syndrome. *Brain Dev.* **1988**, *10*, 181–185. [CrossRef] [PubMed]
25. Tsujino, S.; Suzuki, T.; Azuma, T.; Higa, S.; Sakoda, S.; Kishimoto, S. Hyperornithinemia, hyperammonemia and homocitrullinuria—A case report and study of ornithine metabolism using in vivo deuterium labelling. *Clin. Chim. Acta Int. J. Clin. Chem.* **1991**, *201*, 129–133. [CrossRef] [PubMed]
26. Miyamoto, T.; Kanazawa, N.; Kato, S.; Kawakami, M.; Inoue, Y.; Kuhara, T.; Inoue, T.; Takeshita, K.; Tsujino, S. Diagnosis of Japanese patients with HHH syndrome by molecular genetic analysis: A common mutation, R179X. *J. Hum. Genet.* **2001**, *46*, 260–262. [CrossRef] [PubMed]

27. Salvi, S.; Santorelli, F.M.; Bertini, E.; Boldrini, R.; Meli, C.; Donati, A.; Burlina, A.B.; Rizzo, C.; Di Capua, M.; Fariello, G.; et al. Clinical and molecular findings in hyperornithinemia-hyperammonemia-homocitrullinuria syndrome. *Neurology* **2001**, *57*, 911–914. [CrossRef]
28. Olivieri, G.; Pro, S.; Diodato, D.; Di Capua, M.; Longo, D.; Martinelli, D.; Bertini, E.; Dionisi-Vici, C. Corticospinal tract damage in HHH syndrome: A metabolic cause of hereditary spastic paraplegia. *Orphanet J. Rare Dis.* **2019**, *14*, 208. [CrossRef]

Disclaimer/Publisher's Note: The statements, opinions and data contained in all publications are solely those of the individual author(s) and contributor(s) and not of MDPI and/or the editor(s). MDPI and/or the editor(s) disclaim responsibility for any injury to people or property resulting from any ideas, methods, instructions or products referred to in the content.

Article

Neurometabolic Profile in Obese Patients: A Cerebral Multi-Voxel Magnetic Resonance Spectroscopy Study

Miloš Vuković [1,*], Igor Nosek [1], Johannes Slotboom [2], Milica Medić Stojanoska [1] and Duško Kozić [1]

[1] Faculty of Medicine, University in Novi Sad, 21000 Novi Sad, Serbia; igor.nosek@mf.uns.ac.rs (I.N.); milica.medicstojanoska@kcv.rs (M.M.S.); dusko.kozic@mf.uns.ac.rs (D.K.)
[2] Institute for Diagnostic and Interventional Neuroradiology, University Hospital Bern and Inselspital, 3010 Bern, Switzerland; johannes.slotboom@gmail.com
* Correspondence: milos.vukovic@mf.uns.ac.rs

Abstract: *Background and Objectives:* Obesity-related chronic inflammation may lead to neuroinflammation and neurodegeneration. This study aimed to evaluate the neurometabolic profile of obese patients using cerebral multivoxel magnetic resonance spectroscopy (mvMRS) and assess correlations between brain metabolites and obesity markers, including body mass index (BMI), waist circumference, waist-hip ratio, body fat percentage, and indicators of metabolic syndrome (e.g., triglycerides, HDL cholesterol, fasting blood glucose, insulin, and insulin resistance index (HOMA-IR)). *Materials and Methods:* This prospective study involved 100 participants, stratified into two groups: 50 obese individuals (BMI $\geq$ 30 kg/m^2) and 50 controls (18.5 $\leq$ BMI < 25 kg/m^2). Anthropometric measurements, body fat percentage, and biochemical markers were evaluated. All subjects underwent long- and short-echo mvMRS analysis of the frontal and parietal supracallosal subcortical and deep white matter, as well as the cingulate gyrus, analyzing NAA/Cr, Cho/Cr, and mI/Cr ratios, along with absolute concentrations of NAA and Cho. *Results:* Obese participants exhibited significantly decreased NAA/Cr and Cho/Cr ratios in the deep white matter of the right cerebral hemisphere ($p < 0.001$), while absolute concentrations of NAA and Cho did not differ significantly between groups ($p > 0.05$). NAA levels showed negative correlations with more reliable obesity parameters (waist circumference and waist-to-hip ratio) but not with BMI, particularly in the deep frontal white matter and dorsal anterior cingulate gyrus of the left cerebral hemisphere. Notably, insulin demonstrated a significant negative impact on NAA ($\rho = -0.409$ and $\rho = -0.410$; $p < 0.01$) and Cho levels ($\rho = -0.403$ and $\rho = -0.392$; $p < 0.01$) at these locations in obese individuals. *Conclusions:* Central obesity and hyperinsulinemia negatively affect specific brain regions associated with cognitive and emotional processing, while BMI is not a reliable parameter for assessing brain metabolism.

Keywords: neurodegeneration; brain; neuroinflammation; metabolic syndrome; insulin; magnetic resonance spectroscopy

Citation: Vuković, M.; Nosek, I.; Slotboom, J.; Medić Stojanoska, M.; Kozić, D. Neurometabolic Profile in Obese Patients: A Cerebral Multi-Voxel Magnetic Resonance Spectroscopy Study. *Medicina* **2024**, *60*, 1880. https://doi.org/10.3390/medicina60111880

Academic Editors: Yuzuru Ohshiro, Kunimasa Yagi and Yasuhiro Maeno

Received: 26 September 2024
Revised: 5 November 2024
Accepted: 14 November 2024
Published: 16 November 2024

Copyright: © 2024 by the authors. Published by MDPI on behalf of the Lithuanian University of Health Sciences. Licensee MDPI, Basel, Switzerland. This article is an open access article distributed under the terms and conditions of the Creative Commons Attribution (CC BY) license (https://creativecommons.org/licenses/by/4.0/).

1. Introduction

Obesity presents a significant public health concern, with 300 million individuals classified as obese worldwide and over a billion people falling into the category of being overweight [1,2]. This condition is closely linked to a cluster of disorders known as metabolic syndrome [3], which is believed to be a common factor in various obesity-related illnesses due to its role in causing low-grade systemic inflammation that impacts multiple organs [4] and is connected to a heightened risk of dementia [5]. In cases of obesity, inflammatory reactions within the central nervous system, commonly known as neuroinflammation, have been observed in different areas such as the hypothalamus [6].

A study utilizing population-based MRI scans revealed that obesity and a high waist-to-hip ratio during middle age were linked to an increased likelihood of reduced brain volume and a decline in executive function a decade later. Furthermore, the metabolic obesity profile, marked by increased body fat, visceral adiposity, and systemic inflammation, was linked to a widespread reduction in gray matter volume [7]. Additionally, diabetes was associated with a faster increase in temporal horn volume, which serves as a surrogate marker for accelerated hippocampal atrophy [8]. Previous studies have indicated that greater weight and central obesity are correlated with diminished brain volume [9–12]. Furthermore, a higher body mass index (BMI) and waist circumference have been found to be significantly linked to a thinner cerebral cortex [13].

It is well known that magnetic resonance spectroscopy (MRS) brings additional information to conventional neuroimaging in clinical practice. This method can detect the neurometabolic profile of the lesion and its spatial heterogeneity [14]. Multivoxel MRS (mvMRS) is a sophisticated diagnostic modality in neuroscience that can reveal marked metabolic abnormalities in examinees with no morphological changes on conventional MR imaging, even in mutation carriers of genetic disorders [15]. Ostojic et al. found marked differences among the head, body, and tail of the hippocampus in healthy persons, compatible with different histologic characteristics of the aforementioned segments [16]. Multivoxel MRS can detect significant disturbances in neurometabolic ratios in various disorders, including traumatic brain injury, HIV, and other diseases, although morphologic changes were not evident on routine scanning [17].

It is imperative to comprehend the neurometabolic state of individuals who are obese but show no neurological symptoms. By utilizing MRS, we can non-invasively determine the biochemical profile of brain tissue, which provides insights into both its structural and functional abnormalities. This technique relies on measuring the absolute and relative concentrations of specific metabolites, namely N-acetylaspartate (NAA), choline-containing molecules (Cho), creatine (Cr), myoinositol (mI), glutamate, lactate, and lipid components [18,19].

Brain atrophy and reduced levels of NAA, particularly in the temporal lobes and hippocampus, are considered risk factors for cognitive decline and dementia in older individuals. N-acetylaspartate serves as an established indicator of neuronal viability, only present in mature neurons and their extensions. A decrease in its concentration signifies neuronal loss and dendritic and axonal atrophy. On the other hand, choline-containing compounds primarily play a role in the breakdown and synthesis of cell membranes. Metabolites containing Cr are involved in cellular bioenergetics, while mI is regarded as a marker for the number of glial cells (indicative of neuroinflammation) and an osmoregulator [1].

We analyzed and compared individuals with obesity to those with a normal BMI. We examined the absolute concentrations of NAA and Cho in relation to various anthropometric parameters such as BMI, waist circumference, waist-hip ratio, and body fat percentage in both groups. Additionally, we assessed these concentrations in relation to biochemical markers of metabolic syndrome in the obese participants, including triglycerides, HDL cholesterol, fasting blood glucose and insulin levels, as well as the insulin resistance index (HOMA-IR).

2. Materials and Methods

2.1. Participants in the Study

A total of 100 subjects were included in the study, divided into two groups:

1. The first group consisted of 50 obese persons, with a BMI of ≥ 30 kg/m^2 (25 men and women each), average age 43.32 years (22–62).
2. The second group consisted of 50 control subjects with BMI ≥ 18.5 and <25 kg/m^2 (25 men and women each), average age 43.22 years (23–62).

This study was a prospective cohort study that included participants who did not have any neurological deficits based on the Montreal Cognitive Assessment (MoCA) test [20]. The subjects in both groups were sex- and age-matched, aged between 20 and 65. Exclusion criteria for both groups included acute or chronic neurological disorders, mood disorders, anxiety [21,22], white matter lesions, and the use of medications affecting lipid levels, blood glucose, and insulin (glucocorticoids, statins, fibrates, oral hypoglycemics, and GLP-1 receptor agonists), as well as conditions such as Cushing's syndrome, acromegaly, and uncontrolled hypothyroidism.

Contraindications for an MRI exam included both absolute and relative factors. Absolute contraindications were implanted devices such as pacemakers (unless MRI-compatible), neurostimulators, cochlear implants, and metallic objects near the eyes. Relative contraindications included claustrophobia, first-trimester pregnancy, obesity beyond machine limits, and difficulty remaining still.

The study was approved by the institutional ethical review board, and all participants provided informed consent before joining the study.

2.2. Anthropometric Measurements

The height and weight of all participants in the study were measured, and the BMI was calculated (formula: weight/height^2 in kg/m^2). Additionally, waist circumference (WC) was measured, as well as the waist-hip ratio. Waist circumference was measured halfway between the lower rib and the iliac crest along the midaxillary line, while hip circumference was measured at the widest point over the greater trochanters. Furthermore, body fat percentage was measured by using a bioelectrical impedance body composition analyzer, TBF-300, Tanita, Tokyo, Japan.

2.3. Laboratory Tests and Other Measurements

All obese participants in the study underwent an analysis of biochemical indicators of metabolic syndrome (triglyceride, HDL cholesterol, blood glucose, and insulin with calculation of the HOMA-IR index). For insulin measurement, we used the chemiluminescent microparticle immunoassay (CMIA) method. Blood samples were taken from the cubital vein 12 h after fasting. The HOMA-IR index was calculated according to the formula HOMA-IR = (fasting blood glucose × fasting insulinemia)/22.5.

2.4. Neuroimaging

All subjects underwent conventional MR imaging with additional mvMRS on a 1.5 Tesla magnetic field device (Magnetom Avanto, Siemens, Erlangen, Germany) using a head coil. Conventional MR imaging consisted of T1W sagittal spin echo tomograms, T2W turbo spin echo transverse tomograms, FLAIR transverse tomograms, diffusion imaging (DWI), coronal T2W turbo spin echo tomograms, and 3D T1W MPR sagittal tomograms. Conventional imaging was used to exclude possible focal or diffuse white and gray matter lesions and to correctly position the region of interest for mvMRS (voxel grid).

Multivoxel MR spectroscopy of the supracallosal white and gray matter was performed in all subjects included in the study. Proton MR spectroscopy in the form of the Point RESolved Spectroscopy method (PRESS) was used to obtain data through spectroscopy, with TR/TE for long echo 1690/135 ms and for short echo 1690/30 ms. Section dimensions on CSI spectroscopic imaging were determined by the following parameters: FOV size was 160 mm × 160 mm × 160 mm, the volume of interest (VOI) was 80 mm × 80 mm × 80 mm, and section thickness was 10 mm. The voxel grid was placed directly above the corpus callosum in order to determine the spectra from the subcortical and deep white matter of the centrum semiovale of the frontal and parietal lobes, while the gray matter spectra were determined from the parafalx area of the anterior and posterior cingulate gyrus. The number of phase encoding steps was 16 in all directions (right-left, forward-backward, and up-down). The interpolation resolution was 16 in all directions, obtaining a VOI of 10 mm × 10 mm × 10 mm. To achieve the homogeneity of the magnetic field, automatic, volume-selective shimming was used. Non-water-suppressed CSI data were obtained with the same geometric parameters (1 average) to provide an internal water reference for the absolute quantification of metabolites. The amplitude of the water signal for each processed voxel was assessed from the scan without water suppression and used as an internal reference to calculate the absolute concentration of the metabolites. The SpectrImQMRS program, a tool for the combined analysis of MR spectroscopy and anatomy, was used to calculate the absolute concentrations of the investigated metabolites [23]. To process the spectra of metabolites, pre-processing was carried out, and the obtained values were expressed through the metabolite intensity or the area under the curve (Area) (Figure 1), and these data were further used to quantify the absolute values of the metabolites. A similar pre-processing procedure was also applied to the non-water-suppressed CSI spectra.

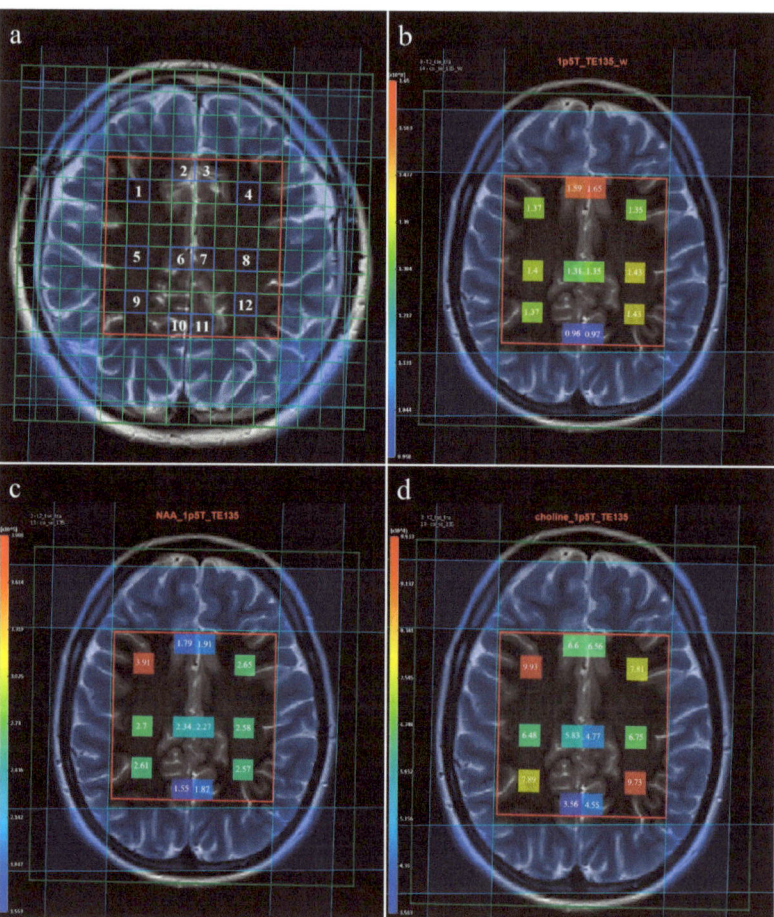

Figure 1. Processing of absolute metabolite concentrations of long-echo mvMRS: (**a**) numbered voxel of interest (VOI); (**b**) intensity of water at VOI; (**c**) intensity of NAA; (**d**) intensity of Cho.

In total, about 4800 spectra were analyzed. We calculated the ratios of metabolites NAA/Cr, Cho/Cr, and mI/Cr from the Leonardo workstation (Figures 2 and 3) and absolute concentrations of NAA at 2.0 parts per million (ppm) and Cho at 3.2 (ppm) using spectral fitting with the appropriate basis set model in the SpectrImQMRS program (Figures 4 and 5) [24]. Monoexponential T1 and T2 relaxation was assumed, and published values of T1 and T2 relaxation times of water and corresponding metabolites measured at 1.5 T in gray and white matter of healthy volunteers were used for relaxation corrections [25]. The obtained absolute concentrations were expressed in mmol/L.

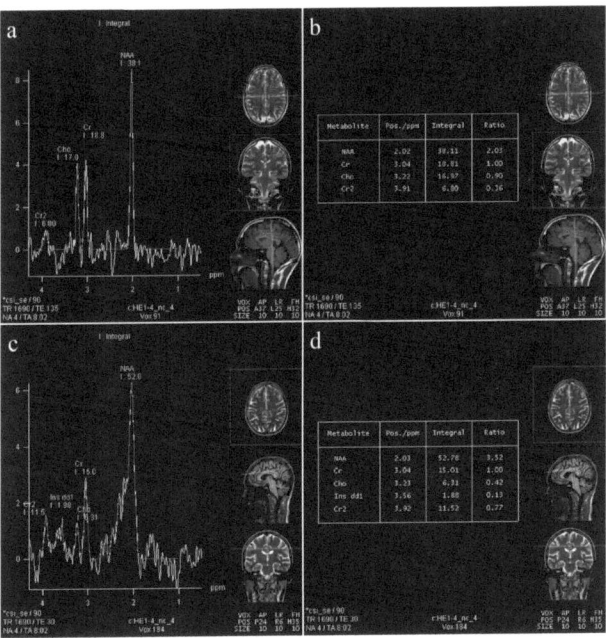

Figure 2. Normal spectrum and ratios in a control subject: (**a,b**) long-echo MRS in the left subcortical FWM (V4); (**c,d**) short-echo MRS of the right PCG (V10).

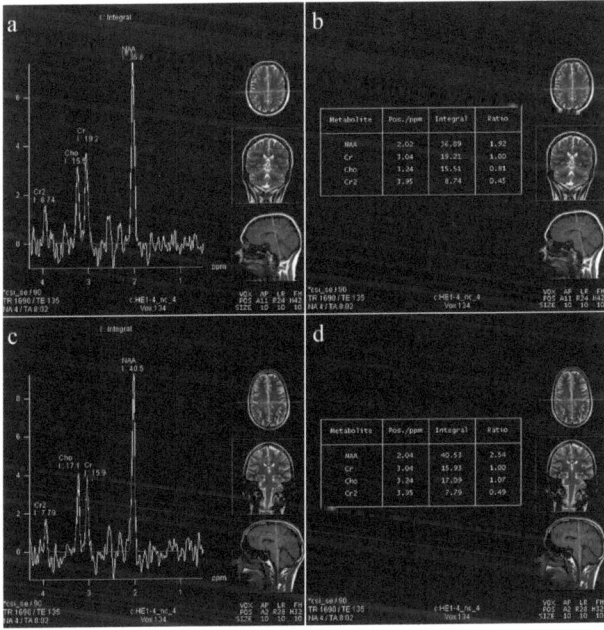

Figure 3. Long-echo mvMRS in the region of the right deep FWM (V5): spectrum (**a**) in an obese subject, showing a slight decrease in the NAA/Cr (1.92) and Cho/Cr ratio (0.81) (**b**); spectrum (**c**) in a control subject, showing a normal ratio of NAA/Cr (2.54) and Cho/Cr (1.07) (**d**).

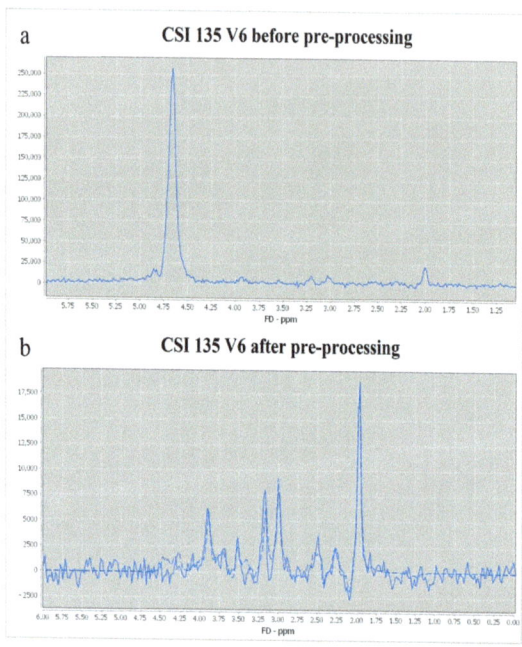

Figure 4. Determination of metabolite intensity through spectrImQMRS: (**a**)—spectrum from the posterior anterior cingulate gyrus (ACG) of the right cerebral hemisphere (V6) before pre-processing; (**b**)—the same spectrum after pre-processing and spectral fitting.

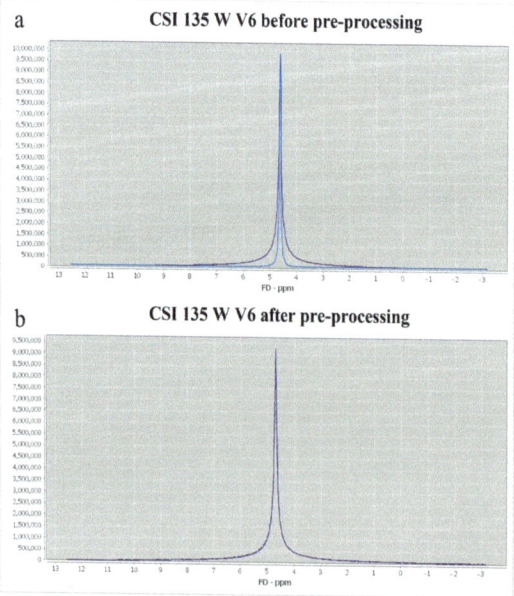

Figure 5. Determination of water intensity through spectrImQMRS: (**a**)—spectrum from the posterior anterior cingulate gyrus (ACG) of the right cerebral hemisphere (V6) without water suppression before pre-processing; (**b**)—the same spectrum after pre-processing and spectral fitting.

We analyzed spectra from 12 individual voxels (Figure 1a), namely: 1. subcortical frontal white matter (FWM) of the right cerebral hemisphere; 2. anterior cingulate gyrus (ACG) of the right cerebral hemisphere; 3. ACG of the left cerebral hemisphere; 4. subcortical FWM of the left cerebral hemisphere; 5. deep FWM of the right cerebral hemisphere; 6. the posterior ACG of the right cerebral hemisphere; 7. the posterior ACG of the left cerebral hemisphere; 8. deep FWM of the left cerebral hemisphere; 9. subcortical parietal white matter (PWM) of the right cerebral hemisphere; 10. posterior cingulate gyrus (PCG) of the right cerebral hemisphere; 11. PCG of the left cerebral hemisphere; 12. subcortical PWM of the left cerebral hemisphere.

2.5. Statistical Analysis

IBM SPSS software version 27.0 (Chicago, IL, USA) was used for statistical data processing. Since all subjects were imaged on the same MR machine, potential scanner-dependent differences between patients did not exist. Descriptive statistics were reported as the mean and standard deviation for variables that exhibited a normal distribution, while for those that did not conform to normality, the statistics were presented as the median and interquartile range. Determining the difference in neurobiochemical profile within the brain parenchyma between obese patients and controls was performed with the t-test or the Mann-Whitney U test depending on the normality of the distribution of the test variables, which was examined with the Kolmogorov Smirnov test. A value of 95% with a statistical significance level of $p < 0.05$ was taken as the confidence interval.

The examination of the relationship between the anthropometric parameters of all subjects, as well as the biochemical parameters of the metabolic syndrome of obese patients with the parameters in the neurobiochemical profile was analyzed with the Spearman correlation coefficient (ρ), because all anthropometric and biochemical parameters have abnormal distribution [26]. The correlation between age and parameters in the neurobiochemical profile of the participants was also examined using the Spearman correlation coefficient.

3. Results

3.1. Demographic, Anthropometric, and Biochemical Data

This prospective cohort study was conducted on 100 participants: 50 obese people with a BMI over 30 kg/m^2 and 50 people with a normal BMI, between 18.5 and 24.9 kg/m^2. An equal number of men and women were recruited in both groups (25 each), matching the participants for age and sex to exclude their influence on examined parameters (Table 1). The level of education can potentially influence the results of the examined parameters, so we recruited subjects with approximately similar educational levels between the groups ($\chi^2 = 1.412$, df = 1, $p = 0.235$).

Table 1. Age and anthropometric characteristics of all participants in the study.

	Groups	Mean (SD)	Median (IQR)	p
Age	Obese	43.32 (11.01)	N/A *	0.950
	Controls	43.22 (11.09)	N/A	
BMI	Obese	36.19 (4.33)	N/A	<0.001
	Controls	22.23 (1.39)	N/A	
Waist circumference (cm)	Obese	115.78 (10.44)	N/A	<0.001
	Controls	N/A	79.00 (8.50)	
Waist-to-hip ratio	Obese	0.94 (0.07)	N/A	<0.001
	Controls	0.85 (0.05)	N/A	
Body fat (%)	Obese	N/A	41.10 (13.02)	<0.001
	Controls	24.81 (6.77)	N/A	

* N/A—not applicable

The anthropometric measurements of all participants in the study are listed in Table 1, while Table 2 presents the biochemical indicators of metabolic syndrome for the obese participants.

Table 2. Biochemical indicators of metabolic syndrome in obese participants.

	Median	Interquartile Range (IQR)
Triglycerides (mmol/L)	1.37	0.90
HDL cholesterol (mmol/L)	0.99	0.33
Glucose (mmol/L)	4.95	0.88
Insulin (μU/mL)	10.30	4.70
HOMA-IR	2.20	1.25

3.2. Correlation Between Age and Neurometabolites

3.2.1. Relative Concentrations

Although the obese and control groups are homogeneous in terms of age, an examination of the association of metabolites obtained by mvMRS long and short echo with age was performed, considering that literature data indicate that the level of certain metabolites changes with age (primarily NAA and Cho).

The results show that NAA/Cr values decrease with age in 6 out of 12 examined voxels, both for long-echo mvMRS and short-echo mvMRS, with a smaller deviation in localizations. The degree of correlation in statistically significant voxels was weak ($\rho < 0.3$), except for V9 (right subcortical PWM) on the short echo, where it was moderate ($\rho = 0.347$, $p < 0.001$).

Regarding the Cho/Cr ratio at almost all observed locations, no statistically significant association with age was found, with the exception of V11 (left PCG) on the short echo, where it can be seen that the level of Cho/Cr increases with age and that this correlation is of moderate intensity ($\rho = 0.327$, $p < 0.001$).

3.2.2. Absolute Concentrations

The correlation of absolute NAA concentrations with age shows a statistically significant negative association at all voxels of interest, which confirms that NAA values decrease with age, independently of Cr, and that in half of the locations there is a strong correlation ($\rho > 0.4$).

Analyzing the correlation between age and absolute concentrations of Cho, a statistically significant negative correlation is observed in only three voxels: at voxels V4 and V11, weak correlation ($\rho < 0.3$), while at the V2 location it is moderate ($0.3 < \rho < 0.3.99$).

3.3. Comparison of Relative Concentrations of Metabolites Between Groups on Long-Echo MRS

3.3.1. NAA/Cr

The results in Table 3 show that there is no difference in the values of the NAA/Cr ratio between the groups in almost all examined locations, with the exception of V5 (right deep FWM), where a statistically significantly lower value of NAA/Cr is observed in obese subjects ($p < 0.001$) (Figure 3a,b) in comparison to controls (Figure 3c,d).

Table 3. Differences in NAA/Cr ratios obtained by the long-echo mvMRS between the groups.

	Groups	Mean (SD)	Median (IQR)	p
NAA/Cr 1	Obese	1.967 (0.431)	N/A	0.972
	Controls	1.970 (0.427)	N/A	
NAA/Cr 2	Obese	1.611 (0.364)	N/A	0.052
	Controls	1.467 (0.367)	N/A	

Table 3. Cont.

	Groups	Mean (SD)	Median (IQR)	p
NAA/Cr 3	Obese	1.532 (0.287)	N/A	0.936
	Controls	1.525 (0.466)	N/A	
NAA/Cr 4	Obese	1.949 (0.378)	N/A	0.127
	Controls	2.064 (0.370)	N/A	
NAA/Cr 5	Obese	2.219 (0.447)	N/A	<0.001
	Controls	2.586 (0.423)	N/A	
NAA/Cr 6	Obese	N/A	1.480 (0.44)	0.250
	Controls	1.609 (0.301)	N/A	
NAA/Cr 7	Obese	1.506 (0.259)	N/A	0.282
	Controls	1.560 (0.246)	N/A	
NAA/Cr 8	Obese	2.184 (0.452)	N/A	0.094
	Controls	2.324 (0.371)	N/A	
NAA/Cr 9	Obese	2.044 (0.306)	N/A	0.114
	Controls	1.949 (0.287)	N/A	
NAA/Cr 10	Obese	1.624 (0.307)	N/A	0.575
	Controls	1.663 (0.375)	N/A	
NAA/Cr 11	Obese	N/A	1.570 (0.33)	0.236
	Controls	1.651 (0.328)	N/A	
NAA/Cr 12	Obese	1.827 (0.366)	N/A	0.535
	Controls	N/A	1.815 (0.55)	

3.3.2. Cho/Cr

The results in Table 4 show the comparison of Cho/Cr ratio between the groups with no statistically significant difference in almost all examined locations with the exception of V5 (right deep FWM), where a statistically significantly lower value of Cho/Cr is observed in obese subjects ($p = 0.009$, $p < 0.01$) (Figure 3a,b).

Table 4. Differences in Cho/Cr ratios obtained by the long-echo mvMRS between the groups.

	Groups	Mean (SD)	Median (IQR)	p
Cho/Cr 1	Obese	N/A	1.210 (0.50)	0.331
	Controls	1.206 (0.239)	N/A	
Cho/Cr 2	Obese	N/A	1.120 (0.32)	0.155
	Controls	N/A	1.030 (0.43)	
Cho/Cr 3	Obese	1.094 (0.234)	N/A	0.942
	Controls	N/A	1.100 (0.31)	
Cho/Cr 4	Obese	1.254 (0.297)	N/A	0.406
	Controls	1.303 (0.294)	N/A	
Cho/Cr 5	Obese	N/A	1.105 (0.30)	0.009
	Controls	1.295 (0.281)	N/A	
Cho/Cr 6	Obese	N/A	1.045 (0.29)	0.735
	Controls	1.034 (0.212)	N/A	
Cho/Cr 7	Obese	1.028 (0.189)	N/A	0.424
	Controls	0.996 (0.209)	N/A	
Cho/Cr 8	Obese	1.260 (0.304)	N/A	0.817
	Controls	1.247 (0.243)	N/A	
Cho/Cr 9	Obese	1.077 (0.232)	N/A	0.169
	Controls	1.015 (0.213)	N/A	

Table 4. Cont.

	Groups	Mean (SD)	Median (IQR)	p
Cho/Cr 10	Obese	N/A	0.750 (0.22)	0.051
	Controls	0.857 (0.238)	N/A	
Cho/Cr 11	Obese	0.780 (0.271)	N/A	0.471
	Controls	0.816 (0.218)	N/A	
Cho/Cr 12	Obese	0.983 (0.231)	N/A	0.294
	Controls	1.036 (0.272)	N/A	

3.4. Comparison of Absolute Concentrations of NAA and Cho Between Groups on Long-Echo MRS

Based on the obtained results, there is no statistically significant difference in absolute concentrations of NAA and Cho between obese subjects and controls at any of the studied locations of interest.

3.5. Comparison of Relative Concentrations of Metabolites Between Groups on Short-Echo MRS

Based on the obtained results, we did not find any difference in the values of the NAA/Cr and mI/Cr ratios between obese subjects and controls in all the examined locations. As for Cho/Cr, the absence of a statistically significant difference is observed in almost all examined locations except for V6 (right posterior ACG), where a statistically significantly lower value of Cho/Cr is observed in obese subjects ($p = 0.037, p < 0.05$).

3.6. Correlation of Absolute Concentration of NAA and Cho with Anthropometric Parameters and Body Fat Percentage in Subjects of Both Groups on Long-Echo MRS

There is a weak negative correlation with waist circumference at V7 and V8 ($p = 0.021$), with waist-hip ratio at V6 ($p = 0.015$), V7 and V8 ($p = 0.017$), as well as with body fat percentage at V4 ($p = 0.023$). The obtained results show us that the parameters that more clearly reflect obesity (waist circumference and waist-hip ratio) have more influence on NAA values than BMI (Table 5).

Table 5. Correlation of absolute concentrations of NAA obtained by long-echo mvMRS with anthropometric parameters and body fat percentage.

Long-Echo		BMI	Waist Circumference	Waist-to-Hip Ratio	Body Fat (%)
NAA 1	ρ	0.079	0.053	−0.051	0.010
	p	0.450	0.617	0.624	0.921
	N	94	94	94	94
NAA 2	ρ	0.050	0.037	−0.023	0.033
	p	0.634	0.727	0.826	0.751
	N	94	94	94	94
NAA 3	ρ	0.023	0.086	0.125	0.086
	p	0.825	0.415	0.232	0.411
	N	94	94	94	94
NAA 4	ρ	−0.072	−0.029	0.071	−0.235 *
	p	0.495	0.783	0.497	0.023
	N	94	94	94	94
NAA 5	ρ	−0.085	−0.086	0.056	−0.188
	p	0.419	0.413	0.595	0.071
	N	94	94	94	94
NAA 6	ρ	−0.102	−0.172	−0.251 *	−0.044
	p	0.330	0.099	0.015	0.676
	N	94	94	94	94

Table 5. Cont.

Long-Echo		BMI	Waist Circumference	Waist-to-Hip Ratio	Body Fat (%)
NAA 7	ρ	−0.187	−0.239 *	−0.246 *	−0.089
	p	0.073	0.021	0.017	0.399
	N	94	94	94	94
NAA 8	ρ	−0.188	−0.240 *	−0.248 *	−0.088
	p	0.072	0.021	0.017	0.399
	N	94	94	94	94
NAA 9	ρ	−0.029	−0.065	−0.085	−0.084
	p	0.786	0.535	0.419	0.422
	N	94	94	94	94
NAA 10	ρ	−0.042	−0.113	−0.127	0.007
	p	0.691	0.282	0.226	0.945
	N	94	94	94	94
NAA 11	ρ	0.021	−0.051	−0.098	0.101
	p	0.843	0.628	0.350	0.337
	N	94	94	94	94
NAA 12	ρ	0.139	0.166	0.121	0.034
	p	0.183	0.111	0.249	0.745
	N	94	94	94	94

* $p = 0.05$

As for Cho, the results show there is a weak negative correlation only with the percentage of body fat at position V4 (left subcortical FWM) ($p = 0.016$).

3.7. Correlation of Absolute Concentrations of NAA and Cho with Biochemical Indicators of Metabolic Syndrome in Obese People on Long-Echo MRS

3.7.1. NAA

The results shown in Table 6 demonstrate that there is a strong negative correlation of the NAA concentrations with the insulin level at the V7 ($p = 0.006$) and V8 ($p = 0.007$) positions, as well as the HOMR-IR index level at the same locations ($p = 0.005$). A slightly contradictory result is the positive correlation of NAA values with insulin at the V4 position, but it is borderline weak to moderate intensity ($p = 0.047$).

Table 6. Correlation of absolute concentrations of NAA obtained by long-echo MRS with biochemical indicators of metabolic syndrome in obese people.

Long-Echo		Triglycerides	HDL Cholesterol	Glucose	Insulin	HOMA-IR
NAA 1	ρ	−0.069	0.092	−0.075	−0.060	−0.083
	p	0.660	0.559	0.634	0.704	0.595
	N	44	44	44	44	44
NAA 2	ρ	0.209	0.056	−0.251	−0.068	−0.119
	p	0.178	0.723	0.105	0.667	0.449
	N	44	44	44	44	44
NAA 3	ρ	0.057	0.005	−0.059	−0.017	−0.046
	p	0.718	0.975	0.708	0.913	0.770
	N	44	44	44	44	44
NAA 4	ρ	−0.058	0.086	0.187	0.305 *	0.300
	p	0.713	0.582	0.230	0.047	0.051
	N	44	44	44	44	44
NAA 5	ρ	0.187	−0.177	0.289	0.118	0.210
	p	0.231	0.257	0.060	0.453	0.176
	N	44	44	44	44	44

Table 6. Cont.

Long-Echo		Triglycerides	HDL Cholesterol	Glucose	Insulin	HOMA-IR
NAA 6	ρ	−0.058	0.011	−0.057	−0.124	−0.111
	p	0.714	0.945	0.718	0.430	0.478
	N	44	44	44	44	44
NAA 7	ρ	0.014	0.091	−0.222	−0.410	−0.422 **
	p	0.930	0.560	0.152	0.006	0.005
	N	44	44	44	44	44
NAA 8	ρ	0.014	0.090	−0.222	−0.409 **	−0.422 **
	p	0.931	0.565	0.152	0.007	0.005
	N	44	44	44	44	44
NAA 9	ρ	0.079	0.070	0.184	0.027	0.091
	p	0.613	0.655	0.238	0.865	0.560
	N	44	44	44	44	44
NAA 10	ρ	0.007	−0.157	0.136	0.023	0.077
	p	0.963	0.313	0.386	0.882	0.621
	N	44	44	44	44	44
NAA 11	ρ	0.087	−0.084	0.040	−0.200	−0.136
	p	0.578	0.592	0.800	0.198	0.385
	N	44	44	44	44	44
NAA 12	ρ	−0.054	−0.004	−0.163	−0.006	−0.061
	p	0.732	0.981	0.297	0.970	0.696
	N	44	44	44	44	44

* $p = 0.05$; ** $p = 0.01$

3.7.2. Cho

The results shown in Table 7 demonstrate that there is a negative correlation of a moderate degree of Cho value with the level of insulin at the position of V7 (left posterior ACG) ($p = 0.009$) and a strong degree at V8 (left deep FWM) ($p = 0.007$), as well as with the level of the HOMR-IR index at the same locations of a moderate degree ($p = 0.023$ and $p = 0.019$). From the previously presented results, it can be observed that NAA and Cho values are negatively correlated with insulin and the HOMA-IR index at V7 and V8, which indicates their influence on degenerative processes at those locations. The result, which is a little contradictory, is a positive correlation of a moderate intensity of Cho with triglycerides at the V5 position ($p = 0.036$).

Table 7. Correlation of absolute concentrations of Cho obtained by long-echo MRS with biochemical indicators of metabolic syndrome in obese people.

Long-Echo		Triglycerides	HDL Cholesterol	Glucose	Insulin	HOMA-IR
Cho 1	ρ	−0.078	−0.104	0.126	−0.055	−0.012
	p	0.619	0.506	0.420	0.726	0.938
	N	44	44	44	44	44
Cho 2	ρ	0.020	0.134	−0.139	−0.018	−0.037
	p	0.898	0.391	0.373	0.910	0.814
	N	44	44	44	44	44
Cho 3	ρ	0.158	0.067	−0.091	−0.001	−0.035
	p	0.310	0.667	0.562	0.993	0.823
	N	44	44	44	44	44
Cho 4	ρ	−0.117	−0.096	0.156	0.246	0.269
	p	0.455	0.539	0.318	0.111	0.081
	N	44	44	44	44	44

Table 7. Cont.

Long-Echo		Triglycerides	HDL Cholesterol	Glucose	Insulin	HOMA-IR
Cho 5	ρ	0.320	−0.252	0.073	−0.008	0.033
	p	0.036	0.102	0.643	0.958	0.836
	N	44	44	44	44	44
Cho 6	ρ	0.222	−0.109	−0.065	−0.072	−0.042
	p	0.152	0.487	0.678	0.647	0.791
	N	44	44	44	44	44
Cho 7	ρ	0.133	0.060	−0.095	−0.392 **	−0.346 *
	p	0.396	0.700	0.547	0.009	0.023
	N	44	44	44	44	44
Cho 8	ρ	0.129	0.074	−0.110	−0.403 **	−0.357 *
	p	0.411	0.638	0.483	0.007	0.019
	N	44	44	44	44	44
Cho 9	ρ	0.112	−0.026	0.046	−0.244	−0.166
	p	0.473	0.867	0.769	0.115	0.287
	N	44	44	44	44	44
Cho 10	ρ	0.078	−0.099	−0.067	0.006	−0.015
	p	0.617	0.527	0.668	0.970	0.923
	N	44	44	44	44	44
Cho 11	ρ	0.227	−0.088	0.026	−0.185	−0.142
	p	0.143	0.574	0.867	0.236	0.364
	N	44	44	44	44	44
Cho 12	ρ	0.017	0.007	−0.268	−0.007	−0.073
	p	0.916	0.966	0.082	0.964	0.640
	N	44	44	44	44	44

* $p = 0.05$; ** $p = 0.01$

4. Discussion

There are no studies in the literature that have examined the association of biochemical indicators of metabolic syndrome (triglyceride, HDL cholesterol, blood glucose, and insulin level, as well as the insulin resistance index (HOMA-IR)) in obese patients with cerebral metabolite concentrations obtained by the mvMRS. There are also no literature data that compare the concentrations of brain metabolites with the percentage of body fat.

There are various mechanisms by which obesity can impact brain health [27,28], particularly regarding the structure of gray matter [29]. First, obesity is associated with hypertension, diabetes, and hyperlipidemia, which are known causes of poor brain health. Second, obesity-associated low-grade chronic inflammation may induce neuroinflammation [30,31]. Different structures of the central nervous system could be affected, including the cerebral cortex, amygdala, cerebellum, and hypothalamus [32,33]. Brain involvement is most likely related to the disturbed blood-brain barrier, with consequent entry of pro-inflammatory cytokines to the brain parenchyma [34] and activation and proliferation of microglia and astrocytes. Microglial proliferation has been previously reported in neurodegenerative diseases and traumatic brain injury [35]. The obesity-related microglial proliferation induces synaptic remodeling, suppressing neurogenesis, leading to cognitive decline [36].

Analyzing the difference in the level of metabolites between obese subjects and controls, we found that only right deep FWM on long-echo MRS has statistically significantly lower values of NAA/Cr and Cho/Cr. Looking at this finding, the obtained results could indicate initial neurodegenerative changes, but a comparison of the absolute values of NAA and Cho between the examined groups did not show the presence of a statistically significant difference. Comparison of mI/Cr ratio between the obese subjects and controls did not find a statistically significant difference. Myoinositol is found mainly in astrocytes, and its high concentrations are interpreted as a reflection of glial proliferation or an increase in the size of glial cells; therefore, it is considered a marker of neuroinflammation [37], which indicates that in our study, there is an absence of a neuroinflammation state between the groups. In another study, elevated concentrations of myoinositol were reported in metabolic syndrome as well as in type 2 diabetes [38].

First of all, it should be noted that there is a negative correlation of NAA values with age, which is in agreement with the results of other studies. This data should be taken into account when analyzing other correlations in order to get a real picture of the relationship between the investigated parameters.

One of the main studies that compared BMI with metabolite values obtained on single-voxel MRS found that higher BMI was significantly associated with lower NAA concentration in frontal, parietal, and temporal white matter, with lower NAA concentration in frontal gray matter, and with lower concentration of choline in the frontal white matter. This study found no association of BMI with regional concentrations of creatine or myoinositol [1]. In contrast to the results of the mentioned study, in our research we did not find that the BMI of participants correlated with the absolute concentrations of NAA in any of the investigated locations. On the other hand, looking at other anthropometric parameters such as waist circumference and waist-hip ratio, we obtained a negative correlation with the level of NAA at positions V7 (left posterior ACG) and V8 (left deep FWM). However, these results were not followed by a correlation of Cho levels. From the above, we can conclude that BMI is not a true parameter that reflects real obesity, that waist circumference and waist-hip ratio better reflect central obesity, and that they correlate better with NAA values in the described locations. Additionally, the waist-hip ratio negatively correlates with NAA values at position V6 (right posterior ACG), and the percentage of body fat correlates negatively with NAA only at position V4 (left subcortical FWM). From these results, we see that the percentage of body fat is an unreliable parameter that reflects obesity, similar to BMI, and perhaps for this reason we did not find a correlation with the values of NAA and Cho.

Analysis of the correlation of biochemical indicators of metabolic syndrome (triglycerides, HDL cholesterol, blood glucose, and insulin with the HOMA-IR index) with absolute concentrations of NAA and Cho in obese individuals showed interesting results. Two locations in the brain parenchyma showed a strong negative correlation between NAA and Cho values with insulin levels and the HOMA-IR index, namely left posterior ACG and left deep FWM. If we look at the correlations with the anthropometric parameters, we see that significant correlations are in the same locations. Changes in NAA levels can be caused by an insulin disorder, i.e., the presence of insulin resistance or hyperinsulinemia, which often exists in obese people [39]. This insulin disorder leads to impaired insulin transport in the brain, which in turn leads to impaired glucose utilization in the brain, which can be associated with lower NAA values [40]. Previous studies have shown low concentrations of NAA in patients with glucose intolerance [41] or diabetes [19,41]. A study examining metabolite concentrations in metabolic syndrome found a decreased ratio of NAA/Cr and an increased ratio of Cho/Cr in the white matter of the frontal lobe in patients with metabolic syndrome compared to patients without it, with the changes being more pronounced in obese patients [42]. In adolescents with metabolic syndrome, the levels of NAA, choline, and myoinositol were significantly reduced in both hippocampi, especially in the right hippocampus [43]. Choline-containing compounds are primarily involved in the breakdown and synthesis of cell membranes, and their concentrations are elevated in type 2 diabetes [44].

One study that examined the reversibility of brain metabolite changes after intragastric balloon insertion found that in a combined group of obese people with and without diabetes, changes in the NAA/Cr ratio during the first 3 months after intragastric balloon insertion were inversely proportional to changes in body weight and BMI. Furthermore, patients with type 2 diabetes had an elevated mI/Cr ratio, which was found to normalize only 3 months after intragastric balloon insertion. This change was accompanied by a decrease in the Cho/Cr ratio. N-acetylaspartate concentrations, but not myoinositol, have been shown to be affected by diet [45,46].

Creatine (Cr) has long been considered a constant in the brain, and metabolite values are expressed through the relationship with Cr. In our study, the analysis of absolute concentrations shows that the only explanation for the mentioned difference in the NAA/Cr and Cho/Cr ratios would be an increase in Cr in obese people in the right deep FWM, which leads to lower metabolic ratios. The possible explanation for this phenomenon could be the generally increased energy and caloric intake in the obese and the consequent higher amount of glucose, i.e., energy in the brain. As Cr is considered to have a role in the storage and transfer of energy in metabolically active tissues, such as the brain, muscles, and heart, it is possible that in obese people there is an excess of Cr in the right deep FWM, i.e., the centrum semiovale, where white matter tracts transit, of which the most important is the corticospinal tract.

One of the important limitations of the study is the use of a 1.5 T MRI scanner, instead of 3 T, which may limit the sensitivity of our study due to the relatively low signal-to-noise ratio (SNR) of MRS at this field strength. This reduced SNR can impair the detection of metabolites and other subtle changes, potentially impacting the accuracy and reliability of the results. Furthermore, an examination of the level of intelligence for the participants in the study was not carried out, which could eventually lead to a difference between the metabolites in the brain, but care was taken to ensure that the obese and control groups did not differ in terms of their level of education. Potential confounding factors, such as hypertension, lifestyle habits, dietary patterns, and levels of physical activity, should be considered too, as they may influence the brain metabolism. Additionally, one of the limitations of the study is the absence of more detailed cognitive and behavioral testing, which would allow for a more comprehensive interpretation of the results, given that obesity has been shown to affect deficits in cognitive control [47].

5. Conclusions

This research is one of the rare prospective studies to observe the impact of obesity on several different locations in the brain, and we concluded that specific changes are noticed in the right deep FWM. However, when analyzing absolute concentrations of neurometabolites, it seems that they are not a reflection of neurodegenerative processes (lower NAA/Cr and Cho/Cr). Instead, there is most likely an increased energy metabolism in the obese primarily due to an increase in Cr in the white matter pathways. Central obesity (a higher waist circumference and waist-to-hip ratio) and hyperinsulinemia have negative effects in the context of neurodegeneration on the deep left FWM and the left posterior ACG, which is responsible for cognitive and emotional processing and which can eventually lead to Alzheimer's disease, while BMI is not a reliable parameter for assessing brain metabolism.

Future research should focus on clarifying the relationship between neurometabolic changes in obesity and a broader range of cognitive and behavioral assessments. Additionally, studies should expand neurometabolic assessments to other brain regions, particularly the hippocampus, and explore the connection between these neurometabolic alterations and potential microstructural and volumetric brain changes in individuals with obesity.

Author Contributions: Conceptualization, M.V. and D.K.; methodology, M.V., I.N., J.S., M.M.S. and D.K.; software, M.V., I.N. and J.S.; validation, M.V., I.N., J.S., M.M.S. and D.K.; formal analysis, M.V., J.S., M.M.S. and D.K.; investigation, M.V. and I.N.; resources, M.V., I.N. and D.K.; data curation, M.V.; writing—original draft preparation, M.V. and D.K.; writing—review and editing, M.V., J.S., M.M.S. and D.K.; visualization, M.V., M.M.S. and D.K.; supervision, M.M.S. and D.K.; project administration, D.K.; funding acquisition, M.V., I.N. and D.K. All authors have read and agreed to the published version of the manuscript.

Funding: This research was funded by the Provincial Secretariat for Higher Education and Scientific Research, Autonomous Province of Vojvodina, Republic of Serbia, under grant number 142-451-3524/2023.

Institutional Review Board Statement: The study was conducted in accordance with the Declaration of Helsinki and approved by the Ethics Committee of the Faculty of Medicine, University of Novi Sad, Serbia (Number: 01-39/259, date 15 September 2020).

Informed Consent Statement: Informed consent was obtained from all subjects involved in the study.

Data Availability Statement: The human data supporting the findings of this study are not openly available due to privacy and sensitivity concerns. They are available from the corresponding author upon request.

Acknowledgments: The authors express their deep gratitude to the late Dušanka Đurić-Nosek for her generous kindness and humanity, which made it possible to carry out this research at her magnetic resonance clinic, Mag-Medica. They also extend their sincere thanks to technician Nemanja Pavlović, whose expertise and constant support were essential in conducting all the examinations.

Conflicts of Interest: The authors declare no conflicts of interest.

References

1. Gazdzinski, S.; Kornak, J.; Weiner, M.W.; Meyerhoff, D.J. Body mass index and magnetic resonance markers of brain integrity in adults. *Ann. Neurol.* **2008**, *63*, 652–657. [CrossRef] [PubMed]
2. Jastreboff, A.M.; Kotz, C.M.; Kahan, S.; Kelly, A.S.; Heymsfield, S.B. Obesity as a disease: The Obesity Society 2018 position statement. *Obesity* **2019**, *27*, 7–9. [CrossRef] [PubMed]
3. Adelantado-Renau, M.; Esteban-Cornejo, I.; Rodriguez-Ayllon, M.; Cadenas-Sanchez, C.; Gil-Cosano, J.; Mora-Gonzalez, J.; Solis-Urra, P.; Verdejo-Román, J.; Aguilera, C.M.; Escolano-Margarit, M.V.; et al. Inflammatory biomarkers and brain health indicators in children with overweight and obesity: The ActiveBrains project. *Brain Behav. Immun.* **2019**, *81*, 588–597. [CrossRef] [PubMed]
4. Dekkers, I.A.; Jansen, P.R.; Lamb, H.J. Obesity, Brain Volume, and White Matter Microstructure at MRI: A Cross-sectional UK Biobank Study. *Radiology* **2019**, *291*, 763–771. [CrossRef] [PubMed]
5. Corlier, F.; Hafzalla, G.; Faskowitz, J.; Kuller, L.H.; Becker, J.T.; Lopez, O.L.; Thompson, P.M.; Braskie, M.N. Systemic inflammation as a predictor of brain aging: Contributions of physical activity, metabolic risk, and genetic risk. *NeuroImage* **2018**, *172*, 118–129. [CrossRef]
6. Parimisetty, A.; Dorsemans, A.-C.; Awada, R.; Ravanan, P.; Diotel, N.; Lefebvre d'Hellencourt, C. Secret talk between adipose tissue and central nervous system via secreted factors-an emerging frontier in the neurodegenerative research. *J. Neuroinflamm.* **2016**, *13*, 67. [CrossRef]
7. Beyer, F.; Kharabian Masouleh, S.; Kratzsch, J.; Schroeter, M.L.; Röhr, S.; Riedel-Heller, S.G.; Villringer, A.; Witte, A.V. A metabolic obesity profile is associated with decreased gray matter volume in cognitively healthy older adults. *Front. Aging Neurosci.* **2019**, *11*, 202. [CrossRef]
8. Debette, S.; Seshadri, S.; Beiser, A.; Au, R.; Himali, J.J.; Palumbo, C.; Wolf, P.; DeCarli, C. Midlife vascular risk factor exposure accelerates structural brain aging and cognitive decline. *Neurology* **2011**, *77*, 461–468. [CrossRef]
9. Gómez-Apo, E.; Mondragón-Maya, A.; Ferrari-Díaz, M.; Silva-Pereyra, J. Structural brain changes associated with over-weight and obesity. *J. Obes.* **2021**, *2021*, 6613385. [CrossRef]
10. Morys, F.; Tremblay, C.; Rahayel, S.; Hansen, J.Y.; Dai, A.; Misic, B.; Dagher, A. Neural correlates of obesity across the lifespan. *Commun. Biol.* **2024**, *7*, 656. [CrossRef]
11. Windham, B.G.; Lirette, S.T.; Fornage, M.; Benjamin, E.J.; Parker, K.G.; Turner, S.T.; Jack, C.R.; Griswold, M.E.; Mosley, T.H. Associations of Brain Structure with Adiposity and Changes in Adiposity in a Middle-Aged and Older Biracial Population. *J. Gerontol. A Biol. Sci. Med. Sci.* **2017**, *72*, 825–831. [CrossRef] [PubMed]
12. Debette, S.; Wolf, C.; Lambert, J.-C.; Crivello, F.; Soumaré, A.; Zhu, Y.-C.; Schilling, S.; Dufouil, C.; Mazoyer, B.; Amouyel, P.; et al. Abdominal obesity and lower gray matter volume: A Mendelian randomization study. *Neurobiol. Aging* **2014**, *35*, 378–386. [CrossRef] [PubMed]
13. Shaw, M.E.; Sachdev, P.S.; Abhayaratna, W.; Anstey, K.J.; Cherbuin, N. Body mass index is associated with cortical thinning with different patterns in mid- and late-life. *Int. J. Obes.* **2018**, *42*, 455–461. [CrossRef]
14. Kozic, D.; Medic-Stojanoska, M.; Ostojic, J.; Popovic, L.; Vuckovic, N. Application of MR spectroscopy and treatment approaches in a patient with extrapituitary growth hormone secreting macroadenoma. *Neuro Endocrinol. Lett.* **2007**, *28*, 560–564. [PubMed]
15. Ostojic, J.; Jancic, J.; Kozic, D.; Semnic, R.; Koprivsek, K.; Prvulovic, M.; Kostic, V. Brain white matter 1 H MRS in Leber optic neuropathy mutation carriers. *Acta Neurol. Belg.* **2009**, *109*, 305–309. [PubMed]
16. Ostojic, J.; Kozic, D.; Konstantinovic, J.; Covickovic-Sternic, N.; Mijajlovic, M.; Koprivsek, K.; Semnic, R. Three-dimensional multivoxel spectroscopy of the healthy hippocampus—Are the metabolic differences related to the location? *Clin. Radiol.* **2010**, *65*, 302–307. [CrossRef]
17. Boban, J.; Kozic, D.; Turkulov, V.; Lendak, D.; Bjelan, M.; Semnic, M.; Brkic, S. Proton chemical shift imaging study of the combined antiretroviral therapy impact on neurometabolic parameters in chronic HIV infection. *AJNR Am. J. Neuroradiol.* **2017**, *38*, 122–129. [CrossRef]

18. Lizarbe, B.; Campillo, B.; Guadilla, I.; López-Larrubia, P.; Cerdán, S. Magnetic resonance assessment of the cerebral al-terations associated with obesity development. *J. Cereb. Blood Flow. Metab.* **2020**, *40*, 2135–2151. [CrossRef]
19. Sinha, S.; Ekka, M.; Sharma, U.; Raghunandan, P.; Pandey, R.M.; Jagannathan, N.R. Assessment of changes in brain metabolites in Indian patients with type-2 diabetes mellitus using proton magnetic resonance spectroscopy. *BMC Res. Notes* **2014**, *7*, 41. [CrossRef]
20. Nasreddine, Z.S.; Phillips, N.A.; Bédirian, V.; Charbonneau, S.; Whitehead, V.; Collin, I.; Cummings, J.L.; Chertkow, H. The Montreal Cognitive Assessment, MoCA: A brief screening tool for mild cognitive impairment. *J. Am. Geriatr. Soc.* **2005**, *53*, 695–699. [CrossRef]
21. Zhang, X.; Han, L.; Lu, C.; McIntyre, R.S.; Teopiz, K.M.; Wang, Y.; Chen, H.; Cao, B. Brain structural and functional alterations in in-dividuals with combined overweight/obesity and mood disorders: A systematic review of neuroimaging studies. *J. Affect. Disord.* **2023**, *334*, 166–179. [CrossRef] [PubMed]
22. van der Kooij, M.A. The impact of chronic stress on energy metabolism. *Mol. Cell Neurosci.* **2020**, *107*, 103525. [CrossRef] [PubMed]
23. Pedrosa de Barros, N.; McKinley, R.; Knecht, U.; Wiest, R.; Slotboom, J. Automatic quality control in clinical ^{1}H MRSI of brain cancer. *NMR Biomed.* **2016**, *29*, 563–575. [CrossRef] [PubMed]
24. Near, J.; Harris, A.D.; Juchem, C.; Kreis, R.; Marjańska, M.; Öz, G.; Slotboom, J.; Wilson, M.; Gasparovic, C. Preprocessing, analysis and quantification in single-voxel magnetic resonance spectroscopy: Experts' consensus recommendations. *NMR Biomed.* **2021**, *34*, e4257. [CrossRef] [PubMed]
25. Minati, L.; Aquino, D.; Bruzzone, M.; Erbetta, A. Quantitation of normal metabolite concentrations in six brain regions by in-vivo1H-MR spectroscopy. *J. Med. Phys.* **2010**, *35*, 154–163. [CrossRef]
26. Dancey, C.P.; Reidy, J. *Statistics Without Maths for Psychology*, 7th ed.; Pearson: Harlow, UK, 2017.
27. Uranga, R.M.; Keller, J.N. The complex interactions between obesity, metabolism and the brain. *Front. Neurosci.* **2019**, *13*, 513. [CrossRef]
28. Karczewski, J.; Zielińska, A.; Staszewski, R.; Eder, P.; Dobrowolska, A.; Souto, E.B. Obesity and the brain. *Int. J. Mol. Sci.* **2022**, *23*, 6145. [CrossRef]
29. Opel, N.; Thalamuthu, A.; Milaneschi, Y.; Grotegerd, D.; Flint, C.; Leenings, R.; Goltermann, J.; Richter, M.; Hahn, T.; Woditsch, G.; et al. Brain structural abnormalities in obesity: Relation to age, genetic risk, and common psychiatric disorders: Evidence through univariate and multivariate mega-analysis including 6420 participants from the ENIGMA MDD working group: Evidence through univariate and multivariate mega-analysis including 6420 participants from the ENIGMA MDD working group. *Mol. Psychiatry* **2021**, *26*, 4839–4852. [CrossRef]
30. Woo, A.; Botta, A.; Shi, S.S.W.; Paus, T.; Pausova, Z. Obesity-related neuroinflammation: Magnetic resonance and mi-croscopy imaging of the brain. *Int. J. Mol. Sci.* **2022**, *23*, 8790. [CrossRef]
31. Ellulu, M.S.; Patimah, I.; Khaza'ai, H.; Rahmat, A.; Abed, Y. Obesity and inflammation: The linking mechanism and the complications. *Arch. Med. Sci.* **2017**, *13*, 851–863. [CrossRef]
32. Guillemot-Legris, O.; Muccioli, G.G. Obesity-induced neuroinflammation: Beyond the hypothalamus. *Trends Neurosci.* **2017**, *40*, 237–253. [CrossRef] [PubMed]
33. Salas-Venegas, V.; Flores-Torres, R.P.; Rodríguez-Cortés, Y.M.; Rodríguez-Retana, D.; Ramírez-Carreto, R.J.; Concepción-Carrillo, L.E.; Pérez-Flores, L.J.; Alarcón-Aguilar, A.; López-Díazguerrero, N.E.; Gómez-González, B.; et al. The obese brain: Mechanisms of systemic and local inflammation, and interventions to reverse the cognitive deficit. *Front. Integr. Neurosci.* **2022**, *16*, 798995. [CrossRef] [PubMed]
34. Van Dyken, P.; Lacoste, B. Impact of metabolic syndrome on neuroinflammation and the blood–brain barrier. *Front. Neurosci.* **2018**, *12*, 930. [CrossRef]
35. Perry, V.H.; Holmes, C. Microglial priming in neurodegenerative disease. *Nat. Rev. Neurol.* **2014**, *10*, 217–224. [CrossRef] [PubMed]
36. Zhou, Q.; Mareljic, N.; Michaelsen, M.; Parhizkar, S.; Heindl, S.; Nuscher, B.; Farny, D.; Czuppa, M.; Schludi, C.; Graf, A.; et al. Active poly-GA vaccination prevents microglia activation and motor deficits in a *C9orf72* mouse model. *EMBO Mol. Med.* **2020**, *12*, e10919. [CrossRef] [PubMed]
37. Gazdzinski, S.; Gaździńska, A.P.; Orzeł, J.; Redlisz-Redlicki, G.; Pietruszka, P.; Mojkowska, A.; Pacho, R.A.; Wylezol, M. Intragastric balloon therapy leads to normalization of brain magnetic resonance spectroscopic markers of diabetes in morbidly obese patients. *NMR Biomed.* **2018**, *31*, e3957. [CrossRef]
38. Haley, A.P.; Gonzales, M.M.; Tarumi, T.; Miles, S.C.; Goudarzi, K.; Tanaka, H. Elevated cerebral glutamate and myo-inositol levels in cognitively normal middle-aged adults with metabolic syndrome. *Metab. Brain Dis.* **2010**, *25*, 397–405. [CrossRef]
39. Craft, S. Insulin resistance and Alzheimer's disease pathogenesis: Potential mechanisms and implications for treatment. *Curr. Alzheimer Res.* **2007**, *4*, 147–152. [CrossRef]
40. O'Neill, J.; Eberling, J.L.; Schuff, N.; Jagust, W.; Reed, B.; Soto, G.; Ezekiel, F.; Klein, G.; Weiner, M.W. Method to correlate 1H MRSI and 18FDG-PET. *Magn. Reson. Med.* **2000**, *43*, 244–250. [CrossRef]
41. Sahin, I.; Alkan, A.; Keskin, L.; Cikim, A.; Karakas, H.M.; Firat, A.K.; Sigirci, A. Evaluation of in vivo cerebral metabolism on proton magnetic resonance spectroscopy in patients with impaired glucose tolerance and type 2 diabetes mellitus. *J. Diabetes Complicat.* **2008**, *22*, 254–260. [CrossRef]

42. El-Mewafy, Z.M.; Razek, A.A.K.A.; El-Eshmawy, M.M.; El-Eneen, N.R.A.; EL-Biaomy, A.A.B. Magnetic resonance spectroscopy of the frontal region in patients with metabolic syndrome: Correlation with anthropometric measurement. *Pol. J. Radiol.* **2018**, *83*, e215–e219. [CrossRef] [PubMed]
43. Sun, J.; Chen, P.; Bi, C. 1H-MRS technique and spectroscopic imaging LCModel based adolescent obese metabolic syndrome research. *Multimed. Tools Appl.* **2016**, *76*, 19491–19505. [CrossRef]
44. Wu, G.; Zhang, Q.; Wu, J.-L.; Jing, L.; Tan, Y.; Qiu, T.-C.; Zhao, J. Changes in cerebral metabolites in type 2 diabetes mellitus: A meta-analysis of proton magnetic resonance spectroscopy. *J. Clin. Neurosci.* **2017**, *45*, 9–13. [CrossRef] [PubMed]
45. Setkowicz, Z.; Gaździńska, A.; Osoba, J.J.; Karwowska, K.; Majka, P.; Orzeł, J.; Kossowski, B.; Bogorodzki, P.; Janeczko, K.; Wyleżoł, M.; et al. Does Long-Term High Fat Diet Always Lead to Smaller Hippocampi Volumes, Metabolite Concentrations, and Worse Learning and Memory? A Magnetic Resonance and Behavioral Study in Wistar Rats. *PLoS ONE* **2015**, *10*, e0139987. [CrossRef] [PubMed]
46. Auer, M.K.; Sack, M.; Lenz, J.N.; Jakovcevski, M.; Biedermann, S.V.; Falfán-Melgoza, C.; Deussing, J.; Steinle, J.; Bielohuby, M.; Bidlingmaier, M.; et al. Effects of a High-Caloric Diet and Physical Exercise on Brain Metabolite Levels: A Combined Proton MRS and Histologic Study. *J. Cereb. Blood Flow Metab.* **2015**, *35*, 554–564. [CrossRef]
47. Zapparoli, L.; Devoto, F.; Giannini, G.; Zonca, S.; Gallo, F.; Paulesu, E. Neural structural abnormalities behind altered brain activation in obesity: Evidence from meta-analyses of brain activation and morphometric data. *NeuroImage Clin.* **2022**, *36*, 103179. [CrossRef]

Disclaimer/Publisher's Note: The statements, opinions and data contained in all publications are solely those of the individual author(s) and contributor(s) and not of MDPI and/or the editor(s). MDPI and/or the editor(s) disclaim responsibility for any injury to people or property resulting from any ideas, methods, instructions or products referred to in the content.

Article

Lower Extremity Amputations Among Patients with Diabetes Mellitus: A Five-Year Analysis in a Clinical Hospital in Bucharest, Romania

Emilia Rusu [1], Eduard Lucian Catrina [2], Iulian Brezean [2], Ana Maria Georgescu [3], Alexandra Vișinescu [1], Daniel Andrei Vlad Georgescu [3], Chivu Anda Mioara [4], Grațiela Maria Dobra [4], Ioana Verde [5,*], Silviu Stanciu [6,7,*], Andrada Coșoreanu [1], Florin Rusu [7], Andra Nica [1], Doina Andrada Mihai [4,8] and Gabriela Radulian [4,8]

1. Department of Diabetes, Nutrition and Metabolic Diseases, "Carol Davila" University of Medicine and Pharmacy, Malaxa Clinical Hospital, 022441 Bucharest, Romania; emilia.rusu@umfcd.ro (E.R.); alexandra.visinescu94@drd.umfcd.ro (A.V.); andrada.cosoreanu@rez.umfcd.ro (A.C.); andra.nica@drd.umfcd.ro (A.N.)
2. Department of General Suregry, "Carol Davila" University of Medicine and Pharmacy, Cantacuzino Clinical Hospital, 030167 Bucharest, Romania; eduard.catrina@umfcd.ro (E.L.C.); iulian.brezean@umfcd.ro (I.B.)
3. Department of Diabetes, Nutrition and Metabolic Diseases, Nicolae Malaxa Clinical Hospital, 022441 Bucharest, Romania; ana-maria.militaru@rez.umfcd.ro (A.M.G.); daniel-andrei-vlad.georgescu@rez.umfcd.ro (D.A.V.G.)
4. Department of Diabetes, Nutrition and Metabolic Diseases, "Prof. Dr. Nicolae Paulescu" National Institute for Diabetes, Nutrition and Metabolic Diseases, 030167 Bucuresti, Romania; anda-mioara.chivu@rez.umfcd.ro (C.A.M.); gratiela-maria.hernest@drd.umfcd.ro (G.M.D.); andrada.mihai@umfcd.ro (D.A.M.); gabriela.radulian@umfcd.ro (G.R.)
5. Department of Internal Medicine, "Carol Davila" University of Medicine and Pharmacy, Theodor Burghele Clinical Hospital, 061344 Bucharest, Romania
6. "Carol Davila" University of Medicine and Pharmacy, 010825 Bucharest, Romania
7. "Doctor Carol Davila" Central Military University Emergency Hospital, 010825 Bucharest, Romania; florinrusumd@yahoo.com
8. Department of Diabetes, Nutrition and Metabolic Diseases, "Carol Davila" University of Medicine and Pharmacy, 030167 Bucharest, Romania
* Correspondence: ioana.verde@umfcd.ro (I.V.); silviu.stanciu@umfcd.ro (S.S.); Tel.: +40-740820004 (I.V.); +40-723590660 (S.S.)

Abstract: *Background and Objectives*: Lower extremity amputations (LEAs) represent a significant health problem. The aim of our study was to analyse the type and trends of diabetes-related LEAs in patients hospitalized in one surgical centre in Bucharest between 2018 and 2021. The second aim was to assess the impact of the COVID-19 pandemic on the trends of LEAs. *Materials and Methods*: We performed a retrospective analysis of all lower limb amputations performed between 01 January 2018 and 31 December 2021 in the Department of Surgery, Dr. I. Cantacuzino Clinical Hospital, Bucharest, Romania. We evaluated demographic parameters, type of LEA, the level, the laterality and trends of the amputations, the main aetiologies leading to amputation, and the length of hospitalization. *Results*: During the study period, 1711 patients underwent an LEA. The mean age was 64.53 ± 9.93 years, 71.6% ($n = 1481$) being over 60. Men outnumbered women by a ratio of 3.62:1. The most frequent interventions were ray amputations in 41.2% ($n = 705$) of patients; then, there were amputations of the toe (20.4%, $n = 349$), transtibial amputations (18.9%, $n = 323$), transfemoral amputations (10.6%, $n = 181$), and midfoot amputations (9%, $n = 154$). Wet gangrene was the most frequent aetiology (40.9%, $n = 699$). The total number of LEAs decreased constantly throughout the analysed period, such that 616 LEAs were performed in 2018 and 323 LEAs in 2021 ($p < 0.001$). There was a statistically significant increase in the rate of major LEAs in the pandemic vs. pre-pandemic period (37% vs. 24.4%, $p < 0.001$). *Conclusions*: In our study, the total number of LEAs decreased throughout the analysed period, but there was an increase in the rate of major LEAs in the pandemic vs. pre-pandemic period. Being over 65 years of age, leucocytosis, sepsis at presentation, and diabetic polyneuropathy were important risk factors for the necessity of LEA in complicated diabetes-related foot disease.

Keywords: lower extremity amputations; diabetes mellitus; diabetes-related foot disease

1. Introduction

Lower extremity amputations (LEAs) represent an important problem in the diabetic population, being associated with lower quality of life, premature mortality, and a significant economic burden for healthcare services. The mortality at 5 years after a diabetes-related LEA speaks for itself; it exceeds 70%. This excess of mortality seems to be caused by a combination of factors, such as associated conditions and complications of advanced diabetes, lack of physical activity, and deconditioning [1].

Diabetes-related foot disease (DFD) encompasses infection, ulceration, or destruction of tissues of the foot induced by neuropathy and/or peripheral artery disease in the lower extremities of a patient with diabetes mellitus. Lower extremity amputation represents the most severe form of this spectrum, which also includes callus or ulcer of the foot or lower limb, peripheral angiopathy with or without gangrene, cellulitis, osteomyelitis, mono/polyneuropathy, or neuropathic arthropathy [2,3].

Indeed, DFD is a major public health problem, as in patients with diabetes, its prevalence varies between 4.6 and 15.1%. Moreover, DFD is the leading cause for nontraumatic lower limb amputation in the developed world, and the global annual incidence of diabetes-related minor and major amputations during the 2010–2020 period was estimated to be, respectively, 139.97 and 94.82 cases/100,000 people with diabetes [4–8].

The most important contributing factors to LEAs in the diabetic population are diabetic foot ulcers and low extremity peripheral arterial disease, both being parts of DFD. At least half of the foot ulcers become infected and, afterward, up to 20% of significantly infected ulcers impose the need for LEA [1].

Prevention and early recognition of aggravation should play a central role in DFD management, being essential to observe early any risk factor for a possible evolution to ulceration or amputation, such as foot deformity, peripheral neuropathy with loss of protective sensation, preulcerative callus or corn, peripheral arterial disease, poor glycaemic control, visual impairment, diabetic nephropathy, cigarette smoking, or an anterior history of ulcers [9]. Increased awareness of possible evolution to amputation would favour timely guidance to a specialized multidisciplinary team in order to perform preventive and curative management of diabetic foot disease aimed at reducing the amputation rate [10]. By bridging the gap between research findings and their clinical application, this study hypothesizes that diabetes complications and metabolic disturbances contribute to the progression of diabetes-related foot disease toward amputations.

Our study aimed to analyse the type and trends of diabetes-related LEAs in patients hospitalized in one surgical centre in Bucharest between 2018 and 2021. The second aim was to assess whether there was a significant change in the number and type of amputations (major vs. minor) during the pandemic period (2020–2021) compared to the pre-pandemic period (2018–2019) and to identify the contributing factors to any observed differences, such as delayed access to healthcare, increased incidence of severe infections, or other pandemic-related barriers.

2. Materials and Methods

2.1. Study Design, Location, and Period

We performed a retrospective analysis of all lower limb amputations performed between 1 January 2018 and 31 December 2021 in the Department of Surgery, Dr. I. Cantacuzino Clinical Hospital, Bucharest, Romania's capital and largest city with approximately 2 million inhabitants. This analysis included all adult patients with diabetes undergoing a lower extremity amputation (LEA). We evaluated demographic parameters, type of LEA, the level and laterality of amputation and trends of the amputations performed, the presence of any chronic complication of diabetes, risk factors (hypertension, obesity, dyslipidaemia), main aetiologies leading to the indication for amputation, and the length of hospitalization. Ethical approval for this study was obtained from the Ethics Committee of the Dr. I. Cantacuzino Clinical Hospital (190/15 Jan 2024).

2.2. Study Population

The inclusion criteria were all patients over 18 years with diabetes (type 1 diabetes mellitus T1DM, type 2 diabetes mellitus T2DM) who underwent an LEA between 2018 and 2021 in the Department of Surgery, Dr. I. Cantacuzino Clinical Hospital. Patients without diabetes who underwent amputations during the analysed period were excluded from the study. All clinical data were collected from the clinical report form each patient's clinical report form.

LEAs were classified as major and minor. The surgical removal of only a part or multiple parts of the lower limb proximal to the ankle joint represents a major lower extremity amputation. A minor lower extremity amputation was described as any LEA distal to the ankle joint.

2.3. Variables

We obtained demographic parameters (e.g., age, gender) and clinical data such as the type of LEA, the level and laterality of amputation (amputation of the toe; amputation of the toe, including the metatarsal bone; mediotarsal amputation; transmetatarsal amputation; ankle disarticulation; transmaleolar amputation of the tibia and fibula; above the knee amputation; below the knee amputation; hip amputation), main aetiologies leading to the indication for amputation, the presence of hypertension, obesity, dyslipidaemia, any chronic complication of diabetes (diabetic polyneuropathy (DPN), peripheral arterial disease (PAD), chronic kidney disease (CKD) or end-stage renal disease (ESRD), diabetic retinopathy) and medical treatment, the associated antibiotic therapy and the postoperative complications, previous vascular intervention, and length of hospitalization recorded from the clinical case report.

Paraclinical laboratory tests at admission, including white blood cell (WBC) counts, C-reactive protein (CRP), erythrocyte sedimentation rate (ESR), fasting plasma glucose (FPG), cholesterol, triglycerides, HDL-cholesterol, and renal function (creatinine, estimated glomerular filtration rate (GFRe)) were analysed.

Hypertension, dyslipidemia (hypercholesterolemia, hypertriglyceridemia, or hypo-HDL cholesterolemia) were diagnosed according to the American Diabetes Association Standards of Care for 2024. The Friedewald equation was applied to estimate the LDL-cholesterol level as follows: estimated LDL-C = [total cholesterol] − [total HDL] − [estimated very low-density lipoprotein (VLDL)]. VLDL level was calculated by dividing the total triglycerides level by 5. There was no direct LDL-C testing performed. Diabetes mellitus (DM) was described using the standard criteria (blood glucose levels, HbA1c, or the use of antidiabetic treatment) [9]. Obesity was defined using a body mass index (BMI) exceeding 30 kg/m^2. The use of tobacco was described as present or absent. Leukocytosis was defined as a number of leukocytes in peripheral blood over >11×10^9/L.

2.4. Statistical Analysis

Qualitative variables are presented as percentages and mean ± standard deviation (SD) for continuous normally distributed data or as median (interquartile range) for continuous abnormally distributed data. We used the Kolmogorov–Smirnov with a Lilliefors significance correction and Shapiro–Wilk statistic to assess the normality of our data. Chi-square tests were used for categorical variables, independent t-tests were used for continuous normally distributed data, and Mann–Whitney U tests were used for nonnormally distributed data. A *p*-value less than 0.05 was considered significant. Data collected were analysed using Statistical Package for Social Software (SPSS) version 19. The cohort was divided according to the year of admission and the type of amputation. All variables with a *p*-value < 0.05 in the univariate analysis were included in the multivariate logistic regression models using the backward stepwise method, and an odds ratio (OR) with a 95% confidence interval was calculated.

3. Results

3.1. General Characteristics of the Population

During the study period, between January 2018 to December 2021, 1711 patients underwent an LEA. The most frequent interventions were ray/rays amputation ($n = 705$); then, in order of frequency, the following were performed: amputation of the toe (20.4%, $n = 349$), transtibial amputation 18.9% ($n = 323$), transfemoral amputation (10.6%, $n = 181$), and transmetatarsal amputation (9%, $n = 154$). Wet gangrene was the most frequent aetiology (40.9%, $n = 699$), followed by superinfected ulcers (21.8%, $n = 373$) and ischemia (11.9%, $n = 204$). The mean age was 64.53 ± 9.93 (range 26–93) years, 71.6% ($n = 1481$) being over 60 (Table 1). Men (78.4%, $n = 1341$) outnumbered women by a ratio of 3.62:1. The majority of patients were with T2DM (98.3%, $n = 1682$), and the median duration of diabetes was 13 years (CI 95% 13.55–15.02). A total of 38.8% ($n = 664$) were current smokers. August to October was the period with the highest LEAs. The most commonly isolated microorganisms were *Staphylococcus aureus* (28.1%, $n = 71$, 17% MSSA ($n = 43$) and 11.1% MRSA ($n = 28$)), Group D Streptococci (14.6%, $n = 37$), and *Proteus Mirabilis* (11.9%, $n = 28$).

Table 1. General characteristics of the population that underwent an LEA.

Variables		2018	2019	2020	2021	Total	p *	p **
Age (years)	Total	64.034 ± 10.35	64.32 ± 10.17	64.88 ± 9.83	65.19 ± 9.43	64.5 ± 9.94	<0.001	ns
	Male	63.400 ± 9.63	63.81 ± 9.73	64.409 ± 9.7	64.510 ± 8.77	63.9 ± 9.53		ns
	Female	66.31 ± 11.14	66.301 ± 11.6	67.029 ± 9.92	67.750 ± 11.25	66.7 ± 11.02		ns
Gender	Male	465 (75.6%)	330 (79.9%)	291 (81.1%)	255 (78.9%)	1341 (78.4%)		ns
	Female	150 (24.4%)	83 (20.1%)	68 (18.9%)	68 (21.1%)	369 (21.6%)		ns
Environment	Rural	227 (37%)	144 (35%)	120 (33.4%)	118 (36.6%)	609 (35.7%)		ns
	Urban	387 (63%)	268 (65%)	239 (66.6%)	204 (63.4%)	1098 (64.3%)		ns
Type of diabetes	1	10 (1.6%)	5 (1.2%)	8 (2.2%)	6 (1.9%)	29 (1.7%)	ns	ns
	2	602 (98.4%)	407 (98.8%)	350 (97.8%)	316 (98.1%)	1675 (98.3%)	ns	ns
Smoking	Total	235 (38.1%)	140 (33.9%)	156 (43.5%)	133 (41.2%)	664 (38.8%)	0.04	0.04
	Male	210 (45.2%)	128 (38.8)	138 (27.4%)	117 (45.9%)	593 (44.2%)	ns	
	Female	25 (16.6%)	12 (14.5%)	18 (26.5%)	16 (23.5%)	71 (19.2%)	ns	
Duration of diabetes (years) #		14.93 ± 9.82 13 (12)	14.27 ± 9.17 13 (12)	13.30 ± 9.91 11.5 (15)	13.64 ± 8.98 12 (13)	14.29 ± 9.54 13 (12)	<0.001	ns
Length of stay (days) #		6 (4)	5 (4)	5 (3)	5 (3)	5 (3)	ns	0.001

The data have been presented as mean ± standard deviation (SD) for continuous normally distributed data or as median and interquartile range (marked with #). p *—between gender, p **—between years; ns, statistically nonsignificant.

A total of 35.4% ($n = 606$) underwent recurrent LEAs on the same limb, and 21.9% on the contralateral limb, 6.3% ($n = 107$) being major LEAs. Only 3.3% ($n = 57$) of participants had a history of revascularization procedures. The length of stay ranged from 1 to 71 days with an average of 6.26 days (CI95% 6.05–6.46), with significantly longer durations in the case of major LEAs (6.61 days (CI95% 6.31–6.9) vs. 6.11 days (5.85–6.38), $p = 0.032$). Upon admission, the patients presented hyperglycaemia, leucocytosis, and inflammatory syndrome (higher C-reactive protein) (Table 2).

The characteristics of the population are described in Table 1.

Table 2. Laboratory parameters at admission.

	2018		2019		2020		2021		p
	Median	Inter-quartile range	Median	Inter-quartile range	Median	Inter-quartile range	Median	Inter-quartile range	
FPG (mg/dL)	189.5	138	174	117	187	166	200	151	ns
Serum creatinine (mg/dL)	0.96	0.64	0.92	0.61	0.95	0.63	1.02	0.66	ns
eGFR (mL/min/1.73 mp)	79	49.75	81	49	81	50	72.5	46.75	ns
Leucocyte (10×10^9 /L)	12	7	12	7	13	8	14	8	<0.001
C-reactive Protein (mg/dL)	147.5	119.5	163.5	191.25	117	151	137.5	135.48	ns

Abbreviations: FPG, fasting plasma glucose; eGFR, estimated glomerular filtration rate. The data have been presented as median and inter-quartile range. ns, statistically nonsignificant.

3.2. Major Amputations

Approximately one-third of all LEAs were major (29.4%, $n = 503$). The total number of LEAs decreased during the period of observation; 616 LEAs were performed in 2018, and 323 LEAs in 2021 ($p < 0.001$) (Table 3). Even though numerically the number of major amputations decreased (Figure 1), the rate of major LEAs increased in 2020 and 2021 (Figure 2). The ratio of major amputations to minor was initially 0.41 and had an upward trend. We also noted a rising trend for the proportion of major amputations in both men ($p < 0.001$) and women ($p = 0.004$).

Table 3. Number of LEAs during the study period, stratified by gender.

Gender	Type of LEAs	2018	2019	2020	2021	Total
Male	Major	115 (24.7%)	69 (20.9%)	96 (33%)	95 (37.3%)	375 (28%)
	Minor	350 (75.3%)	261 (79.1%)	195 (67%)	160 (62.7%)	966 (72%)
Female	Major	48 (31.8%)	19 (22.9%)	27 (39.7%)	34 (50%)	128 (34.6%)
	Minor	103 (68.2%)	64 (77.1%)	41 (60.3%)	34 (50%)	242 (65.4%)
Total	Major	163 (26.5%)	88 (21.3%)	123 (34.3%)	129 (39.9%)	503 (29.4%)
	Minor	453 (73.5%)	325 (78.7%)	236 (65.7%)	194 (60.1%)	1208 (70.6%)
All LEAs		616	413	359	323	1711

Abbreviations: LEA, lower extremity amputation. Variables are presented as percentages.

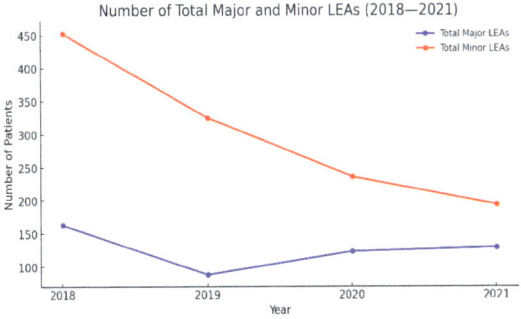

Figure 1. Absolute number of major and minor LEAs.

The most common was transtibial amputation at 18.9% ($n = 323$) (without gender differences). The number of transtibial amputations increased by 1.7 times between 2018 and 2021. This was followed by transfemoral amputations at 10.5% ($n = 180$); the proportion of transfemoral amputations increased by 1.18 times over the years. The majority of patients who underwent major LEAs were male, with an M/F ratio of 2.92, diagnosed with PAD (peripheral arterial disease) (75.1%, *n* = 378), PNP (polyneuropathy) (44.1%, *n* = 222), CKD (chronic kidney disease) (35%, *n* = 176), obesity (46.1%, *n* = 232), hypertension (76.5%, *n* = 385), and 39.1% current smokers (Table 4).

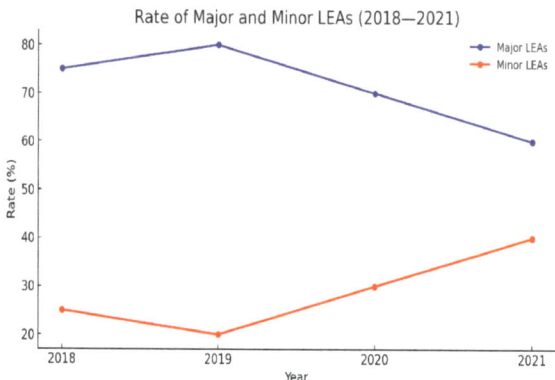

Figure 2. Rate of major and minor LEAs.

Table 4. Characteristics of patients with major and minor LEAs.

	Major LEAs (*n* = 503, 29.4%)	Minor LEAs (*n* = 1208, 70.6%)	*p*
Age (years)	66.57 ± 9.39	63.68 ± 10.03	ns
Gender, male	375 (74.6)	966 (80%)	0.014
Gender, female	128 (25.4)	242 (20%)	0.014
Type 1 diabetes	10 (2%)	19 (1.6%)	0.541
Type 2 diabetes	493 (98%)	1189 (98.4%)	ns
Smoking (*n*, %)	199 (39.6%)	465 (38.5%)	0.703
Hypertension (*n*, %)	385 (76.5%)	899 (74.4%)	0.391
Obesity (*n*, %)	232 (46.1%)	645 (53.4%)	0.001
Dyslipidaemia (*n*, %)	150 (29.8%)	366 (30.3%)	0.862
PAD (*n*, %)	378 (75.1%)	616 (51%)	<0.001
DPN	222 (44.1%)	748 (61.9%)	<0.001
PAD + DPN	160 (31.8%)	361 (29.9%)	0.454
CKD	176 (35%)	359 (29.7%)	0.034
ESRD	53 (10.5%)	51 (4.2%)	0.02
DR	46 (9.1%)	108 (8.9%)	0.926
Charcot foot	32 (6.4%)	26 (2.2%)	0.001
History of revascularisation	13 (2.6%)	44 (3.6%)	0.303
Main causes: Wet gangrene/Ischemia/Infected ulcers	233 (46.3%) 100 (19.9%) 51 (12%)	466 (38.6%) 104 (8.6%) 322 (26.7%)	
SGLT2 inhibitors	8 (1.6%)	37 (3.1%)	0.097

Abbreviations: LEA, lower extremity amputation; PAD, peripheral arterial disease; DPN, diabetic polyneuropathy; CKD, chronic kidney disease; ESRD, end-stage renal disease; DR, diabetic retinopathy. Variables are presented as percentages.

3.3. Minor Amputations

Between 2018 and 2021 were carried out 1208 minor LEAs (70.6%). For minor LEAs, the trend was descending for both genders (Table 3). The most common was ray amputation (including the toe and corresponding metatarsal bone) at 41.2% (*n* = 705). Minor LEAs decreased from 453 procedures in 2018 (73.5% of total LEAs) to 194 procedures in 2021 (60.1% of total LEAs). The patients who underwent minor LEAs were predominantly male (80%, *n* = 966), with an M/F ratio of 3.99, diagnosed with PNP (61.9%, *n* = 748), PAD (51%,

n = 616), CKD (29.7%, n = 359), obesity (53.4%, n = 645), hypertension (74.4%, n = 899), with a high proportion of smoking (38.5%, n = 465) (Table 4).

3.4. Age

The mean age at admission increased from 64.034 ± 10.35 years in 2018 to 65.19 ± 9.43 years in 2021 (p = 0.03), being higher in those with major amputations (66.57 ± 9.39 years versus 63.68 ± 10.03 years, p < 0.001), both in women and men.

Overall, both minor and major amputations belonged to the age group 60–69 years (major LEAs n = 206, 41%; minor LEAs n = 530, 43.9%) (Figure 3); the second most prevalent age group was 70–79 years in women (31.1%, n = 115) and 50–59 years for men (24.2%, n = 324) (Figure 3).

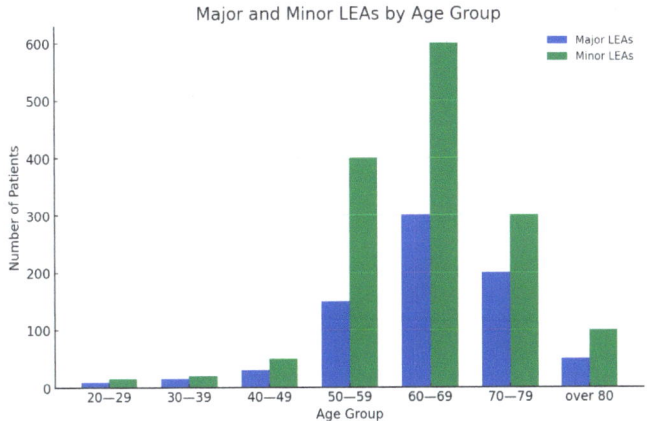

Figure 3. LEAs stratified by age group and type of amputation.

3.5. Smoking

Current smoking status was noticed in 38.8% (n = 664) of the included subjects; the percentage of smokers increased year after year; significantly more men than women were smoking (44.2% versus 19.2%, p = 0.001). Similar proportion of smokers was found both in major (39.6% (n = 199)) and minor amputations (38.5%, n = 465). There were no differences remarked between smokers and nonsmokers regarding the type of intervention.

3.6. Type of Diabetes

There were twenty-nine patients with T1DM, and 34.5% (n = 10) underwent major amputations. In patients with T2DM, 29.3% (n = 493) suffered major amputations, and 70.7% (n = 1208) had minor amputations (p = 0.541). Amputation of the toe and the metatarsal bone was the most common minor LEA in T1DM (n = 11, 57.9%) and T2DM (n = 694, 58.35%). Transtibial amputation was the most frequent major LEA, both in type 1 and type 2 diabetes; in T1DM patients, there were six cases (60%), and in T2DM people, 316 (64.09%).

3.7. The Complications of Diabetes Mellitus

Peripheral arterial disease (PAD) (n = 994, 58.1%) and diabetic polyneuropathy (DPN) (n = 970, 56.7%) were the most prevalent diabetes-related complications in patients who underwent LEAs. Chronic kidney disease (CKD) was present in 31.3% (n = 535), and 9% had end-stage renal disease (ESRD) (n = 154). An increased proportion of patients had associated cardiovascular risk factors such as hypertension (n = 1284, 75%), obesity (n = 877, 51.3%), and dyslipidaemia (n = 516, 30.2%) (Table 5).

Table 5. Complications and comorbidities stratified by year and gender.

Complications and Comorbidity		2018 No. of Patients	(%)	2019 No. of Patients	(%)	2020 No. of Patients	(%)	2021 No. of Patients	(%)	Total No. of Patients	(%)
DPN	Total	343	55.70%	225	54.50%	196	54.60%	206	63.80%	970	56.70%
	Men	270	58.10%	184	55.80%	161	55.30%	167	65.50%	782	58.30%
	Women	73	48.30%	41	49.40%	35	51.50%	39	57.40%	188	50.80%
PAD	Total	361	58.60%	214	51.80%	225	62.70%	194	60.10%	994	58.10%
	Men	266	57.20%	165	50.00%	184	63.20%	152	59.60%	767	57.20%
	Women	95	62.90%	49	59.00%	41	60.30%	42	61.80%	227	61.40%
DPN + PAD	Total	192	31.20%	96	23.20%	116	32.30%	117	36.20%	521	30.50%
	Men	143	30.80%	76	23.00%	98	33.70%	93	36.50%	410	30.60%
	Women	49	32.50%	20	24.10%	18	26.50%	24	35.30%	111	30.00%
CKD	Total	178	29.10%	127	30.80%	118	32.90%	111	34.40%	535	31.30%
	Men	133	28.60%	107	32.40%	93	32.00%	89	34.90%	422	31.50%
	Women	46	30.50%	20	24.10%	25	36.80%	22	32.40%	113	30.50%
ESRD	Total	59	9.60%	41	9.90%	31	8.60%	23	7.10%	154	9.00%
	Men	36	7.70%	29	8.80%	22	7.60%	15	5.90%	102	7.60%
	Women	23	15.20%	12	14.50%	9	13.20%	8	11.80%	52	14.10%
Hypertension	Total	455	73.90%	292	70.70%	280	78.00%	257	79.60%	1284	75.00%
	Men	331	71.20%	227	68.80%	218	74.90%	197	77.30%	973	72.60%
	Women	124	82.10%	65	78.30%	62	91.20%	60	88.20%	311	84.10%
Dyslipidaemia	Total	137	22.20%	111	26.90%	119	33.10%	149	46.10%	516	30.20%
	Men	106	22.80%	92	27.90%	94	32.30%	118	46.30%	410	30.60%
	Women	31	20.50%	19	22.90%	25	36.80%	31	45.60%	106	28.60%
BMI over 30 kg/m^2	Total	294	47.70%	193	46.70%	200	55.70%	190	58.80%	877	51.30%
	Men	229	49.20%	150	45.50%	156	53.60%	152	59.60%	687	51.20%
	Women	65	43.00%	43	51.80%	44	64.70%	38	55.90%	190	51.40%

Abbreviations: LEA, lower extremity amputation; DPN, diabetic polyneuropathy; PAD, peripheral arterial disease; CKD, chronic kidney disease; ESRD, end-stage renal disease; BMI, body mass index. Variables are presented as percentages.

In univariate analysis, factors associated with major LEAs were age over 70, gender, sepsis, obesity, DPN, PAD, CKD, and leucocytosis. In multivariate analysis, patients with diabetes and sepsis, diabetic polyneuropathy, age over 65, and leucocytosis were more likely to undergo major amputation (Table 6). For minor amputation, in multivariate logistic regression, age over 65 years, sepsis, diabetic polyneuropathy, and leucocytosis could be considered independent risk factors (Table 6).

Table 6. Factors associated with LEAs.

	B	S.E.	p	OR	95% C.I. Lower	95% C.I. Upper
Major LEAs						
Sepsis	0.851	0.142	<0.001	2.342	1.772	3.094
DPN	0.597	0.117	<0.001	1.817	1.445	2.285
Leucocytosis	0.081	0.011	<0.001	1.084	1.061	1.107
Age over 65 years	0.032	0.006	<0.001	1.032	1.020	1.045
Minor LEAs						
Sepsis	−1.021	0.139	<0.001	0.360	0.274	0.473
PAD	1.015	0.127	<0.001	2.760	2.154	3.537
DPN	−0.560	0.116	<0.001	0.571	0.455	0.717
Leucocytosis	0.708	0.126	<0.001	2.029	1.586	2.596
Age over 65 years	0.348	0.137	0.011	1.416	1.084	1.851

Abbreviations: LEA, lower extremity amputation; DPN, diabetic polyneuropathy; PAD, peripheral arterial disease. B, standard beta coefficient; SE, standard error; OR, odds ratio; CI, confidence interval.

The total number of LEAs decreased significantly in the pandemic period (682 in pandemic era vs. 1029 in pre-pandemic period). This decrease was obtained mainly because of the reduction of minor LEAs (778 pre-pandemic vs. 430 in the pandemic era), while the number of major LEAs remained approximately constant (251 vs. 252). However, there was a statistically significant increase in the rate of major LEAs in the pandemic vs. pre-pandemic period (37% vs. 24.4%, $p < 0.001$). There were significantly higher percentages of transtibial and transfemoral amputations in the pandemic versus pre-pandemic period (24.9% vs. 14.8% and 12% vs. 9.6%, respectively). The patients who needed amputations in pandemic era had leucocytosis (55.6% to 46.5%, $p < 0.001$) and sepsis (28.9% to 17.5%, $p < 0.001$) more frequently than the patients who needed amputations in the pre-pandemic era (Table 7).

Table 7. Number and type of LEAs in pre-pandemic vs. pandemic period.

	Pre-Pandemic	Pandemic	p
T2DM	1014 (98.5%)	668 (97.9%)	0.347
Minor LEAs	778 (75.6%)	430 (63%)	<0.001
Major LEAs	251 (24.4%)	252 (37%)	<0.001
Total LEAs	1029 (60.1%)	682 (39.9%)	<0.001
Sepsis	180 (17.5%)	197 (28.9%)	<0.001
Leukocytosis	478 (46.5%)	379 (55.6%)	<0.001
Toe amp	225 (21.9%)	124 (18.2%)	ns
TMT and toe	466 (45.3%)	239 (35%)	<0.001
TMT amp	87 (8.5%)	67 (9.8%)	ns
Transtibial amp	152 (14.8%)	170 (24.9%)	<0.001
Transfemoral amp	99 (9.6%)	82 (12%)	<0.001

Abbreviations: T2DM, type 2 diabetes mellitus; LEA, lower extremity amputation; TMT, transmetatarsal; amp, amputation. ns, statistically nonsignificant.

4. Discussion

In our four-year retrospective study, between January 2018 and December 2021, 1711 patients underwent an LEA, and 29.4% of all LEAs were major ($n = 503$). Our data are similar with the statistics published for Romania between 2015 and 2019, when major amputation represented 28.87% of the LEA-affected individuals [11].

Most of our patients were in their 6th decade, with a mean age of 64.53 ± 9.93 (range 26–93) years, similar to other studies [8,11].

Our results showed that men are disproportionately affected by LEAs, with a ratio of 3.62 (1341 vs. 370, $p = 0.014$), consistent with results from extensive studies available in the literature. Ezzatvar et al. published a meta-analysis that provided global estimates of diabetes-related amputation incidence from 2010 to 2020; the incidence of LEAs in males was approximately two-fold higher than in women [8]. In another cohort, men exhibited a higher prevalence of LEAs compared to women, irrespective of whether they had type 1 or type 2 diabetes mellitus [12].

There are several hypotheses for this difference. On the one side, the risk factors for LEA seem to be more prevalent in men: diabetic foot ulcers, peripheral artery disease, or tobacco consumption [8]. Our data likewise showed that more men were smoking (44.2% versus 19.2%, $p = 0.001$). On the other hand, in women of reproductive age, oestrogens could have a vascular and neural protective effect [13].

In our analysis, septic complications were most frequently the reason for LEA, wet gangrene and infected ulcers causing together 62.7% of the indications for LEA. Unlike other large studies that named infected ulcers as the leading cause for LEAs [8,14], in our analysis, wet gangrene (40.9%, $n = 699$) was the most frequent cause, with infected ulcers accounting for only 21.8% of the cases. One explanation may be that our patients were addressed later to the healthcare services, in more advanced stages of the disease.

The risk factors identified in our study are consistent with other results published in the literature [15–17]. In a meta-analysis including 6000 patients with diabetic foot infections, the predictors for LEAs were identified as male gender, smoking habits, a previous amputation, peripheral arterial disease, diabetic retinopathy, osteomyelitis, severe infections, gangrene/necrosis, and advanced scores in the International Working Group on the Diabetic Foot (IWGDF) (grades 3 or 4) and Wagner classifications (grades 4 or 5) [15]. Additionally, inflammatory markers such as leucocytosis, elevated CRP, and erythrocyte sedimentation rate levels were associated with an increased risk of amputation [15].

In another meta-analysis, which included 21 studies with 6505 participants, PAD and infection severity were among the strongest predictors of LEA [16]. Male sex, smoking, and a history of foot ulcers or amputations were also linked to increased risk [16]. Co-morbidities, including osteomyelitis, neuropathy, lower body mass index, and delayed wound healing, were highlighted as contributing factors for LEAs [16]. Early detection and management of PAD and infection, as well as addressing modifiable factors like smoking, could significantly reduce the incidence of amputations and emphasize the importance of targeted interventions and risk stratification in preventing amputations among patients with DFUs [16].

In a more recent meta-analysis that included 9934 subjects, previously identified risk factors for lower extremity amputations (LEAs) were reconfirmed [17]. Male sex and older age were associated with a higher prevalence of LEAs [17]. Although, age and the type of diabetes mellitus were not identified as significant risk factors for lower extremity amputation in individuals with diabetic foot ulcers [17]. Smoking cessation and improved wound care, especially in cases of infection or gangrene, were emphasized as critical for reducing LEA risk [17].

The total number of LEAs decreased constantly throughout the analysed period; 616 LEAs were performed in 2018 and 323 LEAs in 2021 ($p < 0.001$). The total number of LEAs decreased mainly by a reduction in minor LEAs (778 pre-pandemic vs. 430 in the pandemic era), the number of major LEAs remaining approximately constant (251 vs. 252). However, we observed an increasing tendency in the rate of major LEAs among males and females compared to minor LEAs, which was statistically significant ($p < 0.001$). This could support the idea that the access of DFD-affected patients to healthcare services during the pandemic era could have been hampered, as from the patients needing amputation procedures, more individuals needed major LEAs, meaning that they ended up receiving treatment in a more advanced stage of disease. This idea could also be supported by the fact that the patients who needed amputations in the pandemic era had leucocytosis (55.6% to 46.5%, $p < 0.001$) and sepsis (28.9% to 17.5%, $p < 0.001$) more frequently than the amputees in the pre-pandemic era. The hypothesis of impaired access of DFD-affected patients to healthcare services caused by the COVID-19 pandemic is also supported by other authors [18,19].

Disparities seem to be caused not only by impaired access to healthcare services due to the COVID-19 pandemic but also by limited access to complex, expensive medical procedures in our country. For instance, in the Netherlands, the majority (70%) of people with major LEAs had a history of endovascular revascularisation before amputation [20]. On the contrary, only a small proportion of our cohort (3.3%, $n = 57$) had access to revascularization procedures before being subject to LEA. Revascularization, both endovascular and surgical, is a first-class recommendation for limb salvage in patients with chronic limb-threatening ischaemia [21].

A study analysing global variations in amputation rates, using data from 12 countries (primarily European), highlighted that countries with lower gross domestic product (GDP) and healthcare expenditures, such as Hungary and Slovakia, reported the highest rates of major amputations [22,23].

The strengths of this study are the inclusion of a large number of patients treated in a tertiary care centre from the largest city of Romania, a significant analysed period (4 years), including 2 pre-pandemic years and 2 pandemic years, and a direct comparison between the pre-pandemic and pandemic period.

The limitations of our study would be the lack of data regarding quality of life, the impact of different medications in the magnitude of necessary LEAs, possible other disparities significantly involved (such as socioeconomic or ethnic disparities), the severity of diabetes in LEA-affected individuals (possibly quantified by HbA1c, duration of diabetes, or intensity of default antidiabetic treatment), a possible correlation between specific risk factors and repeated need for LEA in the same patient, or the single-centre nature of the study.

This paper brings to attention new data regarding LEAs caused by diabetes mellitus, highlights the risk factors involved in the unfortunate evolution of diabetes-related foot disease, discusses the impact of the COVID-19 pandemic upon diabetes-affected individuals, and could be the basis of future studies aimed at reducing the burden of diabetes.

5. Conclusions

LEAs, in general, and in diabetes-affected individuals in particular, remain an important public health issue, with immense emotional implications for patients and a significant burden for health and social services. This study highlights the ongoing public health challenge posed by lower extremity amputations (LEAs) in patients with diabetes mellitus, exacerbated during the COVID-19 pandemic. In our study, a positive trend was observed, as the total number of LEAs constantly decreased throughout the analysed period. Despite a reduction in the total number of LEAs over the study period, major amputations increased significantly during the pandemic, likely reflecting delays in access to care and advanced disease stages at presentation. Identified key risk factors, including age over 65 years, leucocytosis, sepsis, and diabetic polyneuropathy, underscore the need for early detection and intervention in diabetic foot disease. Further studies, including more years of surveillance and more patients from more centres, could give us a better picture of the trends in amputation rates in diabetic foot disease.

Author Contributions: Conceptualization, E.R., I.B., D.A.M. and G.R.; Methodology, A.M.G., S.S. and A.N.; Software, F.R.; Validation, E.R., E.L.C., I.B., I.V. and A.C.; Formal analysis, A.M.G., C.A.M., G.M.D., A.C. and F.R.; Investigation, A.V. and A.N.; Resources, E.L.C., A.V., D.A.V.G., C.A.M., G.M.D. and F.R.; Data curation, A.M.G., D.A.V.G. and C.A.M.; Writing—original draft, E.R., A.V., D.A.V.G., A.C. and F.R.; Writing—review and editing, E.R., E.L.C., I.V., S.S. and G.R.; Visualization, S.S. and A.N.; Supervision, I.B., I.V., D.A.M. and G.R.; Project administration, I.V., S.S., D.A.M. and G.R. All authors have read and agreed to the published version of the manuscript.

Funding: This research was funded by the University of Medicine and Pharmacy Carol Davila, through the institutional program "Publish not Perish".

Institutional Review Board Statement: The study was conducted in accordance with the Declaration of Helsinki, and approved by the Ethical Commission of the National Institute of Diabetes, Nutrition and Metabolic Diseases (Approval code: 190 and Approval date: 15 January 2024).

Informed Consent Statement: Patient consent was waived due to the fact that this was a retrospective registry study.

Data Availability Statement: The original contributions presented in the study are included in the article; further inquiries can be directed to the corresponding authors.

Conflicts of Interest: The authors declare no conflict of interest.

References

1. Armstrong, D.G.; Boulton, A.J.M.; Bus, S.A. Diabetic Foot Ulcers and Their Recurrence. *N. Engl. J. Med.* **2017**, *376*, 2367–2375. [CrossRef]
2. Ahmed, M.U.; Tannous, W.K.; Agho, K.E.; Henshaw, F.; Turner, D.; Simmons, D. Social determinants of diabetes-related foot disease among older adults in New South Wales, Australia: Evidence from a population-based study. *J. Foot Ankle Res.* **2021**, *14*, 65. [CrossRef]
3. Tekale, S.; Varma, A.; Tekale, S.; Kumbhare, U. A Review on Newer Interventions for the Prevention of Diabetic Foot Disease. Cureus [Internet]. 22 October 2022. Available online: https://www.cureus.com/articles/119931-a-review-on-newer-interventions-for-the-prevention-of-diabetic-foot-disease (accessed on 12 June 2024).

4. Zhang, P.; Lu, J.; Jing, Y.; Tang, S.; Zhu, D.; Bi, Y. Global epidemiology of diabetic foot ulceration: A systematic review and meta-analysis. *Ann. Med.* **2017**, *49*, 106–116. [CrossRef]
5. Zhang, Y.; Lazzarini, P.A.; McPhail, S.M.; Van Netten, J.J.; Armstrong, D.G.; Pacella, R.E. Global Disability Burdens of Diabetes-Related Lower-Extremity Complications in 1990 and 2016. *Diabetes Care* **2020**, *43*, 964–974. [CrossRef] [PubMed]
6. Lazzarini, P.A.; Hurn, S.E.; Kuys, S.S.; Kamp, M.C.; Ng, V.; Thomas, C.; Jen, S.; Kinnear, E.M.; D'Emden, M.C.; Reed, L. Direct inpatient burden caused by foot-related conditions: A multisite point-prevalence study. *BMJ Open* **2016**, *6*, e010811. [CrossRef] [PubMed]
7. Lazzarini, P.A.; Hurn, S.E.; Kuys, S.S.; Kamp, M.C.; Ng, V.; Thomas, C.; Jen, S.; Wills, J.; Kinnear, E.M.; D'Emden, M.C.; et al. The silent overall burden of foot disease in a representative hospitalised population. *Int. Wound J.* **2017**, *14*, 716–728. [CrossRef] [PubMed]
8. Ezzatvar, Y.; García-Hermoso, A. Global estimates of diabetes-related amputations incidence in 2010–2020: A systematic review and meta-analysis. *Diabetes Res. Clin. Pract.* **2023**, *195*, 110194. [CrossRef] [PubMed]
9. American Diabetes Association. Standards of Care in Diabetes–2024. *Diabetes Care* **2024**, *47* (Suppl. S1), S1–S322. [CrossRef]
10. Musuuza, J.; Sutherland, B.L.; Kurter, S.; Balasubramanian, P.; Bartels, C.M.; Brennan, M.B. A systematic review of multi-disciplinary teams to reduce major amputations for patients with diabetic foot ulcers. *J. Vasc. Surg.* **2020**, *71*, 1433–1446.e3. [CrossRef]
11. Rusu, E.; Coman, H.; Coșoreanu, A.; Militaru, A.-M.; Popescu-Vâlceanu, H.-C.; Teodoru, I.; Mihai, D.-A.; Elian, V.; Gavan, N.A.; Radulian, G. Incidence of Lower Extremity Amputation in Romania: A Nationwide 5-Year Cohort Study, 2015–2019. *Medicina* **2023**, *59*, 1199. [CrossRef] [PubMed]
12. Coman, H.; Stancu, B.; Gâvan, N.A.; Bowling, F.L.; Podariu, L.; Bondor, C.I.; Radulian, G. Diabetes-Related Lower Extremity Amputations in Romania: Patterns and Changes between 2015 and 2019. *Int. J. Environ. Res. Public Health* **2022**, *20*, 557. [CrossRef] [PubMed]
13. Somani, Y.B.; Pawelczyk, J.A.; De Souza, M.J.; Kris-Etherton, P.M.; Proctor, D.N. Aging women and their endothelium: Probing the relative role of estrogen on vasodilator function. *Am. J. Physiol. Heart Circ. Physiol.* **2019**, *317*, 395–404. [CrossRef] [PubMed]
14. Perez-Favila, A.; Martinez-Fierro, M.L.; Rodriguez-Lazalde, J.G.; Cid-Baez, M.A.; Zamudio-Osuna, M.D.J.; Martinez-Blanco, M.D.R.; Mollinedo-Montaño, F.E.; Rodriguez-Sanchez, I.P.; Castañeda-Miranda, R.; Garza-Veloz, I. Current Therapeutic Strategies in Diabetic Foot Ulcers. *Medicina* **2019**, *55*, 714. [CrossRef]
15. Sen, P.; Demirdal, T.; Emir, B. Meta-analysis of risk factors for amputation in diabetic foot infections. *Diabetes Metab. Res. Rev.* **2019**, *35*, e3165. [CrossRef]
16. Lin, C.; Liu, J.; Sun, H. Risk factors for lower extremity amputation in patients with diabetic foot ulcers: A meta-analysis. *PLoS ONE* **2020**, *15*, e0239236. [CrossRef]
17. Zhang, H.; Huang, C.; Bai, J.; Wang, J. Effect of diabetic foot ulcers and other risk factors on the prevalence of lower extremity amputation: A meta-analysis. *Int. Wound J.* **2023**, *20*, 3035–3047. [CrossRef]
18. Caruso, P.; Longo, M.; Signoriello, S.; Gicchino, M.; Maiorino, M.I.; Bellastella, G.; Chiodini, P.; Giugliano, D.; Esposito, K. Diabetic Foot Problems During the COVID-19 Pandemic in a Tertiary Care Center: The Emergency Among the Emergencies. *Diabetes Care* **2020**, *43*, e123–e124. [CrossRef]
19. Viswanathan, V.; Nachimuthu, S. Major Lower-Limb Amputation During the COVID Pandemic in South India. *Int. J. Low Extrem. Wounds* **2023**, *22*, 475–479. [CrossRef] [PubMed]
20. Rosien, L.; van Dijk, P.R.; Oskam, J.; Pierie, M.E.; Groenier, K.H.; Gans, R.O.; Bilo, H.J. Lower Extremity Amputation Rates in People with Diabetes Mellitus: A Retrospective Population-Based Cohort Study in Zwolle Region, The Netherlands. *Eur. J. Vasc. Endovasc. Surg.* **2023**, *66*, 229–236. [CrossRef]
21. Kolossváry, E.; Björck, M.; Behrendt, C.A. A Divide between the Western European and the Central and Eastern European Countries in the Peripheral Vascular Field: A Narrative Review of the Literature. *J. Clin. Med.* **2021**, *10*, 3553. [CrossRef]
22. Mazzolai, L.; Teixido-Tura, G.; Lanzi, S.; Boc, V.; Bossone, E.; Brodmann, M.; Bura-Rivière, A.; De Backer, J.; Deglise, S.; Corte, A.D.; et al. ESC Scientific Document Group, 2024 ESC Guidelines for the management of peripheral arterial and aortic diseases: Developed by the task force on the management of peripheral arterial and aortic diseases of the European Society of Cardiology (ESC). *Eur. Heart J.* **2024**, *45*, 3538–3700. [CrossRef] [PubMed]
23. Behrendt, C.A.; Sigvant, B.; Szeberin, Z.; Beiles, B.; Eldrup, N.; Thomson, I.A.; Venermo, M.; Altreuther, M.; Menyhei, G.; Nordanstig, J.; et al. International Variations in Amputation Practice: A VASCUNET Report. *Eur. J. Vasc. Endovasc. Surg.* **2018**, *56*, 391–399. [CrossRef] [PubMed]

Disclaimer/Publisher's Note: The statements, opinions and data contained in all publications are solely those of the individual author(s) and contributor(s) and not of MDPI and/or the editor(s). MDPI and/or the editor(s) disclaim responsibility for any injury to people or property resulting from any ideas, methods, instructions or products referred to in the content.

Article

Robot-Assisted Approach to Diabetes Care Consultations: Enhancing Patient Engagement and Identifying Therapeutic Issues

Yuya Asada [1,†], Tomomi Horiguchi [1,†], Kunimasa Yagi [2,3,*,†], Mako Komatsu [4], Ayaka Yamashita [4], Ren Ueta [4], Naoto Yamaaki [5], Mikifumi Shikida [4], Shuichi Nishio [6] and Michiko Inagaki [1]

[1] Faculty of Health Sciences, Institute of Medical, Pharmaceutical and Health Sciences, Kanazawa University, Kanazawa 920-1192, Japan; y-asada@staff.kanazawa-u.ac.jp (Y.A.); horiguchi@mhs.mp.kanazawa-u.ac.jp (T.H.); ja9xbh@yahoo.co.jp (M.I.)
[2] Department of Internal Medicine, Kanazawa Medical University Hospital, Uchinada 920-0293, Japan
[3] First Department of Internal Medicine, Toyama University Hospital, Toyama 930-0152, Japan
[4] School of Informatics, Kochi University of Technology, Kochi 780-8515, Japan; komatsu.21151@gmail.com (M.K.); 260373c@ugs.kochi-tech.ac.jp (A.Y.); 285089e@gs.kochi-tech.ac.jp (R.U.); shikida.mikifumi@kochi-tech.ac.jp (M.S.)
[5] Isobe Clinic, Ichihara 290-0511, Japan; isobe_dm_clinic@yahoo.co.jp
[6] Institute for Open and Transdisciplinary Research Initiatives, Osaka University, Osaka 565-0871, Japan; nishio@botransfer.org
* Correspondence: yagikuni@icloud.com
† These authors contributed equally to this work.

Abstract: *Background and Objectives*: Diabetes is a rapidly increasing global health challenge compounded by a critical shortage of diabetes care and education specialists. Robot-assisted diabetes care offers a cost-effective and scalable alternative to traditional methods such as training and dispatching human experts. This pilot study aimed to evaluate the feasibility of using robots for diabetes care consultations by examining their ability to elicit meaningful patient feedback, identify therapeutic issues, and assess their potential as substitutes for human specialists. *Materials and Methods*: A robot-assisted consultation programme was developed by selecting an appropriate robot, designing the programme content, and tailoring back-channel communication elements. Experienced diabetes care nurses operated the robot during the consultations. Patient feedback was collected through a 17-item questionnaire using a five-point Likert scale (evaluating functionality, impressions, and effects). Additionally, a five-item questionnaire was used to assess whether the programme helped patients reflect on the key therapeutic domains of diabetes knowledge, diet, exercise, medications, and blood glucose control. *Results*: This study included 32 participants (22 males; mean age, 69.7 ± 12.6 years; mean HbA1c, 7.2 ± 1.0%). None of the participants reported any discomfort during the consultation. Sixteen of the seventeen feedback items scored above the median of 3, as did all five therapeutic reflection items. The interview content analysis revealed the programme's ability to differentiate patients facing issues in treatment compliance from those effectively managing their condition. Robots can elicit valuable patient narratives like human specialists. *Conclusions*: The results of this pilot study support the feasibility of robot-assisted diabetes care to assist human experts. Future research should explore the programme's application with healthcare professionals with limited experience in diabetes care, further demonstrating its scalability and utility in diverse healthcare settings.

Keywords: type 2 diabetes; communication robot; treatment behaviour; diabetes care consultation; response phrase

1. Introduction

The prevalence of diabetes has been increasing rapidly worldwide [1], posing a significant challenge to healthcare systems. This escalating prevalence, combined with the global shortage of diabetes care and education specialists, highlights the urgent need for innovative solutions. Current approaches to addressing this shortage—such as training and dispatching human specialists—are resource-intensive in both time and cost. Therefore, alternative strategies that are cost-effective, scalable, and capable of rapid deployment are imperative.

Robot-assisted diabetes care is a promising approach in this context. Communication robots can be deployed swiftly and economically, potentially alleviating the burden on human specialists. Robots have demonstrated their utility in healthcare-related fields, including promoting physical activity in middle-aged and older adults [2], supporting mental health [3], improving cognitive function [4], rehabilitation for patients with hemiplegia [5], and mental support for patients with blood disorders [6]. Recent reports suggest that some individuals have more comfortable interactions with robots than with human healthcare professionals [7]. However, the applications of communication robots in diabetes care remain unexplored.

To address this gap, we developed a robot-assisted diabetes care consultation programme (hereafter referred to as "the programme") to facilitate patient–professional interactions, identify therapeutic issues, and enhance patient outcomes. Previous studies have shown the potential of such programmes to improve glycaemic control before and after implementation [8]. Unlike conventional consultations, in which patients often experience hesitation or tension, robot-assisted consultations may provide a less intimate environment, thereby encouraging patients to share their concerns more openly.

Given this background, this study aimed to evaluate the feasibility of the programme by documenting its development process, assessing patient feedback, and determining whether healthcare professionals could effectively identify patients' therapeutic issues. By exploring the role of robots in diabetes care, this study seeks to contribute to resolving the global shortage of diabetes care specialists and to improving patient outcomes.

2. Methods

2.1. Study Design and Ethical Issues

All procedures in this cross-sectional observational study followed the ethical standards of the responsible committee on human experimentation and the Helsinki Declaration of 1964, as well as its later amendments. The study protocol was approved by the Ethics Committee of Osaka University (IRB# R4-21). Written informed consent was obtained from all participants after they were informed that they could opt out at any time without penalty.

2.2. Study Participants

A diagnosis of diabetes was established based on the criteria outlined by the American Diabetes Association and the Japan Diabetes Society: HbA1c levels $\geq 6.5\%$ (National Glycohaemoglobin Standardisation Programme), fasting blood glucose concentration ≥ 126 mg/dL (7.0 mmol/L), random blood glucose concentration ≥ 200 mg/dL, or current use of medications for diabetes [9,10]. The exclusion criteria were as follows: (i) type 1 diabetes, (ii) secondary diabetes, (iii) refractory malignant diseases, (iv) dependency on haemodialysis, (v) renal dysfunction with serum creatinine levels > 2.5 mg/dL, (vi) symptomatic coronary artery disease or percutaneous coronary intervention within the past year, (vii) severe hepatic dysfunction (Child–Pugh score ≥ 10), and (viii) patients who had changed their prescriptions within 2 months prior to the interview.

This study was conducted on patients with type 2 diabetes attending a private clinic in Japan. Consecutive cases were approached, and those who provided informed consent participated in this study. Given that this study targeted patients regularly attending a private clinic, it was anticipated that many would have relatively stable glycaemic control and fewer complications. To ensure the homogeneity of the study population, the exclusion criteria above were strictly applied.

Regarding the basic attributes of the subjects, information was collected on age, sex, HbA1c, presence or absence of injectable medication use, and presence or absence of complications.

2.3. Study Period

The study period was from November 2021 to October 2022.

2.4. Programme Creation

2.4.1. Selection of the Robot

In this study, we used SHARP's RoBoHoN-lite (SHARP Corporation, Osaka, Japan). We selected this robot for the following reasons: First, it has a rounded humanoid form, which gives it a charming appearance. Second, its size is convenient for portability because it can fit in one's hand. Third, it is relatively inexpensive for robots, costing approximately JPY 100,000 (USD 154). Fourth, a previous study investigated the possibility of using this robot as a remote-control system [11].

2.4.2. Programme Content

Creation of an Introduction

We created an introductory scenario to reduce the patient's tension and clarify the patient's behavioural role towards the robot.

Creation of Questions to Ascertain Treatment Behaviours

Nightingale [12] stated that humans should be viewed as an integrated entity in which the body, mind, and environment are interrelated. To ensure that these three aspects were depicted as a whole, multiple questions were constructed to capture the overall knowledge, acceptance, implementation level, and support system surrounding a specific health behaviour. Therefore, questions were created to assess whether patients with diabetes engage in the necessary behaviours for diabetes control, such as adjusting their environment, implementing desirable psychological and behavioural practices, and maintaining constructive relationships with their general practitioner (GP) and significant others. These questions were extracted based on guidelines for diabetes care [13]. Additionally, the content implemented by the researcher, a nurse with extensive experience in diabetes consultations, was also referenced. After creation, the questions were tested with three patients to determine whether they were understandable, confirm their relevance to the objectives, and check the appropriateness of the order of the questions, leading to revisions of the scenario.

a. Treatment Behaviours (Diet, Exercise) (Table 1)

The content related to diet was composed of actual dietary intake, its relationship with blood glucose levels, snacking habits, evaluations, satisfaction with the current diet, and the relationship with family and support staff regarding meals.

The content related to exercise was structured to include the type and duration of exercise being performed, the implementation status according to the weather and physical condition, questions regarding exercise, and the status of consultations with GP.

b. Treatment Knowledge (Table 1)

This section comprised content related to foot care, sick day management, and complications.

c. Diabetes Acceptance (Table 1)

Table 1. RoBoHoN interview items and assessment content.

	Interview Items	Assessment Content
Treatment Behaviours	Diet	Whether there are any awkward situations with family regarding meals Satisfaction with meals Relationship between diet and blood glucose levels Self-evaluation of how well things are going Whether family members give compliments Whether you eat anything other than meals
	Exercise	Whether you share details about your exercise, such as the type and duration, with your doctor Selecting and performing exercise based on your physical condition Exercise on rainy days Whether you have any questions about exercise Whether you did any exercise in your youth
Treatment Knowledge	Foot care	Whether you are familiar with the term "foot care" Knowledge of foot care methods, such as nail trimming
	Sick day management	Whether you are familiar with the term "sick day management" Whether you have ever sought a medical consultation on days other than your regular check-up days
	Complications	Whether you want detailed information about complications and how to prevent them The level of ability to explain the complications you know
Diabetes Acceptance	Aetiology	Perception of the cause of diabetes
	Whether there is any disadvantage or burden in daily life	Whether the thought of having diabetes itself feels like a burden Changes in lifestyle following the diagnosis Whether you feel you have lost out because of your diabetes
	Expectations towards the people around you	Whether you have expectations towards healthcare professionals Whether you have expectations towards family Whether you have expectations towards RoBoHoN

This section was structured to include perceptions of the causes of diabetes onset, disadvantages and burdens of living with diabetes, and expectations regarding the surrounding environment (healthcare professionals and family members).

2.4.3. Creation of Response Phrases Emitted by the Robot During Interviews

Response phrases were created to deepen the interviews and ease speech. The creation of these phrases was based on counselling theories [14], interpersonal communication theories [15], and learning theories [16]. To determine the appropriateness of the content, trials were conducted with three patients, and the operator observed and confirmed whether the response phrases interrupted the patient's speech, whether the content developed to a certain depth, and whether the content caused unpleasant reactions. Based on these findings, the response phrases were finalised. The system was designed to allow the robot to select and vocalise responses that the operator judged to be appropriate according to the patient's responses to questions about treatment behaviour and the content of the patient's speech.

2.4.4. Interview Procedure

An interview was conducted with one subject and a robot operator. The robot was operated by nurses experienced in diabetes consultations. The robot operator sat slightly apart from the subject in the same room and operated the robot. The operator did not explicitly inform the subject that the robot was being operated by a person in the same room during the interviews. The interview was recorded using a camera and microphone integrated into the robot, enabling a review of the interview content.

2.5. Evaluation Method of the Programme

2.5.1. Participant Evaluations

After robot-mediated interviews, a questionnaire was conducted to evaluate the robot. The evaluation included three aspects of the robot itself: "Functionality of the RoBoHoN", "Impression of the RoBoHoN", and "Effects of the RoBoHoN". Additionally, participants were asked about the programme content: "Did the programme help you reflect on your diabetes?". The responses were rated on a five-point Likert scale, from "1. Strongly disagree" to "5. Strongly agree". We checked the internal consistency of these items and obtained a Cronbach's alpha coefficient of 0.87. Since a Cronbach's alpha coefficient of ≥ 0.7 is considered acceptable, it was determined that the items in this study were valid.

2.5.2. Criteria for Assessing the Patient's Issues in Diabetes Care

Through conversations with the robot, the patients described three aspects of issues in diabetes care: (1) treatment behaviours (e.g., diet and exercise), (2) treatment knowledge, and (3) diabetes acceptance. Criteria were established to assess the patients' therapeutic issues based on these three aspects. After the interviews, assessments were conducted by two nurses with experience in diabetes consultation who reviewed the interviews.

Treatment Behaviours (e.g., Diet, Exercise) (Table 1)

The criteria for assessing diet and exercise were established from three perspectives: knowledge, behaviour, and support from others. The scores for this section range from 3 to 9 points.

(i) Knowledge: 3 levels (1 point: No knowledge, 2 points: Aware of precautions, 3 points: Have basic knowledge)
(ii) Behaviour: 3 levels (1 point: Not concerned, 2 points: Only pay attention, 3 points: Practising basic principles)
(iii) Support from others: 3 levels (1 point: No concept of cooperation from others, 2 points: Feel that others cooperate but cannot talk about specifics, 3 points: Can discuss specifics regarding cooperation)

Treatment Knowledge (Table 1)

These criteria were based on the presence or absence of treatment knowledge, with three assessment levels: 1 point, never heard of it; 2 points, heard of it but unable to explain; and 3 points, knowledgeable and able to explain. The scores for this section ranged from 3 to 9 points.

Diabetes Acceptance (Table 1)

The criteria for diabetes acceptance were as follows: scores for this section ranged from 3 to 6 points.

(i) Cause of onset: 2 levels (1 point: Do not know, 2 points: Have an idea).
(ii) Burden or disadvantage in daily life: 2 levels (1 point: None, 2 points: Present).
(iii) Expectations from others: 2 levels (1 point: None, 2 points: Present).

2.6. Statistical Analysis

The sample size was determined using Lehr's formula. Assuming an expected difference in questionnaire score of 1 and a standard deviation of 0.7, the effect size was estimated to be 1.4. Consequently, the required sample size was calculated as 32 (16×1.4^2), with a target power of 80% and a significance level of 0.05.

Continuous variables are expressed as mean ± standard deviation (SD) and median values, while categorical variables are reported as counts and percentages. For questionnaire responses, 95% confidence intervals (CIs) were calculated. The Kolmogorov–Smirnov test was used to analyse the cumulative relative frequency distributions.

Statistical analyses were conducted using Microsoft Excel (Microsoft Corporation, Redmond, WA, USA), R version 4.3.0 (R Foundation for Statistical Computing, Vienna, Austria), GUI version 1.79, RStudio version 2024.09.1+394 (Boston, MA, USA) on an Apple Macintosh computer (Apple, Cupertino, CA, USA), and SPSS Statistics version 27 (IBM, Chicago, IL, USA).

3. Results

3.1. Baseline Characteristics of the Participants (Table 2)

In total, interviews were conducted with 32 participants (22 males [68.8%]; mean age, 69.7 ± 12.6 years). All participants were asked a series of questions and responses were recorded for each question. The average HbA1c level was 7.2 ± 1.0%, and a limited number of participants were treated with insulin injections (28.1%) and had diabetic complications (15.6%).

Table 2. Baseline characteristics of the participants. (n = 32).

Variables	Mean ± SD
Male gender, (%)	22 (68.8)
Age, years old	69.7 ± 12.6
HbA1c, %	7.2 ± 1.0
Treated with insulin injections, (%)	9 (28.1)
With diabetic complications, (%)	5 (15.6)

3.2. Created Programme

3.2.1. Introduction

In the introduction, the expected role behaviour was: "I will ask you questions about your diet, exercise, and treatment, so please answer them".

3.2.2. Questions Regarding Treatment Behaviours (Table 3)

Regarding treatment behaviours, the questions about diet included the following: "Are you satisfied with your current diet? Can you tell me if you're satisfied?" and " Do you ever think that your diet is a cause of your fluctuating blood glucose levels?". For family relationships related to meals, the question was: "Do you ever have awkward moments with your family because of meals? For example, are you often told you eat too much?".

Regarding exercise, the questions included the following: "Do you choose and perform exercises based on your condition each day? What do you do when it rains? How do you exercise on rainy days?". The question about exercise-related concerns was "If you have any questions about exercise, please feel free to ask me".

Regarding treatment knowledge, the questions about foot care included the following: "Are you familiar with foot care?" and "If you have any trouble with foot care, such as cutting your nails, please tell me". Regarding sick day management, the questions were:

"Do you know what a 'sick day management' is?" and "If you have ever had the experience of wanting to see a doctor outside of your routine appointments, please share it with me".

Regarding complications, the questions included the following: "Would you like to know more details about the types of complications and their prevention?" and "Can you tell me about any complications you're aware of?". In addition, questions regarding oral medication and insulin therapy were added to the questionnaire.

Regarding diabetes acceptance, the questions were "Could you let me know if you feel burdened just because you have diabetes?" and "Please tell me what you expect from your healthcare professional or family when you go through treatment for diabetes".

All questions and scenarios are presented in Table 3. No participants exhibited unpleasant reactions throughout the programme, and all were able to respond to the questions.

Table 3. Questions regarding treatment behaviours.

Introduction	I will ask you questions about your diet, exercise, and treatment, so please answer them. Please rest assured that your responses will not be shared with anyone without your permission. I will come up with useful advice regarding subsequent treatment based on what you tell me. I will discuss matters with your GP as necessary. Thank you in advance for your time and cooperation.
Treatment Behaviours (Diet and Exercise)	I will ask you about your daily life, so please share your thoughts. First, I will ask about your diet. Do you ever have awkward moments with your family because of meals? For example, are you often told you eat too much? Are you satisfied with your current diet? Can you tell me if you're satisfied? Do you ever think that your diet is a cause of your fluctuating blood glucose levels? Please tell me whether your family praises you for your efforts in following your dietary regimen. Do you think your current diet is working well? Think back to yesterday's meals and tell me what you ate. Is that generally similar to a normal meal for you? Please let me know if you eat anything in addition to your meals. That concludes the questions about your diet.
	Next, I will ask about exercise. Could you let me know if you communicate the types and duration of exercise you do to your primary physician. Do you choose and perform exercises based on your condition each day? What do you do when it rains? How do you exercise on rainy days? If you have any questions about exercise, please feel free to ask me. If there were any sports you played when you were younger, please share that with me. That concludes the questions about exercise. This wraps up the questions about your lifestyle.
Treatment Knowledge	From now on, I will ask questions about your treatment. Are you familiar with foot care? If you have any trouble with foot care, such as cutting your nails, please tell me. Do you know what a 'sick day management' is? If you have ever had the experience of wanting to see a doctor outside of your routine appointments, please share it with me. Would you like to know more details about the types of complications and their prevention? Can you tell me about any complications you're aware of? Do you feel that your insulin injections are going well? If you have any questions about insulin injections, please let me know. If you have any questions about the medication you are currently taking, please let me know.
Diabetes Acceptance	Could you tell me what you think is the cause of your diabetes? We are almost done with the questions. Please do your best to answer. Could you let me know if you feel burdened just because you have diabetes. If you feel that your lifestyle has changed since being diagnosed with diabetes, please share what has changed. If you have ever felt that having diabetes has caused you to lose out on something, please let me know. The following questions are about your expectations for healthcare professionals and your family. Please tell me what you expect from your healthcare professional or family when you go through treatment for diabetes. If you have any requests for me, the robot asking these questions, please let me know.

Table 3. *Cont.*

Conclusion	This is the last question. If there is anything else you would like to ask me, about anything other than the questions I asked today, feel free to ask. That concludes the questions. Thank you very much.

GP, general practitioner.

3.2.3. Response Phrases Issued by the Robot During the Interview (Table 4)

The response phrases were designed as shown in Table 4. These included expressions of empathy such as "I see", "Right", "That's good", and "Yes, indeed". To encourage the participants to elaborate further or reflect on their thoughts, phrases such as "Really?', "Please tell me about your current situation in a little more detail", and "It sounds as though you don't see that as a good thing..." were used. Encouraging phrases included "Thank you for sharing this information with me" and phrases to motivate treatment behaviours, such as "That's wonderful!" and "You've certainly been doing your best". To encourage awareness of the healthcare team, phrases such as "I will discuss this with your GP" and "I will ask your GP in an indirect way" were included.

In total, 22 response phrases were developed. These phrases facilitated the flow of conversation, encouraged participants to provide more detailed responses, and helped maintain their engagement throughout the interviews. As the conversations were originally conducted in Japanese, the original Japanese phrases are presented with their English equivalents in Supplementary Table S1 for reference.

Table 4. Response phrases by the RoBoHoN during Interviews.

I see.	Right.
That's good.	Yes, indeed.
Is that so.	It really is.
I see it's been on your mind as well.	Was this question difficult to answer?
Really?	It sounds as though you don't see that as a good thing...
Please tell me about your current situation in a little more detail.	Oh wow, I never would have thought.
Did you ever wish you had more support?	Could you tell me why?
Indeed, it's not just about you.	How do you cope in those situations?
Please tell me the specific amount, and other details.	Thank you for sharing this information with me.
That is wonderful!	You've certainly been doing your best.
I will discuss this with your GP.	I will ask your GP in an indirect way.

GP, general practitioner.

3.3. Evaluation by the Participants (Table 5)

The average score ranges for the evaluation of the communication robot are as follows: For "Functionality of RoBoHoN" (seven items), the average scores ranged from 3.45 to 4.61 points. For "Impression of RoBoHoN" (five items), the average scores ranged from 3.16 to 4.42 points. For "Effects of RoBoHoN" (five items), the average scores ranged from 2.81 to 3.90 points, with the item 'I thought it would make attending the clinic more enjoyable' being the only one scoring below 3 points. Regarding the average scores for the evaluation of the programme content, the average scores for the five items of "Whether the programme led to a reflection on diabetes" ranged from 3.52 to 3.90.

Table 5. Evaluation of the programme by the participants. (n = 31 *).

	Questionnaire Items	Strongly Disagree	Disagree	Neither Agree Nor Disagree	Somewhat Agree	Strongly Agree	Mean ± S.D.	95% CI
Functionality of RoBoHoN	RoBoHoN's voice was easy to understand.	0 (0.0)	1 (3.2)	2 (6.5)	10 (32.3)	18 (58.1)	4.45 ± 0.77	4.17–4.73
	The speaking pace was just right.	0 (0.0)	0 (0.0)	3 (9.7)	6 (19.4)	22 (71.0)	4.61 ± 0.67	4.37–4.86
	I felt that I was listened to.	0 (0.0)	1 (3.2)	4 (12.9)	11 (35.5)	15 (48.4)	4.29 ± 0.82	3.99–4.59
	The timing of the questions felt natural.	0 (0.0)	0 (0.0)	8 (25.8)	12 (38.7)	11 (35.5)	4.10 ± 0.79	3.81–4.38
	The content of the questions was difficult. [†]	0 (0.0)	5 (16.1)	12 (38.7)	9 (29.0)	5 (16.1)	3.45 ± 0.96	3.10–3.80
	The timing of the responses was natural.	0 (0.0)	0 (0.0)	11 (35.5)	11 (35.5)	9 (29.0)	3.94 ± 0.81	3.64–4.23
	The content of the responses was natural.	0 (0.0)	1 (3.2)	9 (29.0)	12 (38.7)	9 (29.0)	3.94 ± 0.85	3.62–4.25
Impression of RoBoHoN	It was cute.	0 (0.0)	1 (3.2)	6 (19.4)	9 (29.0)	15 (48.4)	4.23 ± 0.88	3.90–4.55
	I had a negative impression (such as fear or coldness). [†]	0 (0.0)	1 (3.2)	3 (9.7)	9 (29.0)	18 (58.1)	4.42 ± 0.81	4.12–4.72
	I developed a sense of attachment while talking.	4 (12.9)	3 (9.7)	13 (41.9)	5 (16.1)	6 (19.4)	3.19 ± 1.25	2.74–3.65
	I felt a sense of familiarity in the way the robot spoke.	2 (6.5)	3 (9.7)	5 (16.1)	11 (35.5)	10 (32.3)	3.77 ± 1.20	3.33–4.22
	I wanted to talk more.	2 (6.5)	3 (9.7)	16 (51.6)	8 (25.8)	2 (6.5)	3.16 ± 0.93	2.82–3.50
Effects of RoBoHoN	I felt that it was easier to talk than when speaking with a healthcare professional.	2 (6.5)	3 (9.7)	16 (51.6)	6 (19.4)	4 (12.9)	3.23 ± 1.02	2.85–3.60
	It was easier to talk about things that are difficult to say compared with speaking with a healthcare professional.	0 (0.0)	6 (19.4)	13 (41.9)	7 (22.6)	5 (16.1)	3.35 ± 0.99	2.99–3.72
	I felt more pressure of being "checked" compared with talking with a healthcare professional. [†]	0 (0.0)	2 (6.5)	8 (25.8)	12 (38.7)	9 (29.0)	3.90 ± 0.91	3.57–4.24
	The childish way of speaking made me feel reluctant to talk about private matters. [†]	1 (3.2)	2 (6.5)	8 (25.8)	8 (25.8)	12 (38.7)	3.90 ± 1.11	3.50–4.31
	I thought it would make attending the clinic more enjoyable.	3 (9.7)	5 (16.1)	20 (64.5)	1 (3.2)	2 (6.5)	2.81 ± 0.91	2.47–3.14
Reflection on diabetes	By having this conversation, I was able to reflect on my knowledge of diabetes.	0 (0.0)	1 (3.2)	8 (25.8)	18 (58.1)	4 (12.9)	3.81 ± 0.70	3.55–4.06
	By having this conversation, I was able to reflect on my diet.	0 (0.0)	2 (6.5)	10 (32.3)	10 (32.3)	9 (29.0)	3.84 ± 0.93	3.50–4.18
	By having this conversation, I was able to reflect on my exercise routine	0 (0.0)	1 (3.2)	10 (32.3)	11 (35.5)	9 (29.0)	3.90 ± 0.87	3.58–4.22
	By having this conversation, I was able to reflect on my medication (oral medication or insulin).	1 (3.2)	3 (9.7)	12 (38.7)	9 (29.0)	6 (19.4)	3.52 ± 1.03	3.14–3.89
	By having this conversation, I was able to reflect on my blood glucose control.	1 (3.2)	1 (3.2)	7 (22.6)	13 (41.9)	9 (29.0)	3.90 ± 0.98	3.54–4.26

* One participant was unable to complete the questionnaire for the robot evaluation; [†] Calculated as a reverse-scored item; CI, confidential interval; S.D., standard deviation.

3.4. Distribution of Scores for Treatment Behaviours (Figure 1a,b)

The total score for diet showed that many participants scored 3 (n = 8) or 5 points (n = 9). A score of 3 points indicated a significant concern with diet; about 25% of the participants fell into this category. Furthermore, because the score range for diet was distributed from 3 to 8 points, this suggests that the assessment was tailored to each participant's individual situation.

For exercise, the most common total score was 3 points (n = 11). A score of 3 points indicated a concern with exercise; approximately 35% of the participants fell into this category. The score range for exercise was from 3 to 9 points, which also enabled an assessment tailored to the individual circumstances of each participant.

3.5. Distribution of Scores for Treatment Knowledge (Figure 1c)

The total score for treatment knowledge was as follows: the most common score was 4 points (n = 15), followed by 3 points (n = 9). The minimum score was 3 points, indicating a lack of knowledge, whereas a score of 4 points suggested that the participant had heard of at least one of the three topics. Therefore, approximately 70% of the patients lacked sufficient knowledge. With scores ranging from 3 to 9 points, this suggests that the assessment of treatment knowledge was also tailored to each participant's individual situation.

3.6. Distribution of Scores for Diabetes Acceptance (Figure 1d)

The most common total score for acceptance of diabetes was 4 points (n = 15). The score for diabetes acceptance ranged from 3 to 6 points, indicating that the assessment of diabetes acceptance was adapted to the individual circumstances of each participant.

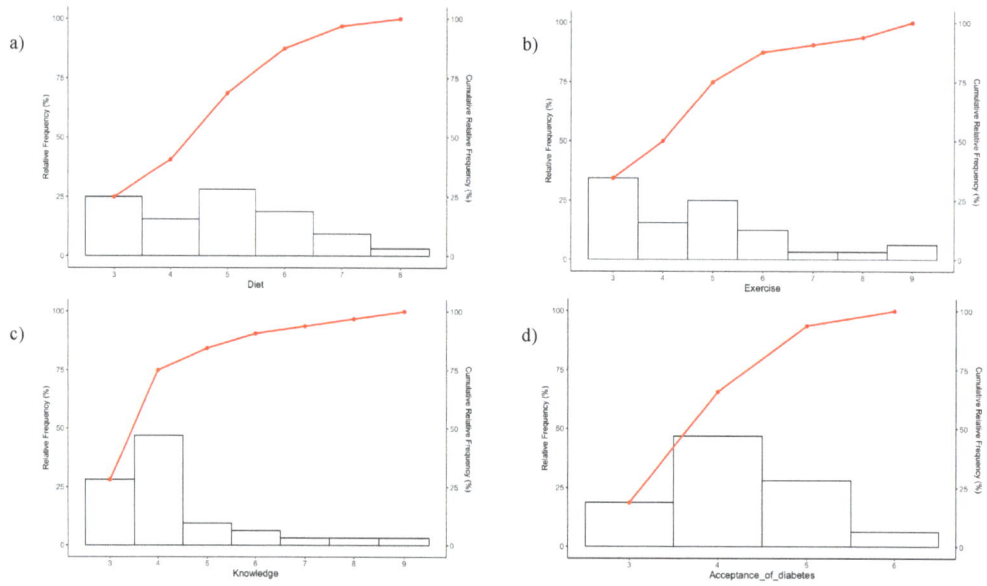

Figure 1. Histograms and cumulative relative frequency graphs depicting the distribution of scores related to treatment behaviours: (**a**) diet, (**b**) exercise, (**c**) treatment knowledge, and (**d**) diabetes acceptance.

3.7. Differences in the Score Distribution of Treatment Behaviours

The Kolmogorov–Smirnov test was used to examine differences in score distributions across treatment behaviours, treatment knowledge, and diabetes acceptance. The analysis revealed a statistically significant difference between diet and treatment knowledge ($p = 0.016$) (Table 6).

Table 6. Kolmogorov–Smirnov tests for the differences in the score distribution of treatment behaviours.

Comparison	KS Statistic	p-Value
Diet behaviour vs. Exercise behaviour	0.093	0.887
Diet behaviour vs. Treatment knowledge	0.34	0.016
Diet behaviour vs. Diabetes acceptance	0.25	0.116
Exercise behaviour vs. Treatment knowledge	0.25	0.118
Exercise behaviour vs. Diabetes acceptance	0.19	0.240
Treatment knowledge vs. Diabetes acceptance	0.094	0.891

KS Statistic, The Kolmogorov–Smirnov statistic (maximum difference between empirical cumulative distribution functions).

4. Discussion

4.1. Evaluating the Feasibility of Robot-Assisted Therapeutic Consultations for Diabetes Care

Effective therapeutic consultations for diabetes management require patients to communicate their experiences openly with healthcare professionals without causing discomfort or hesitation. This study evaluated the feasibility of a robot-assisted consultation programme by examining three key aspects: the suitability of the selected robot model, the design and content of the questions and responses, and the programme's ability to assess treatment behaviours.

4.2. The Selected Robot

This study highlights the significance of human–robot interaction in healthcare, where both entities can complement each other's strengths while enhancing overall engagement. Patient feedback indicated that they found it easier to discuss sensitive topics with the robot than with healthcare professionals. Previous research has explored the potential of tablet-assisted medical guidance [17]. However, given Japan's rapidly ageing population and evolving family dynamics, many individuals now have fewer opportunities for human support and seek alternative means of interaction.

Orem's self-care theory emphasises that "caring" is a fundamental human need [18]. The fact that the patients described the robot as "cute" and developed a sense of attachment suggests that interacting with a robot perceived as non-threatening may help reduce patient anxiety. However, only a small proportion of the participants reported that the robot enhanced their motivation to visit the clinic, indicating that it was not a primary factor in encouraging clinic attendance.

Rather than allowing the robot to operate independently, healthcare professionals actively selected and administered questions and responses. This approach was designed to foster a collaborative dynamic among patients, healthcare professionals, and robots, engaging all parties in meaningful therapeutic discussions. While the extent to which a true sense of camaraderie was achieved remains unclear, this study successfully demonstrated the feasibility of integrating robots into diabetes care consultations as interactive facilitators rather than passive tools.

The financial burden associated with robotic implementation is a recognised concern [19,20]. However, the RoBoHoN used in this study is relatively affordable compared to other humanoid robots. Moreover, its potential for long-term use may help offset initial costs, ultimately leading to cost savings when compared to labour expenses. Additionally, the installed programme allows for intuitive operation, requiring only word selection and a button press. Given these advantages, RoBoHoN demonstrates the potential for widespread adoption and sustainable integration into clinical practice.

4.3. Contents of the Questions and Responses

This programme was designed to facilitate patient-led discussions regarding knowledge, skills, and support systems related to compliance with treatment behaviours. The robot prompted open-ended discussions, asking, "Do you think your current diet is working well?" Multiple questions were developed to assess various dimensions of diabetes care, including knowledge, acceptance, implementation, and support structure. Most participants did not find the questions difficult to answer, and many reported that they could reflect on their understanding of diabetes management. These findings suggest that robot-facilitated dialogue successfully elicited patient narratives, demonstrating its potential as an interactive tool for diabetes care consultation.

An essential feature of this programme is the incorporation of backchannelling, a communication technique that encourages active conversation engagement. The 22 backchannelling phrases used in this study were carefully selected from among those commonly employed by experienced healthcare professionals during consultations. Patient evaluations indicated that very few participants found the timing or content of the backchannelling responses unnatural. In a previous study, a robot was used to confirm knowledge about diabetes, and a questionnaire was used to evaluate the findings [21]. Similarly, our findings highlight the positive reception of robotic interactions, particularly in terms of conversational fluency and engagement. Notably, robot-assisted backchannelling proved effective in sustaining patient narratives without disruption, even within the more complex framework of this study, in which patients were asked to discuss their medical treatment history.

These results indicate that the questions and backchannelling strategies developed in this programme can facilitate meaningful patient discussions from multiple perspectives. These elements are regarded as useful for identifying issues in patient management and for developing solutions.

4.4. Assessment of Issues in Treatment Behaviours

In this study, 30–40% of participants exhibited difficulties complying with dietary and exercise regimens, consistent with previous findings that reported 20–30% non-compliance rates in these behaviours [22].

Histogram analysis indicated that participants scored lower on treatment knowledge than on dietary behaviour. Regarding treatment knowledge, most participants demonstrated limited understanding. This finding is consistent with previous studies, which reported that nearly half of patients did not routinely inspect their feet [23] and that many had little or no awareness of the concept of "sick day management" [24]. The observed lack of recall in our study may be attributed to the participants' relatively stable blood glucose control, absence of foot complications, and limited personal experience with acute illness episodes.

In assessing diabetes acceptance, factors such as understanding the cause of disease onset, perceived burden on daily life, and expectations from others were identified as key elements that could be explored through therapeutic consultations. The distribution of scores suggests that, even with RoBoHoN-assisted consultations, it was possible to facilitate meaningful discussions on these critical aspects of treatment behaviours. Furthermore, the patient narratives elicited through robotic interviews aligned with trends observed in previous studies, indicating that robot-assisted consultations can yield insights comparable to those obtained through clinician-led interactions.

Future research should focus on evaluating long-term changes in patient compliance and behaviour following robot-assisted consultations, as well as assessing the effectiveness of this approach in enhancing diabetes management outcomes.

4.5. Limitations of This Study

This study has several limitations due to its single-centre design and private practice setting. The study population comprised patients with relatively stable blood glucose levels and no severe complications. The findings were based on a single-point interview conducted with a robot. These factors inherently limit the generalisability of the results. Additionally, as the robot in this study was operated by an experienced diabetes care specialist, future research should assess whether similar outcomes can be achieved when operated by healthcare professionals with less experience in diabetes care consultation.

However, these limitations also reflect the reality of diabetes management, where interventions often target asymptomatic patients to promote behavioural changes. This study provides valuable insights into a forward-looking approach to diabetes care in which specialists leverage robotic assistance to enhance therapeutic interventions in real-life settings. Nevertheless, the potential role of robots in the management of patients with established complications remains an important issue and warrants further investigation.

Although the sample size of this study was limited, it was statistically sufficient to draw meaningful conclusions. Unlike large-scale studies where clinically insignificant but statistically significant differences may emerge, this study evaluated the practical benefits and applicability of robot-assisted consultations in diabetes care. While acknowledging its limitations, the findings offer important insights into the feasibility and effectiveness of this approach in a specific healthcare environment.

5. Conclusions

In this study, we developed a consultation programme for diabetes care using a communication robot (RoBoHoN) and evaluated its applicability for patients. The results indicate the feasibility of robot-assisted therapeutic consultations for diabetes care, potentially progressing to assist human experts.

Supplementary Materials: The following supporting information can be downloaded at: https://www.mdpi.com/article/10.3390/medicina61020352/s1, Table S1: Response phrases by the RoBoHoN during interviews and corresponding Japanese words.

Author Contributions: K.Y., M.S., S.N. and M.I. planned the project; Y.A., T.H. and M.I. performed the study as expert Certified Diabetes Care and Education Specialists; K.Y. managed patients as diabetologist; M.K., A.Y., R.U., M.S. and S.N. contributed as experts in information and communications technology and robots; K.Y. and N.Y. performed clinical reviews and as diabetologists; Y.A., T.H. and M.I. interpreted the data and wrote the article. All authors have read and agreed to the published version of the manuscript.

Funding: This work was supported by JSPS KAKENHI [grant numbers JP18K08505 and JP21K10300]. The sponsor of this study was not involved in the study design, data collection, analysis, interpretation of data, report writing, or decision to submit the paper for publication.

Institutional Review Board Statement: This study was conducted in accordance with the Declaration of Helsinki and approved by the Ethics Committee of Osaka University (IRB# R4-21, date of approval: 23 March 2021).

Informed Consent Statement: Written informed consent was obtained from all participants involved in this study. All participants were notified that they could opt out of this study at any time.

Data Availability Statement: The datasets used and analysed in this study are available from the corresponding author upon reasonable request.

Acknowledgments: We would like to express our sincere gratitude to all the patients who participated in this study.

Conflicts of Interest: The authors declare no conflicts of interest.

References

1. World Health Organization. Urgent Action Needed as Global Diabetes Cases Increase Four-Fold over Past Decades. Available online: https://www.who.int/news/item/13-11-2024-urgent-action-needed-as-global-diabetes-cases-increase-four-fold-over-past-decades (accessed on 13 November 2024).
2. Oida, Y.; Kanoh, M.; Yamane, M.; Kasai, T.; Suzuki, T.; Kaga, Y. [Feasibility of a home gymnastic robot to facilitate regular exercise among middle- and old-aged women] Undo shukan no keisei wo shien suru tameno kateiyo taiso robotto no jituyosei no kento. *Jpn. J. Health Educ. Promot.* **2009**, *17*, 184–193. (In Japanese) [CrossRef]
3. Goda, A.; Shimura, T.; Murata, S.; Kodama, T.; Nakano, H.; Ohsugi, H. Effects of robot-assisted activity using a communication robot in elderly individuals. *Jpn J. Appl. Phys.* **2020**, *10*, 131–136. [CrossRef]
4. Taniguchi, Y.; Ishii, C.; Kayanuma, Y. [Effects on physical performance and cognitive functioning of community-dwelling older adults following an exercise program with humanoid robots: A preliminary trial] Hitogata komyunikeshon robotto wo motiita undou puroguramu ga chiiki zaitaku koreisha no sintai kinou oyobi ninti kinou ni oyobosu eikyo. *Jpn. J. Public Health* **2019**, *66*, 267–273. (In Japanese) [CrossRef]
5. Aguirre-Ollinger, G.; Chua, K.S.G.; Ong, P.L.; Kuah, C.W.K.; Plunkett, T.K.; Ng, C.Y.; Khin, L.W.; Goh, K.H.; Chong, W.B.; Low, J.A.M.; et al. Telerehabilitation using a 2-D planar arm rehabilitation robot for hemiparetic stroke: A feasibility study of clinic-to-home exergaming therapy. *J. Neuroeng. Rehabil.* **2024**, *21*, 207. [CrossRef]
6. Yamada, A.; Akahane, D.; Takeuchi, S.; Miyata, K.; Sato, T.; Gotoh, A. Robot therapy aids mental health in patients with hematological malignancy during hematopoietic stem cell transplantation in a protective isolation unit. *Sci. Rep.* **2024**, *14*, 4737. [CrossRef] [PubMed]
7. Silva, J.G.G.d.; Kavanagh, D.j.; Belpaeme, T.; Taylor, L.; Beeson, K.; Andrade, J. Experiences of a motivational interview delivered by a robot: Qualitative study. *J. Med. Internet Res.* **2018**, *20*, e116. [CrossRef] [PubMed]
8. Yagi, K.; Inagaki, M.; Asada, Y.; Komatsu, M.; Ogawa, F.; Horiguchi, T.; Yamaaki, N.; Shikida, M.; Origasa, H.; Nishio, S. Improved glycemic control through robot-assisted remote interview for outpatients with type 2 diabetes: A pilot study. *Medicina* **2024**, *60*, 329. [CrossRef] [PubMed]
9. Araki, E.; Goto, A.; Kondo, T.; Noda, M.; Noto, H.; Origasa, H.; Osawa, H.; Taguchi, A.; Tanizawa, Y.; Tobe, K.; et al. Japanese Clinical Practice Guideline for Diabetes 2019. *Diabetol. Int.* **2020**, *11*, 165–223. [CrossRef]
10. American Diabetes Association Professional Practice Committee; ElSayed, N.A.; Aleppo, G.; Bannuru, R.R.; Bruemmer, D.; Collins, B.S.; Ekhlaspour, L.; Gaglia, J.L.; Hilliard, M.E.; Johnson, E.L.; et al. 2. Diagnosis and Classification of Diabetes: Standards of Care in Diabetes-2024. *Diabetes Care* **2024**, *47*, S20–S42. [CrossRef]
11. Maalouly, E.; Hirano, T.; Yamazaki, R.; Nishio, S.; Ishiguro, H. Encouraging prosocial behavior from older adults through robot teleoperation: A feasibility study. *Front. Comput. Sci.* **2023**, *5*, 1157925. [CrossRef]
12. Nightingale, F. *Florence Nightingale Notes on Nursing: What It Is and What It Is Not*, 8th ed.; Yumaki, M.; Usui, H.; Kodama, K.; Tamura, M.; Kominami, Y., Translators; Gendai-sha: Tokyo, Japan, 2023. (In Japanese)
13. The Japan Diabetes Society. *Treatment Guide for Diabetes 2024*, 1st ed.; Bunko-Do: Tokyo, Japan, 2024. (In Japanese)
14. Rogers, C.R. Empathic: An Unappreciated Way of Being. *Couns. Psychol.* **1975**, *5*, 2–10. [CrossRef]
15. Wiedenbach, E. *Communication: The Key to Effective Nursing*, 1st ed.; Ikeda, A., Translator; Japanese Nursing Association Publishing Company: Tokyo, Japan, 2007. (In Japanese)
16. Knowles, M.S. *The Modern Practice of Adult Education: From Pedagogy to Andragogy*, 1st ed.; Hori, K.; Miwa, K., Translators; Otori Publishing: Tokyo, Japan, 2002. (In Japanese)
17. Sugiyama, N.; Nakato, T.; Uragami, K.; Kitamura, T.; Kawamura, N.; Hiragushi, K.; Watanabe, K.; Itoshima, T. [Implementing diabetes education using tablet computers online storage] Taburetto gata keitai tammatsu to onrain sutorejisabisu womochiita tonyobyo kyoiku shisutemu no kochiku. *J. Jpn. Diab. Soc.* **2011**, *54*, 851–855. (In Japanese) [CrossRef]
18. Orem, D.E. *Nursing-Concepts of Practice 6e by Dorothea, E.Orem*, 4th ed.; Onodera, T., Translator; Igaku-Shoin Ltd.: Tokyo, Japan, 2005. (In Japanese)
19. Bradwell, H.L.; Winnington, R.; Thill, S.; Jones, R.B. Ethical perceptions towards real-world use of companion robots with older people and people with dementia: Survey opinions among younger adults. *BMC Geriatr* **2020**, *20*, 244. [CrossRef] [PubMed]
20. Ficht, G.; Farazi, H.; Rodriguez, D.; Pavlichenko, D.; Allgeuer, P.; Brandenburger, A.; Behnke, S. NimbRo-OP2X: Affordable Adult-sized 3D-printed Open-Source Humanoid Robot for Research. *Int. J. Humanoid Robot.* **2020**, *17*, 2050021. [CrossRef]
21. Chiu, C.J.; Hua, L.C.; Chou, C.Y.; Chiang, J.H. Robot-enhanced diabetes care for middle-aged and older adults living with diabetes in the community: A small sample size mixed-method evaluation. *PLoS ONE* **2022**, *17*, e0265384. [CrossRef]
22. Tanaka, T.; Misaki, M.; Tsuji, M.; Issiki, N.; Ibata, H.; Omoto, Y.; Kitamura, M.; Kato, S.; Shintani, U. [Compliance of exercise and dietary therapy in patients with diabetes mellitus after educational hospitalization] Tounyoubyou kanzya no kyouiku nyuuin go no syokuzi·undou no zissi zyoukyou ni tuite. *IRYO* **2000**, *54*, 136–143. (In Japanese) [CrossRef]

23. Saito, M.; Takakusaki, Y.; Jinguu, A.; Yamaga, R.; Watanuki, M.; Akiyama, S.; Miyazaki, H.; Aoki, T.; Ogiwara, T. [Cognitive function and self-care adherence of patients with diabetes receiving foot-care treatments] Futtokea wo ukete iru tounyoubyou kanzya no nintikinou to asi no ziko kea adohiaransu tono kankei. *Jpn. J. Foot Care* **2019**, *17*, 128–132. (In Japanese) [CrossRef]
24. Diabetes Network. [Diabetes Information BOX & NET for Medical Staff. No. 33.] Iryo Sutaffu No Tame No Tonyobyo Joho Bokkusu Ando Netto. No. 33. 2012. Available online: https://dm-net.co.jp/box/no33.pdf (accessed on 13 March 2024). (In Japanese)

Disclaimer/Publisher's Note: The statements, opinions and data contained in all publications are solely those of the individual author(s) and contributor(s) and not of MDPI and/or the editor(s). MDPI and/or the editor(s) disclaim responsibility for any injury to people or property resulting from any ideas, methods, instructions or products referred to in the content.

Article

Association of Nerve Conduction Study Variables with Hematologic Tests in Patients with Type 2 Diabetes Mellitus

Jung-Eun Han [1], Jun-Hwan Choi [2,*], So-Yeon Yoo [1], Gwan-Pyo Koh [1], Sang-Ah Lee [1], So-Young Lee [2] and Hyun-Jung Lee [2]

[1] Department of Internal Medicine, Jeju National University Hospital, Jeju National University College of Medicine, Jeju 63241, Republic of Korea; heather920@naver.com (J.-E.H.); happyweed@jejunu.ac.kr (S.-Y.Y.); okdom@jejunu.ac.kr (G.-P.K.); sahe7@hanmail.net (S.-A.L.)

[2] Department of Rehabilitation Medicine, Jeju National University Hospital, Jeju National University College of Medicine, Jeju 63241, Republic of Korea; bluelsy900@hanmail.net (S.-Y.L.); sigano@hanmail.net (H.-J.L.)

* Correspondence: miraerojh0728@gmail.com; Tel.: +82-64-717-2711; Fax: +82-64-717-1131

Abstract: *Background and Objective*: Diabetic peripheral neuropathy (DPN) is a prevalent complication of type 2 diabetes mellitus (T2DM), with nerve conduction studies (NCSs) serving as the diagnostic gold standard. Early diagnosis is critical for effective management, yet many cases are detected late due to the gradual onset of symptoms. This study explores the relationship between hematological tests and NCS outcomes in T2DM patients to improve the early detection of DPN. *Material and Methods*: This retrospective study involved T2DM patients exhibiting neuropathic symptoms, and patients were divided based on NCS findings into groups with normal and abnormal results to assess the diagnostic value of various hematological markers, clinical, and demographic data for DPN. *Results*: Among 400 participants, 57% ($n = 228$) had abnormal NCS results indicative of DPN. Significant differences were observed in the abnormal-NCS group, including older age, longer diabetes duration, higher levels of fasting plasma glucose, HbA1c, and apolipoprotein B, along with lower eGFR, HDL-C, and Apo A-I levels. Notably, negative correlations were found between HDL-C, Apo A-I, vitamin B12, and specific NCS measurements, while positive correlations existed with sural sensory nerve amplitudes. Multivariate analysis highlighted the importance of age, diabetes duration, hyperglycemia, and specific hematologic markers in predicting DPN. *Conclusions*: The findings confirm that NCSs, combined with hematologic testing, can effectively identify DPN in T2DM patients. Consistent with prior research, prolonged hyperglycemia and nephropathy progression are strongly linked to DPN development. Additionally, lower levels of HDL-C, Apo A-I, and vitamin B12 are associated with the condition, suggesting their potential utility in early diagnostic protocols.

Keywords: diabetic neuropathy; nerve conduction study; hematologic test

1. Introduction

Diabetic peripheral neuropathy (DPN) is one of the most common microvascular complications of type 2 diabetes mellitus (T2DM), with a prevalence of approximately 50% or more of patients with T2DM [1]. Typical DPN is a symmetrical, length-dependent sensorimotor polyneuropathy caused by metabolic and vascular changes resulting from long-term exposure to hyperglycemia and cardiovascular risk covariates [2]. However, compared with the effect of hyperglycemia treatment on DPN in type 1 diabetes mellitus [3,4], the correlation between DPN and hyperglycemia in patients with T2DM remains unclear. The UK Prospective Diabetes Study found that ten years of glycemic treatment reduced the incidence of overall microvascular complications; however, the reduction in DPN alone was uncertain [5], and the Action to Control Cardiovascular Risk in Diabetes (ACCORD) study showed that intensive glycemic control had no significant effect on the incidence of DPN [6].

The exact pathogenesis is unknown, and experimental studies suggest multifactorial causes and pathogenic pathways driven by oxidative and inflammatory stress, hyperglycemia, and other metabolic factors such as hyperlipidemia and insulin resistance [7,8]. The polyol, glycation, protein kinase C, poly (ADP-ribose) polymerase (PARP), and hexosamine pathways, which are known to cause oxidative stress in neurons and microvessels, are specifically affected by hyperglycemia [9,10]. However, as the previous studies mentioned above could not reduce the occurrence of DPN with blood sugar control alone, many other factors are thought to be involved.

Early diagnosis is vital to prevent the progression of DPN and to manage it properly; however, in the early stages of T2DM, DPN is an insidious and asymptomatically progressing condition [11,12] and no consensus regarding early diagnosis exists [11]. Therefore, therapeutic interventions are often postponed. The diagnosis of DPN usually depends on clinical symptoms and signs of neuropathy, including pain, tingling, and numbness. However, these methods for diagnosing DPN have disadvantages, such as limited sensitivity and high variability [13]. Nerve conduction studies (NCSs) have long been known as the gold standard test for DPN diagnosis and the detection of nerve damage [14]. Although damage to small sensory nerve fibers is one of the earliest manifestations of DPN [13], because it only assesses damage to large fibers, the findings are often separate from the subjective symptoms of DPN. In addition, an NCS requires special equipment and skilled inspectors and, therefore, has moderate reproducibility.

Previous studies have revealed that abnormalities in NCSs are related to glycemic control, DM duration, age, male sex, and height [15–18]. However, there is little evidence of an association between abnormalities in NCSs derived from DPN and blood markers [19,20]. Therefore, this study aimed to investigate the association between hematologic tests and NCS variables in patients with T2DM.

2. Materials and Methods

2.1. Study Design and Participants

This retrospective study was conducted using the medical records of patients with T2DM who visited the Department of Endocrinology at Jeju National University Hospital between March 2011 and June 2019 for evaluation of complications. Among the patients whose medical records were reviewed, those who were referred for NCS examination because of suspected neuropathic symptoms were included.

The exclusion criteria were as follows: (1) focal entrapment results such as carpal and cubital tunnel syndromes, (2) medical conditions (e.g., thyroid disease, malignant neoplasm, and vasculitis), and (3) exposure to toxins or medicines that could cause peripheral neuropathy (e.g., alcohol and neurotoxic chemotherapy).

All patients had NCSs and hematologic tests ordered by the endocrinologists, and both examinations were performed within a week.

2.2. Electrophysiologic Studies

Eligible patients underwent a standard NCS using a Medelec Synergy electromyography machine (Medelec Synergy, Oxford, UK) with surface electrode recording and stimulation. Electrodes were applied to both legs and the right arm. The temperature of the NCS examination room was maintained between 22 °C and 24 °C, and the skin temperature was above 32 °C and 30 °C for the upper and lower limbs, respectively. All the NCS assessments were performed by a well-trained technician.

The conduction velocity, amplitude (from baseline to peak), and distal latency of compound muscle action potentials (CMAPs) were obtained in the median, ulnar, peroneal, and tibial nerves using the orthodromic technique. The CMAPs were measured using a 3–10,000 Hz filter, 5 ms/division sweep speed, and 5 mV/division sensitivity. The distal latency and amplitude (from baseline to peak) of the sensory nerve action potentials (SNAPs) in the median, ulnar, sural, and superficial peroneal nerves were measured using an antidromic technique. The SNAPs were performed using a 20–2000 Hz filter, 1 ms/division sweep speed, and 20 µV/division sensitivity. The latency of the F-waves in the median, ulnar, peroneal, and tibial nerves and the Hoffmann reflex (H-reflex) of the tibial nerve were recorded. The F-wave and H-reflex were measured by a 20–10,000 Hz filter, 5 ms/division sweep speed, and 500 µV/division sensitivity, and a 30–10,000 Hz filter, 10 ms/division sweep speed, and 500 µV/division sensitivity, respectively.

DPN diagnosis was determined based on the presence of at least two abnormalities in the aforementioned NCSs. DPN severity was classified into the following stages: mild (two or more abnormalities in the H-reflex [21], tibial F-wave [22], or sural SNAP results [23]), moderate (mild stage with two or more abnormalities on SNAP in the upper extremity nerves or CMAP in the lower extremity nerves), and severe (two or more abnormalities on CMAP or F-waves in the upper extremities).

2.3. Hematologic Tests

All blood samples were collected in the morning, following a fast beginning at midnight. Plasma glucose was measured by the hexokinase-glucose-6-phosphate method, total cholesterol (TC) was analyzed by the enzymatic colorimetric method, triglyceride (TG) was analyzed by the glycerol-3-phosphate (GPO)–peroxidase (POD) chromogenic method, high-density lipoprotein cholesterol (HDL-C) and low-density lipoprotein cholesterol (LDL-C) were analyzed by the direct enzymatic method, apolipoprotein A1 (Apo A1) and B (Apo B) were measured by an immunoassay (TIA) using an automatic chemical analyzer (TBA-FX8, Toshiba, Tokyo, Japan). Vitamin B12 levels were analyzed using a Chemiluminescent Microparticle Immunoassay (CMIA).

The urine albumin–creatinine ratio (UACR) was measured using an automatic chemical analyzer (Hitachi 7600-110, High-Tech, Tokyo, Japan).

The following formula was used to determine the glycemic exposure (GE) index using HbA1c and diabetes duration [24]:

$$\text{Glycemic exposure (GE) index} = (\text{HbA1c})^{1/2} \times (\text{duration of DM in years})^{1/8}$$

2.4. Statistical Analysis

All variables were analyzed using descriptive statistics. Differences in demographics, disease-related traits, and laboratory investigations were compared using Student's t-test. Pearson correlation coefficients were used to evaluate the relationships between laboratory studies and the latencies of the H-reflex, tibial F-wave, and amplitude of sural SNAP. Multivariate analysis of variance (MANOVA) was used to determine the relationship between DPN severity and laboratory results. Independent factors and the presence of DPN were determined using multivariate analysis and binary logistic regression. The optimal cut-point selection of variables was determined by the maximum value of the Youden index in the ROC curve analysis among the significant variables in MANOVA. All statistical analyses were performed using IBM SPSS Statistics for Windows version 22.0 (SPSS Inc., Armonk, NY, USA) and MedCalc Version 22.023 (Medcalc software Ltd., Ostend, Belgium). A p-value of <0.05 was considered statistically significant.

3. Results

A total of 494 patients were referred for DPN evaluation during the study period. Among these patients, 94 met the exclusion criteria (Figure 1). The final analysis included 400 patients (173 male and 227 female) with a mean age of 58.7 ± 14.8 and a mean diabetes duration of 11.5 ± 8.3 years. The NCSs revealed abnormal values in 57% (n = 228) of the patients, and 125 (25.3%), 123 (25.3%), and 74 (15%) patients were in the mild, moderate, and severe stages, respectively.

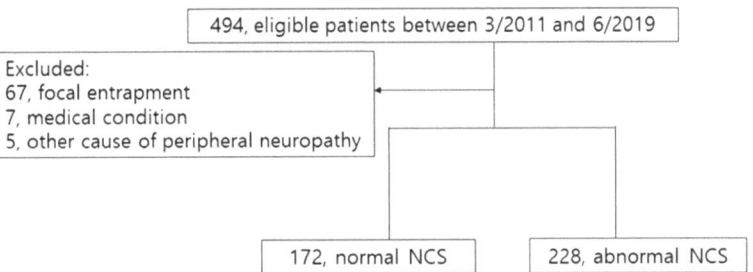

Figure 1. Flow chart of this study.

The demographic and disease-related characteristics and hematological test results of the patients are presented in Table 1. The mean age, DM duration, HbA1c level, GE index, fasting plasma glucose (FPG) level, Apo B level, and UACR were higher in the abnormal-NCS group. HDL-C, Apo A1, estimated glomerular filtration ratio (eGFR), and vitamin B12 levels were higher in the normal-NCS group.

Table 1. Demographic, disease-related characteristics, and laboratory studies of the subjects (n = 400).

Variables	Normal NCS (n = 172)	Abnormal NCS (n = 228)	p-Value
Age (years)	54.5 ± 14.3	62.1 ± 13.7	<0.001 **
Sex, males/females	81 (47.1)/91 (52.9)	92 (40.4)/136 (59.6)	0.186
BMI	25.1 ± 4.4	25.6 ± 4.7	0.299
DM duration (years)	8.6 ± 6.6	13.6 ± 9.0	<0.001 **
HbA1c (%)	8.4 ± 2.3	9.2 ± 2.6	<0.001 **
GE Index	4.4 ± 3.4	7.7 ± 5.8	<0.001 **
FPG (mg/dL)	172.9 ± 83.9	198.8 ± 116.4	0.011 *
C-peptide (nmol/L)	2.18 ± 1.4	2.36 ± 1.8	0.265
TC (mg/dL)	156.4 ± 64.0	164.9 ± 44.7	0.135
TG (mg/dL)	148.1 ± 141.7	158.1 ± 232.2	0.445
HDL-C (mg/dL)	50.4 ± 19.9	44.1 ± 16.6	<0.001 **
Apo A1 (mg/dL)	63.7 ± 70.7	40.2 ± 63.3	0.001 **
LDL-C (mg/dL)	84.3 ± 38.7	88.4 ± 34.3	0.239
Apo B (mg/dL)	31.5 ± 50.5	49.9 ± 57.9	0.001 **
eGFR	95.8 ± 69.7	79.6 ± 30.5	0.009 *
UACR	62.5 ± 171.3	299.0 ± 828.8	<0.001 **
Vitamin B12	752.1 ± 380.7	632.2 ± 358.3	0.003 **

Values represent mean ± standard deviation or number (%) of cases. Abbreviations: NCS, nerve conduction study; DM, diabetes mellitus; FPG, fasting plasma glucose; HbA1c, glycosylated hemoglobin; GE Index, glycemic exposure index; TC, total cholesterol; TG, triglyceride; HDL-C, high-density lipoprotein cholesterol; Apo, apolipoprotein; LDL-C, low-density lipoprotein cholesterol; eGFR, estimated glomerular filtration rate; UACR, urine albumin–creatinine ratio; * < 0.05, ** < 0.01.

Table 2 shows the correlation between the demographic, disease-related characteristics, laboratory results, H-reflex latency, tibial F-wave latency, and sural SNAP amplitude. Older age and long-term exposure to hyperglycemia were positively correlated with delayed H-reflex and tibial F-wave latencies and negatively correlated with sural SNAP amplitude. HDL-C, Apo A1, and vitamin B12 levels had significant negative correlations with the latencies of the H-reflex and tibial F-wave and negative correlations with the amplitude of the sural SNAP.

Table 2. Correlation between latency of NCS and disease-related characteristics in the laboratory studies.

Variables	Coefficients (r) H-Reflex (Rt/Lt)	Coefficients (r) Tibial F-Wave (Rt/Lt)	Coefficients (r) Sural Amp (Rt/Lt)
Age (years)	0.196 **/0.217 **	0.201 **/0.131 **	−0.397 **/−0.376 **
BMI	0.118/0.147	0.052/0.015	−0.028/0.043
DM duration	0.217 **/0.179 *	0.223 **/0.163 **	−0.299 **/−0.309 **
GE Index	0.221 **/0.177 **	0.264 **/0.195 **	−0.318 **/−0.319 **
HbA1c (%)	0.093/0.076	0.204 **/0.230 **	−0.105/−0.084
FPG (mg/dL)	0.146/0.157	0.120 */0.146 **	−0.099/−0.058
C-peptide (nlmol/L)	0.064/0.089	−0.002/0.049	−0.065/−0.046
TC (mg/dL)	−0.058/−0.021	−0.065/−0.009	0.163 */0.159 *
TG (mg/dL)	0.096/0.118	0.003/0.108	0.055/0.055
HDL-C (mg/dL)	−0.152 **/−0.180 **	−0.202 **/−0.183 **	0.141 */0.160 *
Apo A1 (mg/dL)	−0.373 **/−0.195 *	−0.231 **/−0.238 **	0.214 */0.278 **
LDL-C (mg/dL)	−0.040/−0.001	0.034/−0.014	0.103/0.106
Apo B (mg/dL)	−0.030/−0.121	−0.004/−0.009	−0.033/0.041
eGFR	−0.057/0.022	−0.088/−0.027	0.262 **/0.216 *
UACR	0.132/0.195 *	0.173 **/0.205 **	−0.163 */−0.204 *
Vitamin B12	−0.205 **/−0.246 **	−0.113 */−0.125 *	0.261 **/0.168 *

NCS, nerve conduction study; Rt, right; Lt, left; DM, diabetes mellitus; GE Index, glycemic exposure index; FPG, fasting plasma glucose; HbA1c, glycosylated hemoglobin; TC, total cholesterol; TG, triglyceride; HDL-C, high-density lipoprotein cholesterol; Apo, apolipoprotein; LDL-C, low-density lipoprotein cholesterol; eGFR, estimated glomerular filtration rate; UACR, urine albumin–creatinine ratio. * < 0.05, ** < 0.01.

In the multivariate binary logistic regression analysis, older age, higher GE index, HbA1c, and UACR, and lower Apo A1 and eGFR were significantly associated with DPN (Table 3).

Table 3. Multivariate analysis using binary logistic regression of laboratory studies and disease related characteristics for DPN.

Variables	Logistic Regression		
	Odds Ratio	95% CI	p-Value
Age > 62.5	2.582	1.708–3.902	<0.001
GE index > 4.78	3.629	2.386–5.519	<0.001
HbA1c > 7.75	2.195	1.461–3.299	<0.001
ApoA1 < 100	2.003	1.315–3.050	0.001
eGFR < 62.65	7.969	3.457–18.370	<0.001
UACR > 37.465	4.250	2.700–6.691	<0.001
Vitamin B12 < 520	1.684	1.041–2.722	0.034

GE Index, glycemic exposure index; HbA1c, glycosylated hemoglobin; Apo, apolipoprotein; eGFR, estimated glomerular filtration rate; UACR, urine albumin–creatinine ratio.

In this study, we conducted a subgroup analysis of the correlations based on the NCS results. In patients with normal NCS results, HbA1c showed a significant positive correlation with bilateral tibial F-wave latency (r = 0.190/0.248, $p < 0.001$), HDL-C was negatively correlated with bilateral tibial F-wave latency (r = −0.284/−0.254, $p < 0.001$), and Apo A1 was negatively correlated with bilateral H-reflex latency (r = −0.732/−0.382, r < 0.001) and tibial F-wave latency (r = −0.229/−0.261, $p < 0.001$). The eGFR and UACR were positively correlated with the bilateral amplitude of sural SNAP (r = 0.256/211 and

0.329/0.291, respectively; all $p < 0.001$). In the subgroup analysis of abnormal NCS results, DPN severity was related to the UACR in the MANOVA.

4. Discussion

This study investigated the relationship between hematological tests and NCS results in patients with clinical DPN. The abnormal-NCS group showed poor glycemic control, dyslipidemia, and decreased renal function compared to the patients with normal electrophysiological studies.

The results of this study showed that poor blood sugar control and the period of exposure to hyperglycemia worsened DPN risk and severity. These results are similar to those of hyperglycemia, which was the main cause of neuropathy in patients with type 2 diabetes in previous studies [25,26]. Although hyperglycemia is known to be an important etiology of DPN, glycemic control did not reduce DPN risk in patients with T2DM in large-scale studies such as the ACCORD trial [6]. However, this delayed its occurrence. Therefore, it is thought that the occurrence of DPN in T2DM patients is not only caused by hyperglycemia but also involves various metabolic factors. Previous studies have indeed identified dyslipidemia and insulin resistance as relevant hematological findings [20,25]. However, there are limited studies on other hematological markers. Therefore, it is necessary to determine the relationship between hematological tests for T2DM and neuropathy. Therefore, prolonged hyperglycemia is still considered one of the most critical etiological factors of DPN.

The American Academy of Neurology, American Association of Electrodiagnostic Medicine, and American Academy of Physical and Rehabilitation defined DPN as a symmetrical, length-dependent sensorimotor polyneuropathy caused by metabolic causes such as hyperglycemia and microvessel alteration [14]. In addition, an NCS is the gold standard for confirming nerve damage, and the NCS for diagnosing DPN is the most objective and widely used worldwide [2]. In this study, assuming that diabetic neuropathy is length-dependent, we analyzed the H-reflex, F-wave, and sural SNAP test results, which are thought to be the first abnormal findings in NCS DPN results. The H-reflex is a monosynaptic reflex arc arising from the Ia afferent activation of muscles to the alpha motor neurons [27]. Therefore, this is one of the evaluations of the longest nerve pathways in the NCS. The H-reflex could have a predictive value in early DPN diagnosis [28] and the F-wave latency of the tibial nerve is one of the most sensitive measures in an NCS to detect subclinical or overt DPN [22,29]. Sural nerve conduction study amplitude is a reliable method for diagnosing mild DPN [12].

Based on its reproducibility, reasonable sensitivity, and specificity, NCSs have been used as an endpoint in clinical trials on human diabetic neuropathy to measure large myelinated nerve fiber function [30]. Therefore, this study analyzed the correlation between changes in the H-reflex, tibial F-wave, and sural nerve amplitude, which are the initial findings of NCS abnormalities in patients with DPN among those with normal NCS test results. However, this study was based on the NCS results of patients clinically thought to have DPN, and it cannot be concluded that abnormal findings in the NCS results indicate that the neuropathy has progressed further. Since the subjects were patients clinically suspected of having neuropathy, it cannot be said that neuropathy does not exist even if the NCS results are normal. Because approximately 80–90% of peripheral nerves are small fibers [12,31], several studies have shown that small sensory nerve fibers are one of the earliest manifestations of DPN [32–34]. However, when referring to studies showing that DPN progresses from damage to small fibers to damage to large fibers, abnormal findings on the NCS can be considered to be slightly more advanced DPN. Although the results of the NCS are within the normal range of the diagnostic criteria, it would be helpful for early diagnosis if changes in the NCS could be identified more quickly.

Dyslipidemia, especially low HDL-C levels, is a noteworthy phenomenon. In this study, HDL-C and its precursor Apo A1 were statistically correlated with the latency of the H-reflex, tibial F-wave, and sural nerve amplitude. Additionally, the logistic regression analysis demonstrated that the likelihood of receiving a DPN diagnosis in the NCS increased with decreasing Apo A1 levels, a precursor to HDL. Additionally, logistic regression analysis showed that the probability of DPN diagnosis in the NCS increased with decreasing Apo A1 levels. However, no statistically significant results were observed for LDL or its precursor Apo B. Considering that all patients in this study had DPN, HDL may play a role in preventing exacerbations. A possible explanation is that in diabetes animal models, HDL-C and Apo A1 improve insulin sensitivity and pancreatic β-cell survival and function, which in turn improves glycemic control [35].

In addition to treating hyperlipidemia with drugs, a strategy to increase HDL levels is needed. These results showed no significant differences in the TC, TG, and LDL levels between the two groups, which is thought to be because most patients were receiving treatment for hyperlipidemia. Current study showed no discernible changes in TC, TG, or LDL levels between the two groups, which is assumed to be because most patients were receiving treatment for hyperlipidemia. Drugs that lower LDL and TG levels, such as niacins, fibrates, and statins, occasionally increase HDL levels. However, large-scale randomized controlled trials of interventions for HDL-C levels have not responded favorably to epidemiological evidence of an inverse relationship between HDL-C levels and the risk of atherosclerotic cardiovascular disease [36]. Therefore, physical activity, weight loss, diet, and smoking cessation are important factors that increase HDL-C levels.

Significant negative correlations were found between H-reflex latency, tibial F-wave, kidney function, and vitamin B12. Diabetic nephropathy is a major complication of diabetes and is associated with long-term exposure to high blood sugar levels and other metabolic or hemodynamic problems [37]. In this study, increased microalbumin levels and decreased renal function, which are early indicators of nephropathy, appeared to be associated with DPN occurrence. This suggests that neuropathy must be identified in the early stages. However, the prevalence of nephropathy and neuropathic complications is not well established. In addition, the relationship between the severity of nephropathy and neuropathy, as well as that between the degree of nephropathy progression and neuropathy occurrence, is not well known.

In addition, several studies have shown that the administration of vitamin B12 improves DPN symptoms [38,39], and studies have shown that a decrease in vitamin B12 increases DPN occurrence [40]. Similar to the aforementioned studies, vitamin B12 levels were lower in patients with abnormal NCS results who were thought to have more advanced neuropathy. In the NCS results, lower vitamin B12 levels increased H-reflex and F-wave latencies and decreased the sural nerve amplitude. Binary logistic regression analysis showed that lower vitamin B12 levels were associated with an increased risk of DPN diagnosis based on NCS.

This study has some limitations. First, there is no established NCS protocol for diagnosing DPN. Although it was conducted based on the literature, additional research is needed to determine whether the changes in the H-reflex, F-wave, and Sural nerve, which were assumed to be initial changes in NCS in this study, can truly represent changes in the initial DPN. Second, this was a retrospective study, and the participants were clinically presumed to be patients. Therefore, participants could not be evaluated using screening or assessment tools. Future studies should use additional quantitative assessment tools. Finally, this study only included patients with T2DM who had been diagnosed with diabetes for a relatively long period. Therefore, a prospective study that can confirm these changes in patients with early-stage diabetes is necessary.

5. Conclusions

This study demonstrated that NCS can be used to identify hematological indicators helpful in diagnosing DPN. Patients with T2DM with long-term exposure to hyperglycemia and progression of nephropathy are thought to be associated with the development of DPN. It is necessary not only to control hyperglycemia but also to assess metabolic factors such as HDL-C, Apo A-I, and vitamin b12 to prevent DPN. Glycemic control alone is insufficient to prevent the progression and worsening of DPN. Additionally, lower levels of HDL-C, Apo A-I, and vitamin B12 are associated with the condition, suggesting their potential utility in early diagnostic protocol. Therefore, this study is clinically meaningful in identifying the risk of developing DPN early through hematological testing and confirming it through NCS.

Author Contributions: Conceptualization, J.-H.C. and J.-E.H.; methodology, J.-H.C. and S.-Y.Y.; formal analysis, J.-H.C. and J.-E.H.; investigation, J.-H.C., S.-Y.Y., G.-P.K., S.-A.L., S.-Y.L. and H.-J.L.; data curation, J.-H.C., S.-Y.Y., G.-P.K., S.-A.L., S.-Y.L. and H.-J.L.; writing, J.-H.C. and J.-E.H.; original draft preparation, J.-H.C. and J.-E.H.; project administration, J.-H.C.; Supervision, J.-H.C. All authors have read and agreed to the published version of the manuscript.

Funding: This work was supported by a research grant from Jeju National University Hospital in 2023 (202300620001).

Institutional Review Board Statement: This study was conducted in accordance with the Declaration of Helsinki and approved by the Ethics Committee of Jeju National University Hospital (JNUH 2023-11-014, 5 January 2024).

Informed Consent Statement: Patients did not have to provide informed consent due to the retrospective nature of the study.

Data Availability Statement: Data are available upon request.

Conflicts of Interest: The authors declare that there are no conflicts of interest.

References

1. Pop-Busui, R.; Ang, L.; Boulton, A.J.; Feldman, E.L.; Marcus, R.L.; Mizokami-Stout, K.; Singleton, J.R.; Ziegler, D. *Diagnosis and Treatment of Painful Diabetic Peripheral Neuropathy*; American Diabetes Association: Arlington, VA, USA, 2022.
2. Tesfaye, S.; Boulton, A.J.; Dyck, P.J.; Freeman, R.; Horowitz, M.; Kempler, P.; Lauria, G.; Malik, R.A.; Spallone, V.; Vinik, A. Diabetic neuropathies: Update on definitions, diagnostic criteria, estimation of severity, and treatments. *Diabetes Care* **2010**, *33*, 2285. [CrossRef]
3. Control, D.; Group, C.T.R. The effect of intensive treatment of diabetes on the development and progression of long-term complications in insulin-dependent diabetes mellitus. *N. Engl. J. Med.* **1993**, *329*, 977–986.
4. Galosi, E.; Hu, X.; Michael, N.; Nyengaard, J.R.; Truini, A.; Karlsson, P. Redefining distal symmetrical polyneuropathy features in type 1 diabetes: A systematic review. *Acta Diabetol.* **2022**, *59*, 1–19. [CrossRef] [PubMed]
5. UK Prospective Diabetes Study (UKPDS) Group. Intensive blood-glucose control with sulphonylureas or insulin compared with conventional treatment and risk of complications in patients with type 2 diabetes (UKPDS 33). *Lancet* **1998**, *352*, 837–853. [CrossRef]
6. Ismail-Beigi, F.; Craven, T.; Banerji, M.A.; Basile, J.; Calles, J.; Cohen, R.M.; Cuddihy, R.; Cushman, W.C.; Genuth, S.; Grimm, R.H. Effect of intensive treatment of hyperglycaemia on microvascular outcomes in type 2 diabetes: An analysis of the ACCORD randomised trial. *Lancet* **2010**, *376*, 419–430. [CrossRef]
7. Biessels, G.; Bril, V.; Calcutt, N.; Cameron, N.; Cotter, M.; Dobrowsky, R.; Feldman, E.; Fernyhough, P.; Jakobsen, J.; Malik, R. Phenotyping animal models of diabetic neuropathy: A consensus statement of the diabetic neuropathy study group of the EASD (Neurodiab). *J. Peripher. Nerv. Syst.* **2014**, *19*, 77–87. [CrossRef]
8. Vincent, A.M.; Callaghan, B.C.; Smith, A.L.; Feldman, E.L. Diabetic neuropathy: Cellular mechanisms as therapeutic targets. *Nat. Rev. Neurol.* **2011**, *7*, 573–583. [CrossRef] [PubMed]
9. Feldman, E.L.; Callaghan, B.C.; Pop-Busui, R.; Zochodne, D.W.; Wright, D.E.; Bennett, D.L.; Bril, V.; Russell, J.W.; Viswanathan, V. Diabetic neuropathy. *Nat. Rev. Dis. Prim.* **2019**, *5*, 41. [CrossRef]

10. Sloan, G.; Selvarajah, D.; Tesfaye, S. Pathogenesis, diagnosis and clinical management of diabetic sensorimotor peripheral neuropathy. *Nat. Rev. Endocrinol.* **2021**, *17*, 400–420. [CrossRef]
11. Himeno, T.; Kamiya, H.; Nakamura, J. Diabetic polyneuropathy: Progress in diagnostic strategy and novel target discovery, but stagnation in drug development. *J. Diabetes Investig.* **2020**, *11*, 25–27. [CrossRef] [PubMed]
12. Malik, R.; Tesfaye, S.; Newrick, P.; Walker, D.; Rajbhandari, S.; Siddique, I.; Sharma, A.; Boulton, A.; King, R.; Thomas, P. Sural nerve pathology in diabetic patients with minimal but progressive neuropathy. *Diabetologia* **2005**, *48*, 578–585. [CrossRef] [PubMed]
13. Petropoulos, I.N.; Ponirakis, G.; Khan, A.; Almuhannadi, H.; Gad, H.; Malik, R.A. Diagnosing diabetic neuropathy: Something old, something new. *Diabetes Metab. J.* **2018**, *42*, 255. [CrossRef] [PubMed]
14. England, J.; Gronseth, G.; Franklin, G.; Miller, R.; Asbury, A.; Carter, G.; Cohen, J.; Fisher, M.; Howard, J.; Kinsella, L. Distal symmetric polyneuropathy: A definition for clinical research: Report of the American Academy of Neurology, the American Association of Electrodiagnostic Medicine, and the American Academy of Physical Medicine and Rehabilitation. *Neurology* **2005**, *64*, 199–207. [CrossRef] [PubMed]
15. Sosenko, J.M.; Gadia, M.T.; Fournier, A.M.; O'Connell, M.T.; Aguiar, M.C.; Skyler, J.S. Body stature as a risk factor for diabetic sensory neuropathy. *Am. J. Med.* **1986**, *80*, 1031–1034. [CrossRef] [PubMed]
16. Tesfaye, S.; Chaturvedi, N.; Eaton, S.E.; Ward, J.D.; Manes, C.; Ionescu-Tirgoviste, C.; Witte, D.R.; Fuller, J.H. Vascular risk factors and diabetic neuropathy. *N. Engl. J. Med.* **2005**, *352*, 341–350. [CrossRef]
17. Marshall, A.; Alam, U.; Themistocleous, A.; Calcutt, N.; Marshall, A. Novel and emerging electrophysiological biomarkers of diabetic neuropathy and painful diabetic neuropathy. *Clin. Ther.* **2021**, *43*, 1441–1456. [CrossRef]
18. Dunker, Ø.; Nilsen, K.B.; Olsen, S.E.; Åsvold, B.O.; Bjørgaas, M.R.R.; Sand, T. Which combined nerve conduction study scores are best suited for polyneuropathy in diabetic patients? *Muscle Nerve* **2022**, *65*, 171–179. [CrossRef]
19. Park, J.H.; Park, J.H.; Won, J.C. Associations of nerve conduction study variables with clinical symptom scores in patients with type 2 diabetes. *Ann. Clin. Neurophysiol.* **2019**, *21*, 36–43. [CrossRef]
20. Cho, Y.N.; Lee, K.O.; Jeong, J.; Park, H.J.; Kim, S.-M.; Shin, H.Y.; Hong, J.-M.; Ahn, C.W.; Choi, Y.-C. The role of insulin resistance in diabetic neuropathy in Koreans with type 2 diabetes mellitus: A 6-year follow-up study. *Yonsei Med. J.* **2014**, *55*, 700–708. [CrossRef]
21. Troni, W. Analysis of conduction velocity in the H pathway: Part 2. An electrophysiological study in diabetic polyneuropathy. *J. Neurol. Sci.* **1981**, *51*, 235–246. [CrossRef] [PubMed]
22. Andersen, H.; Stålberg, E.; Falck, B. F-wave latency, the most sensitive nerve conduction parameter in patients with diabetes mellitus. *Muscle & Nerve* **1997**, *20*, 1296–1302.
23. Group, D.R. Factors in development of diabetic neuropathy: Baseline analysis of neuropathy in feasibility phase of Diabetes Control and Complications Trial (DCCT). *Diabetes* **1988**, *37*, 476–481. [CrossRef]
24. Dyck, P.J.; Davies, J.L.; Clark, V.M.; Litchy, W.J.; Dyck, P.J.B.; Klein, C.J.; Rizza, R.A.; Pach, J.M.; Klein, R.; Larson, T.S. Modeling chronic glycemic exposure variables as correlates and predictors of microvascular complications of diabetes. *Diabetes Care* **2006**, *29*, 2282–2288. [CrossRef]
25. Pop-Busui, R.; Boulton, A.J.; Feldman, E.L.; Bril, V.; Freeman, R.; Malik, R.A.; Sosenko, J.M.; Ziegler, D. Diabetic neuropathy: A position statement by the American Diabetes Association. *Diabetes Care* **2017**, *40*, 136. [CrossRef] [PubMed]
26. Smith, S.; Normahani, P.; Lane, T.; Hohenschurz-Schmidt, D.; Oliver, N.; Davies, A.H. Prevention and management strategies for diabetic neuropathy. *Life* **2022**, *12*, 1185. [CrossRef] [PubMed]
27. JW, M. Electrophysiological studies of nerve and reflex activity in normal man. I. Identification of H-reflex of certain reflexes in the electromyogram and the conduction velocity of peripheral nerve fibers. *Bull. Johns Hopkins Hosp.* **1950**, *86*, 265–290.
28. Millan-Guerrero, R.; Trujillo-Hernandez, B.; Isais-Millan, S.; Prieto-Diaz-Chavez, E.; Vasquez, C.; Caballero-Hoyos, J.; Garcia-Magana, J. H-reflex and clinical examination in the diagnosis of diabetic polyneuropathy. *J. Int. Med. Res.* **2012**, *40*, 694–700. [CrossRef]
29. Pan, H.; Jian, F.; Lin, J.; Chen, N.; Zhang, C.; Zhang, Z.; Ding, Z.; Wang, Y.; Cui, L.; Kimura, J. F-wave latencies in patients with diabetes mellitus. *Muscle Nerve* **2014**, *49*, 804–808. [CrossRef] [PubMed]
30. Dyck, P.J.; Overland, C.J.; Low, P.A.; Litchy, W.J.; Davies, J.L.; Dyck, P.J.B.; O'Brien, P.C.; Cl, vs. NPhys Trial Investigators. Signs and symptoms versus nerve conduction studies to diagnose diabetic sensorimotor polyneuropathy: Cl vs. NPhys trial. *Muscle Nerve* **2010**, *42*, 157–164. [CrossRef] [PubMed]
31. Said, G.; Baudoin, D.; Toyooka, K. Sensory loss, pains, motor deficit and axonal regeneration in length-dependent diabetic polyneuropathy. *J. Neurol.* **2008**, *255*, 1693–1702. [CrossRef] [PubMed]
32. Baron, R.; Binder, A.; Wasner, G. Neuropathic pain: Diagnosis, pathophysiological mechanisms, and treatment. *Lancet Neurol.* **2010**, *9*, 807–819. [CrossRef] [PubMed]

33. Malik, R.; Veves, A.; Tesfaye, S.; Smith, G.; Cameron, N.; Zochodne, D.; Lauria, G.; Toronto Consensus Panel on Diabetic Neuropathy. Small fibre neuropathy: Role in the diagnosis of diabetic sensorimotor polyneuropathy. *Diabetes/Metab. Res. Rev.* **2011**, *27*, 678–684. [CrossRef]
34. Ainslie, P.N.; Murrell, C.; Peebles, K.; Swart, M.; Skinner, M.A.; Williams, M.J.; Taylor, R.D. Early morning impairment in cerebral autoregulation and cerebrovascular CO2 reactivity in healthy humans: Relation to endothelial function. *Exp. Physiol.* **2007**, *92*, 769–777. [CrossRef] [PubMed]
35. Manandhar, B.; Cochran, B.J.; Rye, K.A. Role of High-Density Lipoproteins in Cholesterol Homeostasis and Glycemic Control. *J. Am. Heart Assoc.* **2020**, *9*, e013531. [CrossRef] [PubMed]
36. Madsen, C.M.; Varbo, A.; Nordestgaard, B.G. Novel insights from human studies on the role of high-density lipoprotein in mortality and noncardiovascular disease. *Arterioscler. Thromb. Vasc. Biol.* **2021**, *41*, 128–140. [CrossRef] [PubMed]
37. Fowler, M.J. Microvascular and macrovascular complications of diabetes. *Clin. Diabetes* **2011**, *29*, 116–122. [CrossRef]
38. Talaei, A.; Siavash, M.; Majidi, H.; Chehrei, A. Vitamin B12 may be more effective than nortriptyline in improving painful diabetic neuropathy. *Int. J. Food Sci. Nutr.* **2009**, *60* (Suppl. S5), 71–76. [CrossRef] [PubMed]
39. Didangelos, T.; Karlafti, E.; Kotzakioulafi, E.; Margariti, E.; Giannoulaki, P.; Batanis, G.; Tesfaye, S.; Kantartzis, K. Vitamin B12 supplementation in diabetic neuropathy: A 1-year, randomized, double-blind, placebo-controlled trial. *Nutrients* **2021**, *13*, 395. [CrossRef] [PubMed]
40. Alvarez, M.; Sierra, O.R.; Saavedra, G.; Moreno, S. Vitamin B12 deficiency and diabetic neuropathy in patients taking metformin: A cross-sectional study. *Endocr. Connect.* **2019**, *8*, 1324–1329. [CrossRef] [PubMed]

Disclaimer/Publisher's Note: The statements, opinions and data contained in all publications are solely those of the individual author(s) and contributor(s) and not of MDPI and/or the editor(s). MDPI and/or the editor(s) disclaim responsibility for any injury to people or property resulting from any ideas, methods, instructions or products referred to in the content.

MDPI AG
Grosspeteranlage 5
4052 Basel
Switzerland
Tel.: +41 61 683 77 34

Medicina Editorial Office
E-mail: medicina@mdpi.com
www.mdpi.com/journal/medicina

Disclaimer/Publisher's Note: The title and front matter of this reprint are at the discretion of the Guest Editors. The publisher is not responsible for their content or any associated concerns. The statements, opinions and data contained in all individual articles are solely those of the individual Editors and contributors and not of MDPI. MDPI disclaims responsibility for any injury to people or property resulting from any ideas, methods, instructions or products referred to in the content.

www.ingramcontent.com/pod-product-compliance
Lightning Source LLC
LaVergne TN
LVHW072346090526
838202LV00019B/2486